# Edexcel AS/A level
## CHEMISTRY 1

**Cliff Curtis**
**Jason Murgatroyd**
**David Scott**

ALWAYS LEARNING                                            PEARSON

Published by Pearson Education Limited, 80 Strand, London WC2R 0RL.

www.pearsonschoolsandfecolleges.co.uk

Copies of official specifications for all Edexcel qualifications may be found on the website: www.edexcel.com

Text © Pearson Education Limited 2015
Edited by Tracey Cowell and Ashley Craig
Designed by Elizabeth Arnoux for Pearson Education Limited
Typeset by Techset Ltd, Gateshead
Original illustrations © Pearson Education Limited 2008
Illustrated by Techset Ltd and Peter Bull
Cover design by Elizabeth Arnoux for Pearson Education Limited
Picture research by Susie Prescott
Cover photo/illustration © Science Photo Library/Alfred Paseika

The rights of Cliff Curtis, Jason Murgatroyd and David Scott to be identified as authors of this work have been asserted by them in accordance with the Copyright, Designs and Patents Act 1988.

First edition published 2008
Second edition published 2015

19 18 17 16 15
10 9 8 7 6 5 4 3 2 1

**British Library Cataloguing in Publication Data**
A catalogue record for this book is available from the British Library

ISBN 978 1 447 97582 3

**Copyright notice**
All rights reserved. No part of this publication may be reproduced in any form or by any means (including photocopying or storing it in any medium by electronic means and whether or not transiently or incidentally to some other use of this publication) without the written permission of the copyright owner, except in accordance with the provisions of the Copyright, Designs and Patents Act 1988 or under the terms of a licence issued by the Copyright Licensing Agency, Saffron House, 6–10 Kirby Street, London EC1N 8TS (www.cla.co.uk). Applications for the copyright owner's written permission should be addressed to the publisher.

Printed in the UK by CPI

**Acknowledgements**
Every effort has been made to contact copyright holders of material reproduced in this book. Any omissions will be rectified in subsequent printings if notice is given to the publishers.

**A note from the publisher**
In order to ensure that this resource offers high-quality support for the associated Pearson qualification, it has been through a review process by the awarding body. This process confirms that this resource fully covers the teaching and learning content of the specification or part of a specification at which it is aimed. It also confirms that it demonstrates an appropriate balance between the development of subject skills, knowledge and understanding, in addition to preparation for assessment.

Endorsement does not cover any guidance on assessment activities or processes (e.g. practice questions or advice on how to answer assessment questions), included in the resource nor does it prescribe any particular approach to the teaching or delivery of a related course.

While the publishers have made every attempt to ensure that advice on the qualification and its assessment is accurate, the official specification and associated assessment guidance materials are the only authoritative source of information and should always be referred to for definitive guidance.

Pearson examiners have not contributed to any sections in this resource relevant to examination papers for which they have responsibility.

Examiners will not use endorsed resources as a source of material for any assessment set by Pearson.

Endorsement of a resource does not mean that the resource is required to achieve this Pearson qualification, nor does it mean that it is the only suitable material available to support the qualification, and any resource lists produced by the awarding body shall include this and other appropriate resources.

## Picture credits

The publisher would like to thank the following for their kind permission to reproduce their photographs:

(Key: b-bottom; c-centre; l-left; r-right; t-top)

**123RF.com:** preecha bamrungrai 178l (fig B)
**Alamy Images:** age fotostock Spain, S.L. 10tc, Brian North Gardens 97cl (fig C), D. Hurst 183tl (fig B), David Taylor 101cr (fig A), David Zanzinger 182br (fig A), E.R. Degginger 95bl, Extreme Sports Photo 264, G.L Archive 10tl, Gordon Shoosmith 34–35, INSADCO Photography 216, Leslie Garland Picture Library 209tr (fig A1), Libby Welch 71c (fig A), MaximImages 67bl (fig H), Nigel Cattlin 67tl (fig F), Paul Fleet 66bl (fig C), Peter Hermes Furian 222c (fig A), photodreams2 132cl (fig A), RGB Ventures/Superstock 10tr, Sara RIchards 276cl (fig A), sciencephotos 155tl (fig A)
**Armin Kubelbeck:** 153t (fig A)
**Bridgeman Art Library Ltd:** De Agostini Picture Library/Chomon 125br (fig B)
**Brown University Library:** Wang Lab 70 (fig A)
**Corbis:** Andrew Brookes 8–9, Ann Johansson 226cl (fig A), David Muench 250
**Courtesy of Omega:** 96bl (fig B)
**Getty Images:** A & F Michler/Photolibrary 66br (fig E), David Caudery/PhotoPlus Magazine 230, Jeff Foot/Discovery Channel 86, Katsiaryna Kapusta/iStock 67cl (fig G), Photos.com/360 166tl (fig A)
**NASA:** JPL-Caltech 71b (fig B), NASA/ESA/S. Beckwith (STScI) and The HUDF Team 31bl, NASA Images/JPL-Caltech 71b (fig C)
**Pearson Education Ltd:** Miguel Domínguez Muñoz 107cl (fig B), Trevor Clifford 136bl (fig A), Coleman Yuen 121br (fig C)
**Science Photo Library Ltd:** 20, 90, 96cl (fig A), 97cr (fig D), 100c (fig A), 156tl (fig A), Andrew Lambert Photography 92tl (fig A), 94tl (fig Ab), 106cr (fig A), 112bl, 141cr (fig C), 171c (fig A), 175cl (fig A), 188bl (fig A), 200c, Carlos Goldin 181bl (fig A), Charles D. Winters 94tl (fig Aa), 94bc, 125cr (fig A), Martyn F. Chillmaid 95tl, 113tr (fig B), 139bl (fig A), 139bc (fig A), 139br (fig A), 204tl (fig A), Sinclair Stammers 198cl (fig A)
**Shutterstock.com:** Aleksey Stemmer 103cl (fig C), Christian Lagerek 131br (fig A), Federico Rostagno 164, gosphotodesign 198cr (fig A), janprachal 149cl (fig B), Jerome Whittingham 159c (fig A), Minerva Studio 178tl (fig A), PhotoSGH 150tl (fig A), Pitsanu Kraichana 155bl (fig B), Sementer 160, Sergio Ponomarev 149cr (fig A)
**Veer/Corbis:** Aja 74, alexraths 166b (fig B), Edward_Foto 185tr (fig A), Only Fabrizio 87, Rudyanto Wijaya 118

**Cover images:** *Front:* **Science Photo Library Ltd:** Alfred Paseika

All other images © Pearson Education

Every effort has been made to trace the copyright holders and we apologise in advance for any unintentional omissions. We would be pleased to insert the appropriate acknowledgement in any subsequent edition of this publication.

*We are grateful to the following for permission to reproduce copyright material:*

### Figures

Figure on page 10 from 'Mass Number and Isotope', http://www.shimadzu.com, copyright © 2014 Shimadzu Corporation. All rights reserved; Figures on page 207 from Education Scotland © Crown copyright 2012; Figures on page 212 from 'Catalysts for a green industry' *Education in Chemistry* (Tony Hargreaves), July 2009, http://www.rsc.org/education/eic/issues/2009July/catalyst-green-chemistry-research-industry.asp, copyright © Royal Society of Chemistry; Figures on page 226 from 'Five rings good, four rings bad', *Education in Chemistry* (Dr Simon Cotton), March 2010, http://www.rsc.org/education/eic/issues/2010Mar/FiveRingsGoodFourRingsBad.asp, copyright © Royal Society of Chemistry.

### Text

Extract on page 30 from *From Stars to Stalagmites: How Everything Connects* by Paul Braterman, World Scientific Publishing Co., 2012, p.76, copyright © 2012 World Scientific Publishing Co. Pte Ltd; Quote on page 58 from Professor Gautam Desiraju of the University of Hyderabad, 1999. Reproduced with kind permission; Extracts on pages 86 and 114 from 'Some of our selenium is missing', *New Scientist Magazine*, 18/11/1995, Issue 2004 (Michelle Knott), copyright © 2002, 2004 Reed Business Information and 'Science: Nitrate disappears in "impossible" reaction', *New Scientist Magazine*, 16/07/1994, Issue 1934 (John Emsley) – UK. All rights reserved. Distributed by Tribune Content Agency; Extracts on page 160 from 'Ancient coins', http://www.rsc.org/Education/EiC/issues/2006Nov/AncientCoins.asp, copyright © Royal Society of Chemistry; Extracts on pages 212 and 226 from 'Catalysts for a green industry' http://www.rsc.org/education/eic/issues/2009July/catalyst-green-chemistry-research-industry.asp, and 'Five rings good, four rings bad', http://www.rsc.org/education/eic/issues/2010Mar/FiveRingsGoodFourRingsBad.asp, copyright © Royal Society of Chemistry; Extract on page 227 from 'How athletics is still scarred by the reign of the chemical sisters', *The Daily Mail*, 06/08/2012 (Matt Lawton), copyright © Solo Syndication, 2012; Extract on page 246 from 'Organic Chemists Contribute to Renewable Energy', http://www.rsc.org/Membership/Networking/InterestGroups/OrganicDivision/organic-chemistry-case-studies/organic-chemistry-biofuels.asp, copyright © Royal Society of Chemistry; Extracts on pages 260 and 276 from 'Tiny tubes set chemical reactions racing', *New Scientist Magazine*, 23/05/2007 (Tom Simonite), and 'Something in the water...', *New Scientist Magazine*, 13/04/2002, Issue 2338 (Emma Young), copyright © 2002, 2007 Reed Business Information - UK. All rights reserved. Distributed by Tribune Content Agency.

The Publisher would like to thank Chris Curtis, Chris Ryan and Adelene Cogill for their contributions to the Maths skills and Exam preparation sections of this book.

Every effort has been made to contact copyright holders of material reproduced in this book. Any omissions will be rectified in subsequent printings if notice is given to the publishers.

# Contents

How to use this book ............................................. 6

## TOPIC 1 Atomic structure and the Periodic Table

**1.1 Atomic structure** ............................................. 8
   1 Structure of the atom and isotopes ............................ 10
   2 Mass spectrometry and relative masses of atoms, isotopes and molecules ............................ 12
   3 Atomic orbitals and electronic configurations ............................ 16
   4 Ionisation energies ............................ 20

**1.2 The Periodic Table** ............................ 24
   1 The Periodic Table ............................ 24
   2 Periodicity ............................ 26
   Thinking Bigger ............................ 30
   Exam-style questions ............................ 32

## TOPIC 2 Chemical bonding and structure

**2.1 Giant structures** ............................ 34
   1 Metallic bonding ............................ 36
   2 Ionic bonding ............................ 38
   3 Covalent bonding ............................ 42
   4 Electronegativity and bond polarity ............................ 45

**2.2 Discrete molecules** ............................ 48
   1 Bonding in discrete molecules ............................ 48
   2 Dative covalent bonds ............................ 50
   3 Shapes of molecules and ions ............................ 51
   4 Non-polar and polar molecules ............................ 53
   5 Intermolecular interactions ............................ 55
   6 Intermolecular interactions and physical properties ............................ 59

**2.3 Physical properties related to structure and bonding** ............................ 65
   1 Solid lattices ............................ 65
   2 Structure and properties ............................ 68
   Thinking Bigger ............................ 70
   Exam-style questions ............................ 72

## TOPIC 3 Redox reactions

**3.1 Oxidation and reduction in terms of electrons** ............................ 74
   1 Electron loss and gain ............................ 76

**3.2 Oxidising agents and reducing agents** ............................ 78
   1 Calculating oxidation numbers ............................ 78
   2 Recognising reactions using oxidation numbers ............................ 80
   3 Use of oxidation numbers in nomenclature ............................ 82
   4 Writing full equations from ionic half-equations ............................ 84
   Thinking Bigger ............................ 86
   Exam-style questions ............................ 88

## TOPIC 4 Inorganic chemistry and the Periodic Table

**4.1 Group 2** ............................ 90
   1 Trends in the Group 2 elements ............................ 92
   2 Reactions of the Group 2 elements ............................ 94
   3 Reactions of the Group 2 oxides and hydroxides, and trends in solubility ............................ 96
   4 Thermal stability of Group 2 compounds, and the comparison with Group 1 ............................ 98
   5 Flame tests and the test for ammonium ions ............................ 101

**4.2 Group 7** ............................ 104
   1 General trends in Group 7 ............................ 104
   2 Redox reactions in Group 7 ............................ 106
   3 Reactions of halides with sulfuric acid ............................ 109
   4 Other reactions of halides ............................ 112
   Thinking Bigger ............................ 114
   Exam-style questions ............................ 116

## TOPIC 5 Formulae, equations and amounts of substance

**5.1 Empirical and molecular formulae** ............................ 118
   1 Empirical formulae ............................ 120
   2 Molecular formulae ............................ 122

**5.2 Amount of substance** ............................ 125
   1 Calculations using moles and the Avogadro constant ............................ 125
   2 Writing chemical equations ............................ 127
   3 Calculations using reacting masses ............................ 130
   4 Avogadro's law and gas volume calculations ............................ 132
   5 Molar volume calculations ............................ 134

**5.3 Equations and calculations** ............................ 136
   1 Concentrations of solutions ............................ 136
   2 Making standard solutions ............................ 138
   3 Doing titrations ............................ 140
   4 Calculations from titrations ............................ 142

**5.4 Errors and uncertainties** ............................ 144
   1 Mistakes, errors, accuracy and precision ............................ 144
   2 Measurement errors and measurement uncertainties ............................ 146
   3 Percentage measurement uncertainty ............................ 148

**5.5 Yield and atom economy** ............................ 150
   1 The yield of a reaction ............................ 150
   2 Atom economy ............................ 152

| 5.6 Types of reaction | 154 |
|---|---|
| 1 Displacement reactions | 154 |
| 2 Precipitation reactions | 156 |
| 3 Reactions of acids | 158 |
| Thinking Bigger | 160 |
| Exam-style questions | 162 |

## TOPIC 6 Organic chemistry

| 6.1 Introduction to organic chemistry | 164 |
|---|---|
| 1 What is organic chemistry? | 166 |
| 2 Different types of formulae | 168 |
| 3 Functional groups and homologous series | 170 |
| 4 Nomenclature | 172 |
| 5 Isomerism | 175 |

| 6.2 Hydrocarbons | 178 |
|---|---|
| 1 Alkanes from crude oil | 178 |
| 2 Alkanes as fuels | 180 |
| 3 Alternative fuels | 182 |
| 4 Substitution reactions of alkanes | 184 |
| 5 Alkenes and their bonding | 186 |
| 6 Addition reactions of alkenes | 188 |
| 7 The mechanisms of addition reactions | 191 |
| 8 Polymerisation reactions | 194 |
| 9 Dealing with polymer waste | 196 |

| 6.3 Halogenoalkanes | 198 |
|---|---|
| 1 Halogenoalkanes and hydrolysis reactions | 198 |
| 2 Comparing the rates of hydrolysis reactions | 200 |
| 3 Halogenoalkane reactions and mechanisms | 202 |

| 6.4 Alcohols | 204 |
|---|---|
| 1 Alcohols and some of their reactions | 204 |
| 2 Oxidation reactions of alcohols | 206 |
| 3 Purifying an organic liquid | 208 |
| Thinking Bigger | 212 |
| Exam-style questions | 214 |

## TOPIC 7 Modern analytical techniques

| 7.1 Mass spectrometry | 216 |
|---|---|
| 1 Mass spectrometry in organic compounds | 218 |
| 2 Deducing structures from mass spectra | 220 |

| 7.2 Infrared spectroscopy | 222 |
|---|---|
| 1 Infrared spectroscopy | 222 |
| 2 Using infrared spectra | 224 |
| Thinking Bigger | 226 |
| Exam-style questions | 228 |

## TOPIC 8 Chemical energetics

| 8.1 Heat energy and enthalpy | 230 |
|---|---|
| 1 Introducing enthalpy and enthalpy change | 232 |
| 2 Enthalpy level diagrams | 234 |
| 3 Standard enthalpy change of combustion | 235 |
| 4 Standard enthalpy change of neutralisation | 237 |
| 5 Standard enthalpy change of formation and Hess's Law | 239 |

| 8.2 Bond enthalpy | 242 |
|---|---|
| 1 Bond enthalpy and mean bond enthalpy | 242 |
| 2 Using mean bond enthalpies | 244 |
| Thinking Bigger | 246 |
| Exam-style questions | 248 |

## TOPIC 9 Reaction kinetics

| 9.1 Reaction rate | 250 |
|---|---|
| 1 Reaction rate, collision theory and activation energy | 252 |
| 2 Making a reaction go faster – Part 1 | 254 |
| 3 Making a reaction go faster – Part 2 | 256 |
| 4 Making a reaction go faster – Part 3 | 258 |
| Thinking Bigger | 260 |
| Exam-style questions | 262 |

## TOPIC 10 Chemical equilibrium

| 10.1 Reversible reactions and dynamic equilibrium | 264 |
|---|---|
| 1 Reversible reactions and dynamic equilibrium | 266 |
| 2 The effect of changes in conditions on equilibrium composition | 270 |

| 10.2 Equilibrium position | 272 |
|---|---|
| 1 The equilibrium constant | 272 |
| 2 Reversible reactions in industry | 270 |
| Thinking Bigger | 276 |
| Exam-style questions | 278 |

| Maths skills | 280 |
|---|---|
| Exam preparation | 286 |
| Glossary | 292 |
| Periodic Table | 296 |
| Index | 297 |

# How to use this book

Welcome to your Edexcel AS/A level Chemistry course. In this book, which covers the full AS course and the first year of the A level course, you will find a number of features designed to support your learning.

## Topic openers

Each topic starts by setting the context for that topic's learning:

- Links to other areas of Chemistry are shown, including previous knowledge that is built on in the topic, and future learning that you will cover later in your course.
- The **All the maths you need** checklist helps you to know what maths skills will be required.

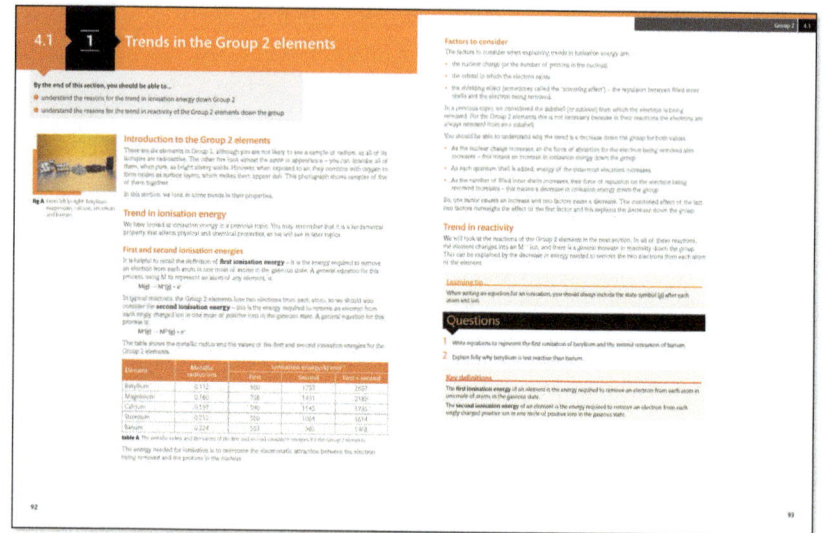

## Main content

The main part of each topic covers all the points from the specification that you need to learn. The text is supported by diagrams and photos that will help you understand the concepts.

Within each section, you will find the following features:

- **Learning objectives** at the beginning of each section, highlighting what you need to know and understand.
- **Key definitions** shown in bold and collated at the end of each section for easy reference.
- **Worked examples** showing you how to work through questions, and how your calculations should be set out.
- **Learning tips** to help you focus your learning and avoid common errors.
- **Did you know?** boxes featuring interesting facts to help you remember the key concepts.
- **Questions** to help you check whether you have understood what you have just read, and whether there is anything that you need to look at again. Answers to the questions can be found on Pearson's website as a free resource.

## Thinking Bigger

At the end of each topic there is an opportunity to read and work with real-life research and writing about science.
The timeline at the bottom of the spreads highlights which other topics the material relates to. These spreads will help you to:

- read real-life material that's relevant to your course
- analyse how scientists write
- think critically and consider the issues
- develop your own writing
- understand how different aspects of your learning piece together.

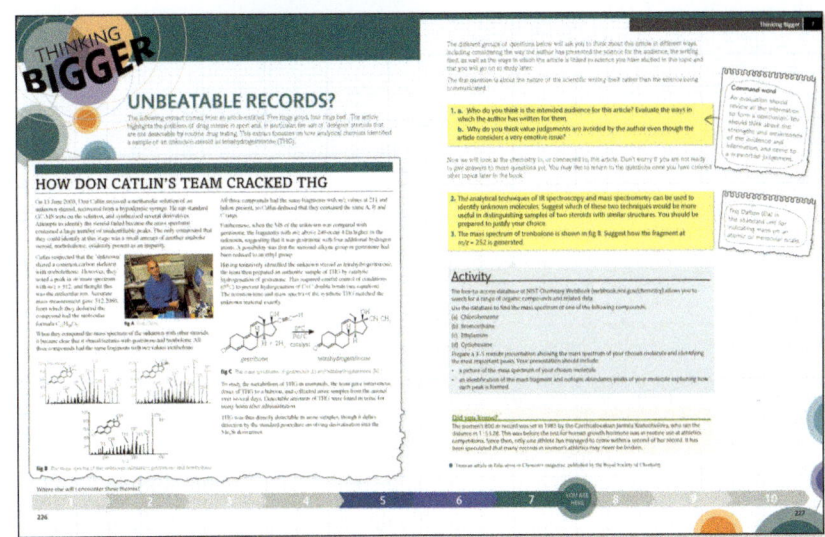

## Exam-style questions

At the end of each topic there are also **exam-style questions** to help you to:

- test how fully you have understood the learning
- practise for your exams.

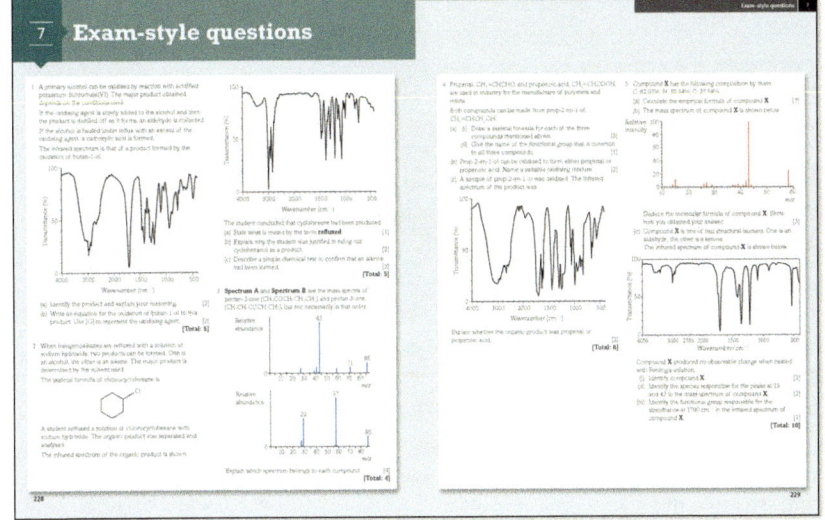

## Getting the most from your online ActiveBook

This book comes with 3 years' access to ActiveBook* – an online, digital version of your textbook. Follow the instructions printed on the inside front cover to start using your ActiveBook.

Your ActiveBook is the perfect way to personalise your learning as you progress through your Edexcel AS/A level Chemistry course. You can:

- access your content online, anytime, anywhere
- use the inbuilt highlighting and annotation tools to personalise the content and make it really relevant to you
- search the content quickly using the index.

### Highlight tool
Use this to pick out key terms or topics so you are ready and prepared for revision.

### Annotations tool
Use this to add your own notes, for example links to your wider reading, such as websites or other files. Or make a note to remind yourself about work that you need to do.

*For new purchases only. If this access code has already been revealed, it may no longer be valid. If you have bought this textbook secondhand, the code may already have been used by the first owner of the book.

# TOPIC 1

# Atomic structure and the Periodic Table

## Introduction

Chemistry is the study of matter and the changes that can be made to matter. Matter is anything that has mass and takes up space. This includes all solids, liquids and gases. The ancient Greeks thought that all matter was made from four elements: Air, Earth, Fire and Water. This idea was gradually abandoned as scientists began to experiment and apply logical scientific thinking to the observations they made.

In 1803, John Dalton put forward his atomic theory, which was based on the premise that the atoms of different elements could be distinguished by differences in their weights. Some parts of his theory are still considered to have a grain of truth, but scientists have gradually refined the theory into one that more precisely explains the observations we are able to make today. Science works by scientists continuing to experiment and make new observations. If those observations cannot be explained satisfactorily using the existing theory, then the theory has to be refined or even possibly replaced.

Atoms, and the particles of which they are made, govern how everything works in the world around us. The continuing study and understanding of atoms has allowed us to push forward with developing new medicines to combat illnesses; the materials and sophisticated technology for use in communications and computing; and newer and better materials. In fact, there is probably not a single area of modern life unaffected by the study of atoms.

## All the maths you need

- Recognise and use expressions in decimal and ordinary form
- Use ratios, fractions and percentages
- Use an appropriate number of significant figures
- Find arithmetic means

## What have I studied before?

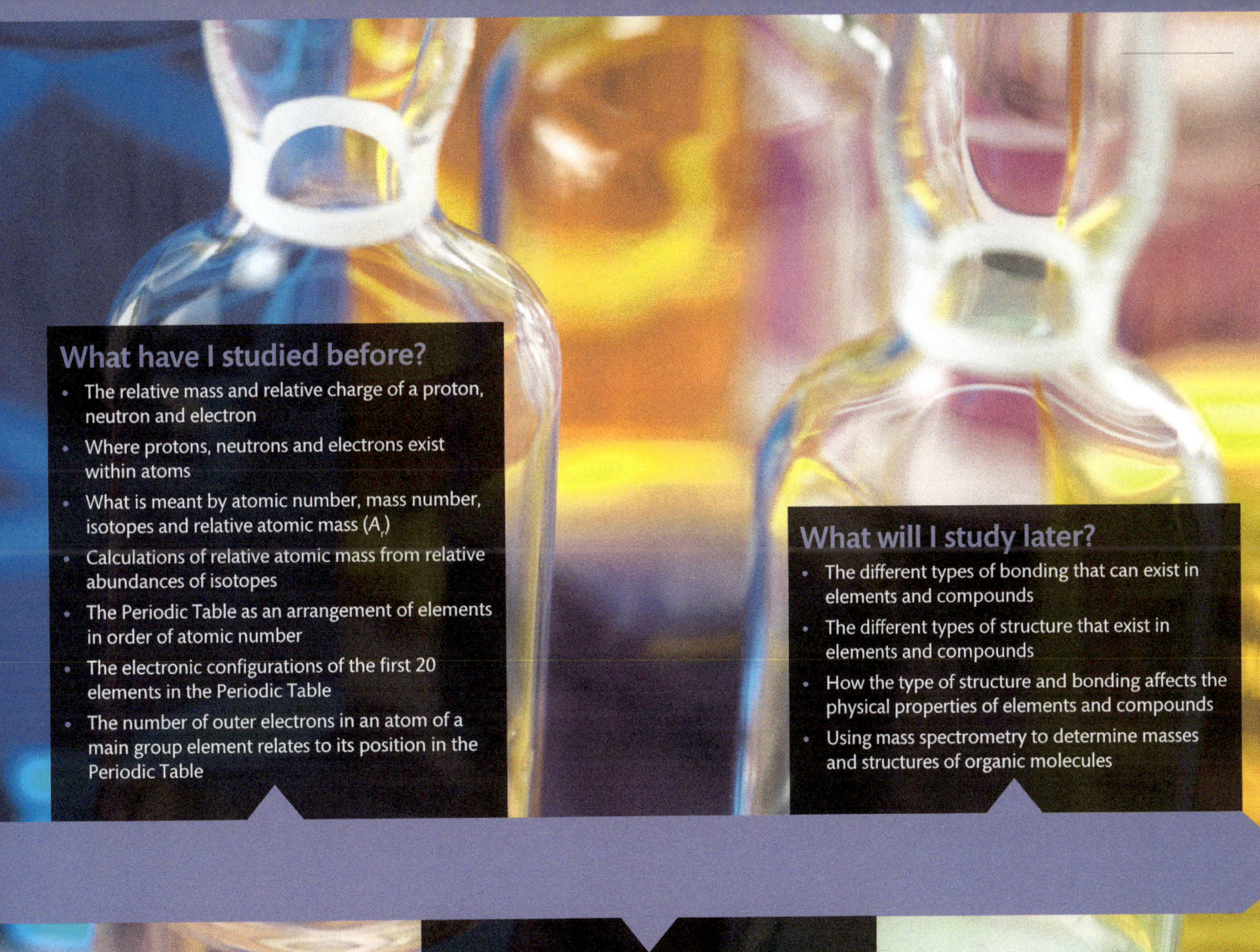

- The relative mass and relative charge of a proton, neutron and electron
- Where protons, neutrons and electrons exist within atoms
- What is meant by atomic number, mass number, isotopes and relative atomic mass ($A_r$)
- Calculations of relative atomic mass from relative abundances of isotopes
- The Periodic Table as an arrangement of elements in order of atomic number
- The electronic configurations of the first 20 elements in the Periodic Table
- The number of outer electrons in an atom of a main group element relates to its position in the Periodic Table

## What will I study later?

- The different types of bonding that can exist in elements and compounds
- The different types of structure that exist in elements and compounds
- How the type of structure and bonding affects the physical properties of elements and compounds
- Using mass spectrometry to determine masses and structures of organic molecules

## What will I study in this topic?

- How our understanding of the structure of the atom has developed over time
- How the masses of atoms and molecules can be determined
- The evidence for the existence of quantum shells, subshells and electron orbitals
- The electronic configurations of the first 36 elements in the Periodic Table
- Periodicity of properties in Periods 2 and 3 of the Periodic Table

# 1.1 1 Structure of the atom and isotopes

**By the end of this section, you should be able to...**

- understand the structure of an atom in terms of electrons, protons and neutrons
- know the relative mass and relative charge of protons, neutrons and electrons
- determine the number of each type of subatomic particle in an atom, molecule or ion from the atomic number and mass number
- understand the term 'isotopes'

## Who discovered electrons, protons and neutrons?

Our current understanding of the structure of atoms is influenced by the theories put forward by scientists such as J.J. Thomson, Ernest Rutherford and James Chadwick.

J.J. Thomson discovered the electron in 1897; Ernest Rutherford discovered the proton in 1917; James Chadwick discovered the neutron in 1932.

## Structure of an atom

Although many other subatomic particles such as quarks, leptons and bosons have since been discovered, chemistry is concerned solely with electrons, protons and neutrons.

The structure of the atom in terms of these three subatomic particles can be summarised as follows.

**fig A** J.J. Thomson, Ernest Rutherford and James Chadwick.

| Particle | Symbol | Relative mass | Relative charge | Position in the atom |
|---|---|---|---|---|
| proton | p | 1 | +1 | nucleus |
| neutron | n | 1 | 0 | nucleus |
| electron | e⁻ | $\frac{1}{1840}$ | −1 | energy levels surrounding the nucleus |

**table A** The structure of the atom in terms of protons, neutrons and electrons.

As well as discovering the proton, Rutherford is credited for first suggesting that the atom has a very small core containing the bulk of the mass of the atom. This core is called the nucleus and contains all of the protons and neutrons that an atom has.

The energy levels surrounding the nucleus in which electrons exist are called quantum shells. We will develop this concept later in this chapter.

### Learning tip

Never state that the mass of an electron is zero.

You do not need to know the exact masses in grams, or the exact charges in coulombs, of the subatomic particles. It is the relative values you need to know.

## Atomic number, mass number and isotopes

The diagram below shows the **atomic number**, **mass number** and **isotopes** of carbon.

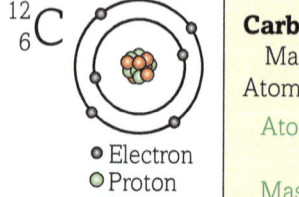

Carbon atoms

Mass number → $^{12}_{6}C$, $^{13}_{6}C$
Atomic number →

Atomic number = number of protons
= (number of electrons)
Mass number = number of protons + number of neutrons

**fig B** Atomic number, mass number and isotopes of carbon.

10

Two isotopes of carbon are $^{12}C$ and $^{13}C$. They have the same atomic number, 6, so they both show the same chemical properties. However, the mass of $^{13}C$ is larger than $^{12}C$. This is because $^{13}C$ has one more neutron.

## Isotopes

In 1919, Francis Aston invented an instrument called a mass spectrometer. He used it to discover that not all the atoms of an element have the same mass. This was eventually attributed to differing numbers of neutrons in the nuclei of the atoms. These different atoms are called isotopes.

We can illustrate this point by considering two isotopes of chlorine: chlorine-35 and chlorine-37. The table shows the number of protons, neutrons and electrons in an atom of each isotope.

| Isotope | Symbol for isotope | Number of protons | Number of neutrons | Number of electrons |
|---|---|---|---|---|
| chlorine-35 | $^{35}_{17}Cl$ | 17 | 18 | 17 |
| chlorine-37 | $^{37}_{17}Cl$ | 17 | 20 | 17 |

**table B** Isotopes of chlorine.

### Did you know?
All atoms of a given element have the same atomic number, which is different from the atomic number of any other element.

The number of electrons in a neutral atom of an element is equal to the number of protons in the nucleus of that atom.

The electrons surround the nucleus in well-defined energy levels called quantum shells.

### Did you know?
Isotopes of the same element have identical chemical properties because they have identical electronic configurations.

## Questions

1. Use the information in the table below to answer the following questions about particles **A** to **F**.
   (a) Which two particles are isotopes of the same element?
   (b) Which two particles are positive ions?
   (c) Which two particles are negative ions?
   (d) Which two particles have the same mass number?

| Particle | Number of protons | Number of neutrons | Number of electrons |
|---|---|---|---|
| A | 12 | 13 | 12 |
| B | 17 | 18 | 18 |
| C | 11 | 14 | 10 |
| D | 12 | 12 | 12 |
| E | 35 | 44 | 36 |
| F | 19 | 21 | 18 |

2. Complete the table to show the numbers of protons, neutrons and electrons in these atoms, molecules and ions.

| Symbol for atom or ion | Number of protons | Number of neutrons | Number of electrons |
|---|---|---|---|
| $^{3}_{1}H$ | | | |
| $^{18}_{8}O^{2-}$ | | | |
| $^{24}_{12}Mg^{2+}$ | | | |
| $^{14}_{7}N^{1}_{1}H_{3}$ | | | |
| $^{14}_{7}N^{2}_{1}H^{+}_{4}$ | | | |

### Key definitions

The **atomic number** ($Z$) of an element is the number of protons in the nucleus of an atom of that element.

The **mass number** of an atom is the sum of the number of protons and the number of neutrons in the nucleus of that atom.

**Isotopes** are atoms of the same element with different masses.

# 1.1 2 Mass spectrometry and relative masses of atoms, isotopes and molecules

By the end of this section, you should be able to...

- define the terms relative atomic mass and relative isotopic mass, based on the $^{12}C$ scale
- analyse and interpret data from mass spectrometry to calculate relative atomic mass from the relative abundance of isotopes, and vice versa
- predict the mass spectra for diatomic molecules, including chlorine
- understand how mass spectrometry can be used to determine the relative molecular mass of a molecule
- understand how to use the terms relative molecular mass and relative formula mass, including calculating these values from relative atomic masses

### Learning tip
Remember that:
- the mass number and relative isotopic mass are not the same
- the mass number is always a whole number since it is the sum of two whole numbers (the number of protons and the number of neutrons in the nucleus of an atom)
- relative isotopic mass is relative to the mass of a carbon-12 atom and is not likely to be a whole number
- relative isotopic masses should be used to calculate relative atomic masses, but sometimes mass numbers are used to make the arithmetic easier.

### Did you know?
The relative atomic mass of an element is the weighted mean mass of an atom of the element compared to $\frac{1}{12}$ the mass of an atom of carbon-12, which has a mass of 12.

The relative isotopic mass is the mass of an atom of an isotope of the element compared to $\frac{1}{12}$ mass of an atom of carbon-12, which has a mass of 12.

## Relative atomic mass and relative isotopic mass

### Relative atomic mass

Chemists working in the nineteenth century established the chemical formulae of some common compounds and calculated the masses of atoms from which they are made. What they had measured were not the actual masses of the atoms but how the masses of atoms compared with one another, in other words **relative atomic masses**.

The sulfur atom was twice as heavy as the oxygen atom, and the oxygen atom was sixteen times heavier than the hydrogen atom. The hydrogen atom was the lightest atom so its mass was given a value of 1. The masses of the atoms of the other elements were stated relative to this value. So the mass of an atom of oxygen on this scale was 16, and sulfur was 32.

The discovery of isotopes complicated the situation. It was eventually decided, in 1961, to adopt an isotope of carbon, carbon-12, as the standard.

### Relative isotopic mass

**Relative isotopic masses** and their percentage abundance in a sample of an element are used to calculate relative atomic masses.

**WORKED EXAMPLE**

Lithium, in its naturally occurring compounds, has two isotopes of relative isotopic masses 6.015 and 7.016. The percentage abundance of each isotope is 7.59 and 92.41 respectively.
Here is how you can calculate the relative atomic mass of lithium. There are two stages.
Stage 1: $(6.015 \times 7.59) + (7.016 \times 92.41) = 694.00241$
Stage 2: $\frac{694.00241}{100} = 6.9400241$
The relative atomic mass of lithium, to three significant figures, is 6.94.

## Using data obtained from a mass spectrometer

### What is a mass spectrometer?

A mass spectrometer measures the masses of atoms and molecules.

It produces positive ions that are deflected by a magnetic field according to their mass-to-charge ratio ($m/z$). It also calculates the relative abundance of each positive ion and displays this as a percentage.

The positive ions could be positively charged atoms, positively charged molecules or positively charged fragments of molecules.

Dealing with fragments of molecules is particularly useful when you are analysing organic compounds. We will discuss this further in **Topic 6**, Organic chemistry, of this book.

### Determining relative isotopic and atomic masses

You can determine the *exact* values of the relative masses of isotopes from a mass spectrum of an element, together with the percentage abundance of each isotope.

You can use this information to calculate the relative atomic mass of the element.

# Determining relative molecular mass ($M_r$) of diatomic molecules

Some elements and compounds contain two or more atoms covalently bonded together. If these substances are analysed by mass spectrometry, you can obtain the relative molecular mass of the element or compound by observing the peaks with the largest $m/z$ ratios (assuming a value of $z = 1$).

The diagram shows the mass spectrum of chlorine, which exists as diatomic molecules: $Cl_2$.

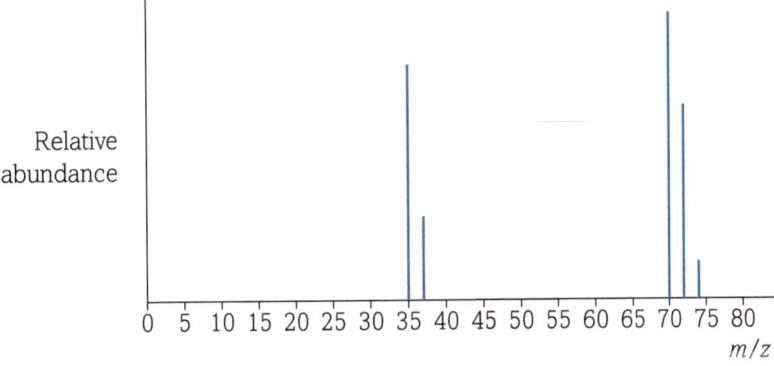

**fig A** The mass spectrum of chlorine.

To make calculations easier, in this example the relative molecular and isotopic masses are quoted whole numbers.

There are two peaks corresponding to isotopic masses of 35 and 37. The approximate relative peak heights for each isotope are 3 : 1 for chlorine-35 to chlorine-37.

This gives an approximate relative atomic mass of this sample of chlorine of $((3 \times 35) + (1 \times 37)) \div 4 = 35.5$.

You will notice in the diagram that there are three peaks corresponding to molecular masses of 70, 72 and 74. How can we explain the relative heights of these peaks?

The table shows the atomic composition of each molecule of chlorine.

| Mass of molecule | Formula of molecule |
|---|---|
| 70 | $^{35}Cl^{35}Cl$ |
| 72 | $^{35}Cl^{37}Cl$ |
| 74 | $^{37}Cl^{37}Cl$ |

**table A** Atomic composition of each molecule of chlorine.

The ratio of $^{35}Cl$ to $^{37}Cl$ in this sample is 3 : 1. The molecular composition is shown in the table below.

| Formula of molecule | Ratio of molecules |
|---|---|
| $^{35}Cl^{35}Cl$ | 9 |
| $^{35}Cl^{37}Cl$ | 6 |
| $^{37}Cl^{37}Cl$ | 1 |

**table B** Molecular composition of sample.

Let's make sense of this.

- There is a 3 in 4 chance of selecting a $^{35}Cl$ atom from a sample of chlorine atoms. So the total chance of two $^{35}Cl$ atoms combining together is $\frac{3}{4} \times \frac{3}{4} = \frac{9}{16}$.
- There is a 1 in 4 chance of selecting a $^{37}Cl$ atom. So the chance of $^{35}Cl$ combining with a $^{37}Cl$ is $\frac{1}{4} \times \frac{3}{4} = \frac{3}{16}$. The chance of a $^{37}Cl$ atom combining with a $^{35}Cl$ atom is also $\frac{3}{16}$. So the total chance of $^{35}Cl$ and $^{37}Cl$ combining together in any order is $2 \times \frac{3}{16}$ or $\frac{6}{16}$.
- The chance of two $^{37}Cl$ atoms combining together is $\frac{1}{4} \times \frac{1}{4} = \frac{1}{16}$.

$\frac{9}{16} : \frac{6}{16} : \frac{1}{16}$ in whole number ratios is 9 : 6 : 1

This corresponds approximately to the peak heights of 70, 72 and 74.

### Learning tip

The peak heights will not be exactly 9 : 6 : 1 because the relative isotopic masses are not whole numbers (as mentioned above).

You can then determine the relative molecular mass ($M_r$) of chlorine by calculating the weighted mean of the various molecules present:

$$M_r(Cl_2) = \frac{(9 \times 70) + (6 \times 72) + (1 \times 74)}{16} = 71$$

You will usually be asked to calculate relative molecular masses by simply adding together the relative atomic masses of the elements in the molecule.

If we use this method, the relative molecular mass of chlorine is $2 \times 35.5 = 71$.

### Learning tip

Make sure you always use the relative atomic masses you are given. For example, the value for magnesium given in the Periodic Table at the end of this book is 24.3. However, you may occasionally be asked to use a more accurate value, such as 24.305.

## Determining relative molecular mass of a polyatomic molecule

It is quite tricky to use the data supplied by a mass spectrum of a compound to determine its exact relative molecular mass. You need to know the exact relative isotopic masses of all the atoms present, and the relative composition of all the different molecules.

You are more likely to be asked to work out the relative molecular mass of a compound by considering what is called the **molecular ion peak**.

You have to be careful when analysing organic compounds. This is because there is always a small percentage of the carbon-13 isotope present in the compound, which can lead to what is often referred to as an $M + 1$ peak. This peak can often be seen in molecules with large masses, where the percentage of carbon-13 becomes significant. The peak is often missing, or insignificant, in molecules of small mass.

The following examples illustrate this point.

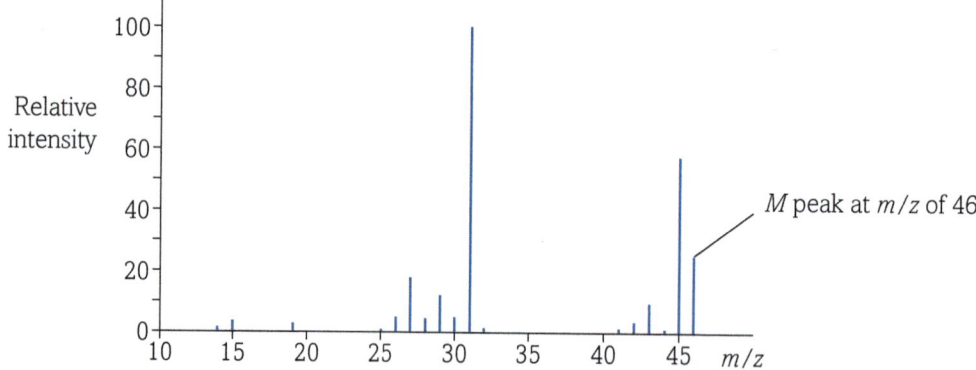

**fig B** Mass spectrum of ethanol ($M_r = 46$).

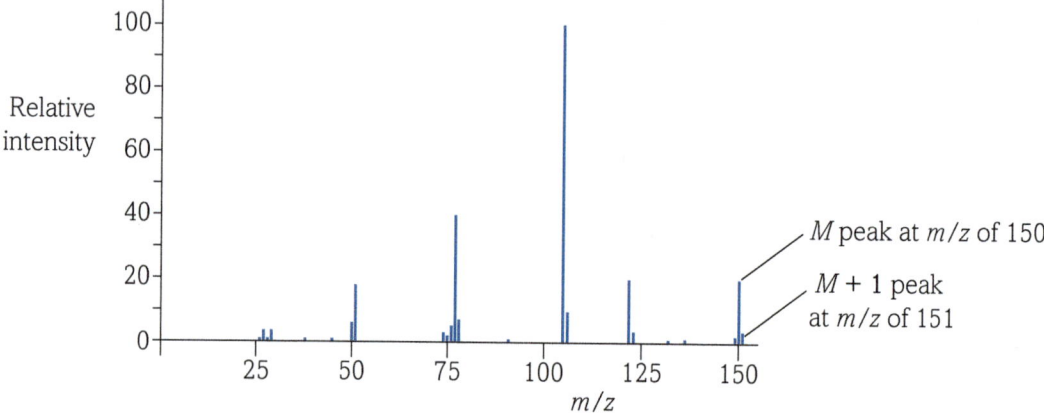

**fig C** Mass spectrum of ethyl benzoate ($M_r = 150$).

# Questions

1. Describe the difference between relative atomic mass and relative isotopic mass.

2. Explain why the relative atomic masses of many elements are not exact whole numbers.

3. Calculate the relative atomic mass of a sample of magnesium that has the following isotopic composition:
   - magnesium-24: 78.6%
   - magnesium-25: 10.1%
   - magnesium-26: 11.3%
   
   Give your answer to three significant figures.

4. A sample of copper contains two isotopes of relative isotopic mass 63.0 and 65.0. If the relative atomic mass of copper is 63.5, calculate the relative abundance of each isotope.

5. The mass spectrum of bromine vapour, $Br_2$, is shown below:

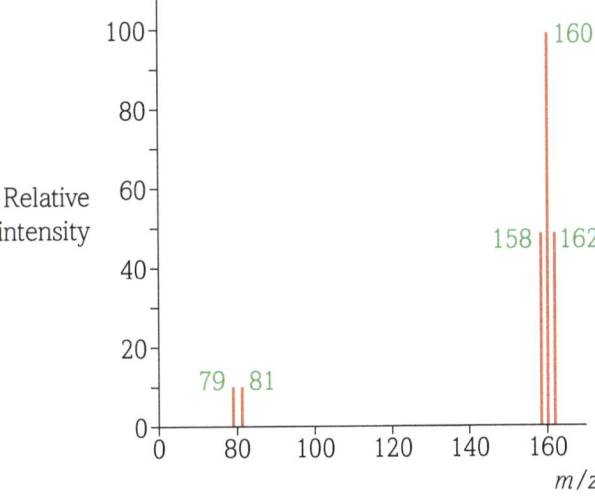

**fig D** Mass spectrum of bromine vapour.

   (a) What are the relative isotopic masses of the two isotopes present in bromine?
   (b) Identify the particles responsible for the peaks at $m/z$ 158, 160 and 162.
   (c) Deduce the relative abundance of the two isotopes and explain the relative heights of the three peaks at $m/z$ 158, 160 and 162.

## Key definitions

The weighted mean (average) mass of an atom of an element compared to $\frac{1}{12}$ of the mass of an atom of carbon-12 is called the **relative atomic mass** ($A_r$) of the element.

The mass of an individual atom of a particular isotope relative to $\frac{1}{12}$ of the mass of an atom of carbon-12 is called the **relative isotopic mass**.

The **molecular ion peak** is the peak with the highest $m/z$ ratio in the mass spectrum, the M peak.

# 1.1 3 Atomic orbitals and electronic configurations

By the end of this section, you should be able to...

- know the number of electrons that can fill the first four quantum shells
- know that an orbital is a region within an atom that can hold up to two electrons with opposite spins
- know the shape of an s orbital and a p orbital
- state the number of electrons that occupy s, p and d subshells
- predict the electronic configurations, using 1s notation and 'electrons-in-boxes' notation, of atoms, given the atomic number, Z, for up to Z = 36

## Quantum shells

Max Planck first put forward the quantum theory, which is used to describe the arrangement of electrons around the nuclei of atoms.

According to this theory, electrons can only exist in certain well-defined energy levels called **quantum shells**.

All electrons in a given quantum shell have similar, but not identical, energies.

## Electrons in the first four quantum shells

Electrons in the first quantum shell have the lowest energy for that element. The first quantum shell is located in the region closest to the nucleus. The second quantum shell is in a region outside the first; the third is outside the second, and so on.

Each quantum shell, apart from the first, is further divided into subshells of slightly different energy levels.

- There is only one subshell in the first quantum shell. This is labelled the 1s subshell.
- The second quantum shell has two subshells, labelled 2s and 2p. Electrons in the 2p subshell have a slightly higher energy level than those in the 2s subshell.
- The third quantum shell is divided into three subshells, labelled 3s, 3p and 3d. Electrons in the 3p subshell have slightly higher energy than those in 3s, and those in the 3d subshell have slightly higher energy than those in 3p.
- The fourth quantum shell is divided into four subshells, labelled 4s, 4p, 4d and 4f. Once again, the electron energies increase in the order 4s < 4p < 4d < 4f.

## s orbitals

Each subshell contains **orbitals**. If you could see a single electron in a hydrogen atom and map its position at regular intervals, then it would be possible to build up a three-dimensional map of places where the electron is likely to be found. Most of the time, the electron is located within a fairly easily defined region of space close to the nucleus.

**Fig A** shows a cross-section through a spherical s orbital.

Since the electron of the hydrogen atom is in an s subshell, the orbital is described as a *1s orbital*.

Electrons in the 2s subshell also exist in an s orbital called the 2s orbital. The 2s orbital has the same shape as the 1s orbital but is larger. **Fig B** shows the shape of typical 1s and 2s orbitals. Note that $x$, $y$ and $z$ are 3D Cartesian axes (i.e. axes at mutual right angles).

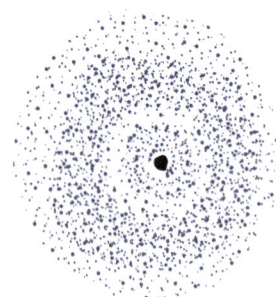

**fig A** Cross-section through an s orbital.

The sizes and shapes of the orbitals are such that there is a 90% probability of finding the electron within their boundaries.

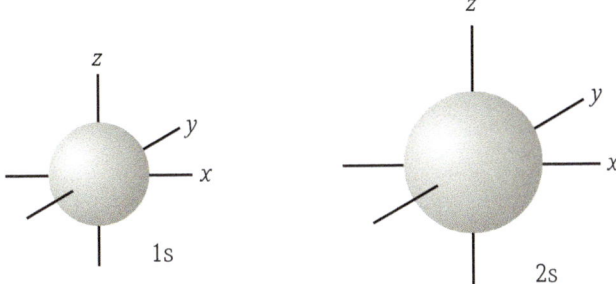

**fig B** The shape of typical 1s and 2s orbitals.

## p orbitals

The 2p subshell contains three separate p orbitals. These orbitals are more complex and harder to visualise. They have an elongated dumbbell shape and a variable charge density. The only difference between the three orbitals is their orientation in space, as shown in **fig C**.

Again, the size and shape of the orbitals means that there is a 90% probability of finding the electron within their boundaries. The three orbitals are arranged at mutual right angles.

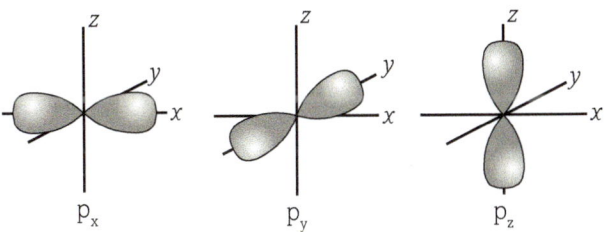

**fig C** Orientation of the three p orbitals.

## d orbitals

The d orbitals are very complicated and you do not need to know their shapes for AS or A level studies. There are five d orbitals in the d subshell.

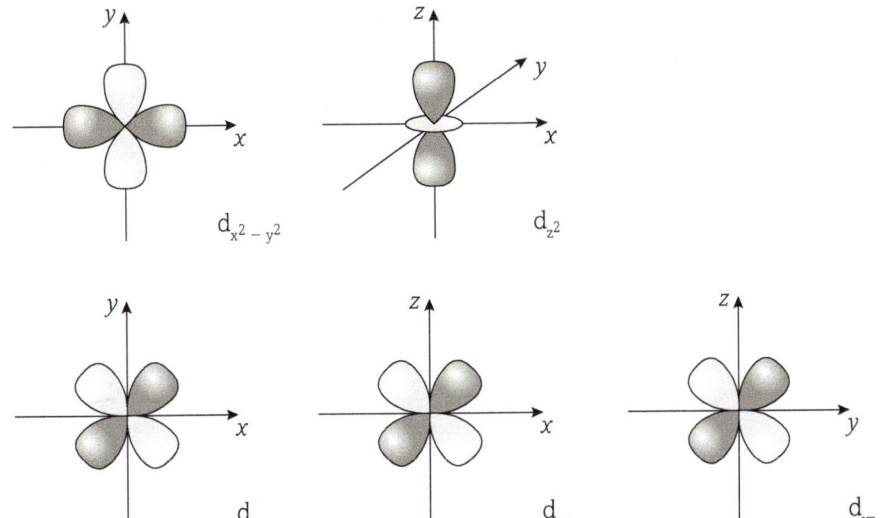

**fig D** The five d orbitals.

There can be up to two electrons in each orbital. With this in mind, the number of electrons that can exist in each subshell and each quantum shell is shown in the table below.

| | Number of electrons | | Number of electrons |
|---|---|---|---|
| s subshell | 2 (1 × 2) | first quantum shell | 2 |
| p subshell | 6 (3 × 2) | second quantum shell | 8 |
| d subshell | 10 (5 × 2) | third quantum shell | 18 |
| f subshell | 14 (7 × 2) | fourth quantum shell | 32 |

**table A** The number of electrons in each subshell and each quantum shell.

## Electronic configurations

The **electronic configuration** of an atom of an element is the distribution of electrons among atomic orbitals.

Here are the electronic configurations for the first 36 elements in the Periodic Table.

| Atomic number | Symbol | 1s | 2s | 2p | 3s | 3p | 3d | 4s | 4p |
|---|---|---|---|---|---|---|---|---|---|
| 1 | H | 1 | | | | | | | |
| 2 | He | 2 | | | | | | | |
| 3 | Li | 2 | 1 | | | | | | |
| 4 | Be | 2 | 2 | | | | | | |
| 5 | B | 2 | 2 | 1 | | | | | |
| 6 | C | 2 | 2 | 2 | | | | | |
| 7 | N | 2 | 2 | 3 | | | | | |
| 8 | O | 2 | 2 | 4 | | | | | |
| 9 | F | 2 | 2 | 5 | | | | | |
| 10 | Ne | 2 | 2 | 6 | | | | | |
| 11 | Na | 2 | 2 | 6 | 1 | | | | |
| 12 | Mg | 2 | 2 | 6 | 2 | | | | |
| 13 | Al | 2 | 2 | 6 | 2 | 1 | | | |
| 14 | Si | 2 | 2 | 6 | 2 | 2 | | | |
| 15 | P | 2 | 2 | 6 | 2 | 3 | | | |
| 16 | S | 2 | 2 | 6 | 2 | 4 | | | |
| 17 | Cl | 2 | 2 | 6 | 2 | 5 | | | |
| 18 | Ar | 2 | 2 | 6 | 2 | 6 | | | |
| 19 | K | 2 | 2 | 6 | 2 | 6 | | 1 | |
| 20 | Ca | 2 | 2 | 6 | 2 | 6 | | 2 | |
| 21 | Sc | 2 | 2 | 6 | 2 | 6 | 1 | 2 | |
| 22 | Ti | 2 | 2 | 6 | 2 | 6 | 2 | 2 | |
| 23 | V | 2 | 2 | 6 | 2 | 6 | 3 | 2 | |
| 24 | Cr | 2 | 2 | 6 | 2 | 6 | 5 | 1 | |
| 25 | Mn | 2 | 2 | 6 | 2 | 6 | 5 | 2 | |
| 26 | Fe | 2 | 2 | 6 | 2 | 6 | 6 | 2 | |
| 27 | Co | 2 | 2 | 6 | 2 | 6 | 7 | 2 | |
| 28 | Ni | 2 | 2 | 6 | 2 | 6 | 8 | 2 | |
| 29 | Cu | 2 | 2 | 6 | 2 | 6 | 10 | 1 | |
| 30 | Zn | 2 | 2 | 6 | 2 | 6 | 10 | 2 | |
| 31 | Ga | 2 | 2 | 6 | 2 | 6 | 10 | 2 | 1 |
| 32 | Ge | 2 | 2 | 6 | 2 | 6 | 10 | 2 | 2 |
| 33 | As | 2 | 2 | 6 | 2 | 6 | 10 | 2 | 3 |
| 34 | Se | 2 | 2 | 6 | 2 | 6 | 10 | 2 | 4 |
| 35 | Br | 2 | 2 | 6 | 2 | 6 | 10 | 2 | 5 |
| 36 | Kr | 2 | 2 | 6 | 2 | 6 | 10 | 2 | 6 |

**table B**

As you can see, in general the lowest energy orbitals are occupied and filled first.

For example, the electronic configuration of helium (He) is $1s^2$ not $1s^1 2s^1$. This is because the 1s orbital can accommodate two electrons, and electrons in the 1s orbital have a lower energy than those in the 2s orbital.

## Predicting electronic configurations

However, as you can see from the table, there are some exceptions to this general rule.

For example, the electronic configuration of potassium (K) is $1s^2 2s^2 2p^6 3s^2 3p^6 4s^1$ and not, as you may expect, $1s^2 2s^2 2p^6 3s^2 3p^6 3d^1$.

This is because the exact energies of the electrons in the orbitals are determined by the number of protons in the nucleus of the atom and the repulsion between all of the electrons present in the atom. For both the potassium atom and the calcium atom, the energy of the 4s orbital is lower than that of 3d orbitals.

So the electronic configuration of calcium (Ca) is $1s^2 2s^2 2p^6 3s^2 3p^6 4s^2$ and *not* $1s^2 2s^2 2p^6 3s^2 3p^6 3d^2$.

### Did you know?

Interestingly, for all the elements after calcium in the Periodic Table, the energy of 3d orbitals is *less* than that of 4s orbital.

So you would expect the electronic configuration of scandium (Sc) to be $1s^2 2s^2 2p^6 3s^2 3p^6 3d^3$.

However, it is $1s^2 2s^2 2p^6 3s^2 3p^6 3d^1 4s^2$.

As Professor Eric Scerri of the University of California has pointed out, the 3d orbitals accept electrons *before* the 4s orbitals: so why are two electrons pushed into the higher-energy 4s orbital? The reason is that 3d orbitals are more compact than 4s ones and, as a result, any electrons entering the 3d orbital will experience greater mutual repulsion than they would in the 4s orbital. So when scandium ionises, it is the 4s electrons, i.e. those with the greater energy, that are lost first. This is what happens to the elements from scandium to zinc. We will discuss this in more detail in **Book 2**.

However, other atoms do not always act in this way. The nuclear charge increases as we move from scandium to zinc. This complicates the interactions between the electrons and the nucleus, and between the electrons themselves. This is what ultimately produces an electronic configuration. Unfortunately, there is no one simple rule that can cope with this complicated situation.

For example, you might think, incorrectly, that the most stable electronic configuration for chromium (Cr) is $3d^4 4s^2$ and for copper (Cu) is $3d^9 4s^2$. This is not the case, as shown in the table of electronic configurations on the previous page.

### Learning tip

You may have come across a principle called the 'Aufbau principle'. This principle is often quoted as a method for working out the electronic configuration of the atoms. According to the Aufbau principle, as protons are added to the nucleus, the electrons are successively added to orbitals of increasing energy, beginning with the lowest energy orbitals, and continue doing so until all electrons are accommodated. It can be a useful rule of thumb for working out the configurations of the first 36 elements, with the exceptions of chromium and copper. However, as you study more elements you will come across more exceptions. By all means use the Aufbau principle to work out electronic configurations, but do not assume that it offers an explanation for them.

## Spin-spin pairing and electronic configurations

**Hund's rule** and the **Pauli Exclusion Principle** govern the way in which electrons fill the p and the d subshells.

If we apply Hund's rule to boron (B), carbon (C) and nitrogen (N), we obtain the following electronic configurations:

B  $1s^2 2s^2 2p_x^1$

C  $1s^2 2s^2 2p_x^1 2p_y^1$

N  $1s^2 2s^2 2p_x^1 2p_y^1 2p_z^1$

If we apply the Pauli Exclusion Principle to oxygen (O), fluorine (F) and neon (Ne), the electronic configurations are displayed as boxes.

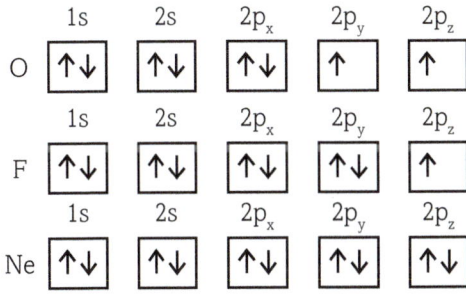

**fig E** Electronic configurations for oxygen, fluorine and neon, displayed as boxes.

# Questions

1  The way that electrons of an atom are located in atomic orbitals is subject to Hund's rule and the Pauli Exclusion Principle.

  (a) State what you understand by Hund's rule and the Pauli Exclusion Principle.

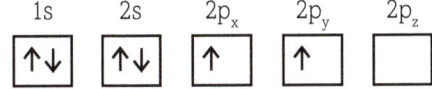

  **fig F** Box notation diagram for Question 1 (b) and (c).

  (b) (i) State the significance of the arrows and of the letters *x*, *y* and *z*.
  
  (ii) Suggest why the electron in the $2p_y$ orbital is not located in the $2p_x$ orbital.
  
  (iii) Describe the shapes of s and p orbitals.

  (c) Using the box notation shown above, give the electronic configuration of:
  
  (i) sulfur
  
  (ii) phosphorus
  
  (iii) chromium.

### Key definitions

A **quantum shell** defines the energy level of an electron.

An **orbital** is a region within an atom that can hold up to two electrons with opposite spins.

The **electronic configuration** of an atom shows the number of electrons in each sublevel in each energy level of the atom.

**Hund's rule** states that electrons will occupy the orbitals singly before pairing takes place.

The **Pauli Exclusion Principle** states that two electrons cannot occupy the same orbital unless they have opposite spins. Electron spin is usually shown by the use of upward and downward arrows: ↑ and ↓.

# 1.1 4 Ionisation energies

By the end of this section, you should be able to...

- know what is meant by first ionisation energy and successive ionisation energies
- understand how ideas about electronic configuration developed
- know how ionisation energies are influenced by the number of protons, the electron shielding and the electron subshell/orbital from which the electron is removed
- explain the general increase in first ionisation energy across a period
- explain the decrease in first ionisation energy down a group

When you studied chemistry at GCSE level, you probably learned that electrons exist in shells surrounding the nucleus. You may not have been told how we know this.

There is lots of experimental evidence to support this theory, including atomic emission spectra and data from considering the ionisation energies of elements.

## Energy levels and emission spectra

Evidence for the different energy levels in which electrons can exist within atoms came initially from atomic emission spectra. When atoms in the gaseous state are given energy by heating them or by passing an electric current through them, the electrons move to higher energy levels. Eventually they return back to the lower energy levels and emit electromagnetic radiation as they do so. We can analyse the electromagnetic radiation emitted using a device called a spectroscope. This shows us that atoms do not emit radiation across the whole of the electromagnetic spectrum. Only specific frequencies are emitted, and these are unique to an individual element. All atoms of a particular element radiate the same set of frequencies. The emission spectrum produced is called a line spectrum. The line emission spectra of hydrogen, helium and mercury are shown in **fig A**.

The fact that only certain frequencies of electromagnetic radiation are emitted, rather than a continuous spectrum, is compelling evidence that the energy of electrons in atoms can only have certain, fixed values and not a continuous range of values. This is the basis of the quantum theory in which the energy of electrons is said to be quantised (rather than continuous).

## First ionisation energy

Ionisation energy is a measure of the energy required to completely remove an electron from an atom of an element.

We can represent the **first ionisation energy** of an element, A, by the equation:

$$A(g) \rightarrow A^+(g) + e^-$$

We can represent the **second ionisation energy** of A by the equation:

$$A^+(g) \rightarrow A^{2+}(g) + e^-$$

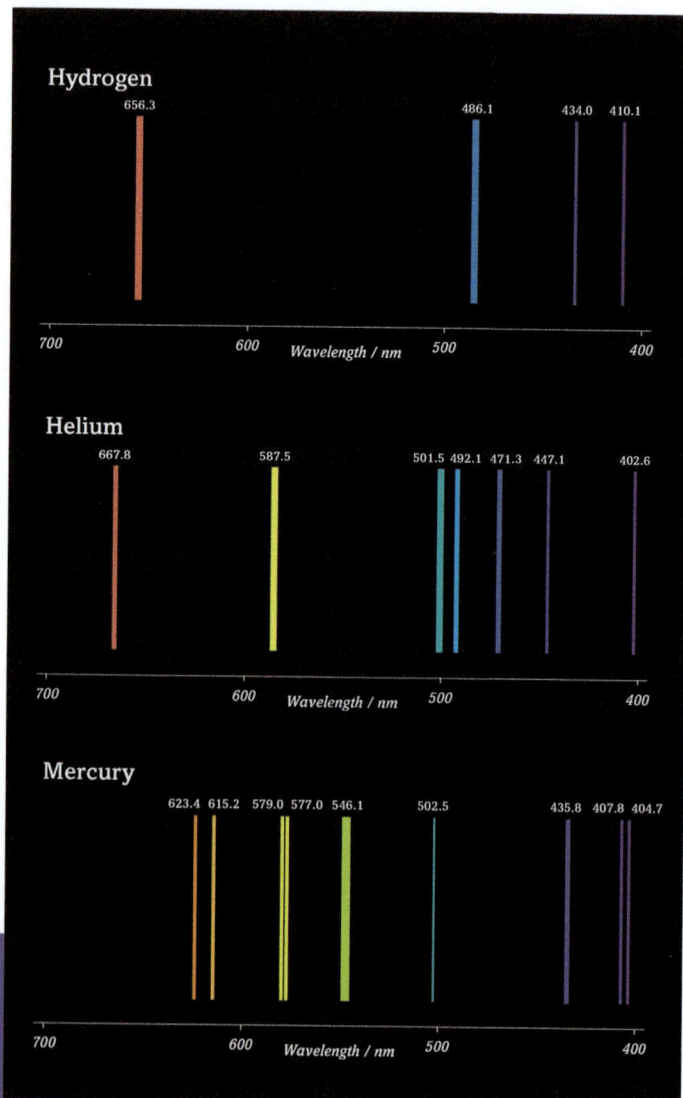

**fig A** The line emission spectra of hydrogen, helium and mercury.

## Successive ionisation energies

When successive energies of an element are listed there are steady increases, and big jumps occur at defined places. This is one piece of evidence for the existence of quantum shells.

A sodium (Na) atom has 11 electrons. Its successive ionisation energies are shown in the table below. The places where the ionisation energy has jumped significantly are shown in red.

| 1st | 2nd | 3rd | 4th | 5th | 6th |
|---|---|---|---|---|---|
| 496 | 4563 | 6913 | 9544 | 13 352 | 16 611 |

| 7th | 8th | 9th | 10th | 11th |
|---|---|---|---|---|
| 20 115 | 25 491 | 28 934 | 141 367 | 159 079 |

**table A** Successive ionisation energies, in kJ mol$^{-1}$, of sodium.

The first electron is considerably easier to remove than the second. There is a steady rise in ionisation energy for the next eight electrons, but there is a big jump in energy from the ninth electron to the tenth. The last two electrons are much harder to remove than the previous eight.

The explanation for this trend is that the last two electrons to be removed are in the first quantum shell, the one that has the lowest energy. The next eight electrons are in the second quantum shell, which is of lower energy than the third. The first electron to be removed is in the third quantum shell of highest energy. So the basic electronic configuration of a sodium atom is 2,8,1.

The trend in successive ionisation energies can be seen when the logarithm of the ionisation energy is plotted against the order of electron removal. The graph shown below is for a sodium atom.

It is easy to see from the graph that there are three distinct quantum shells containing electrons. The first quantum shell contains two electrons, the second quantum shell contains eight electrons and the third quantum shell contains one electron.

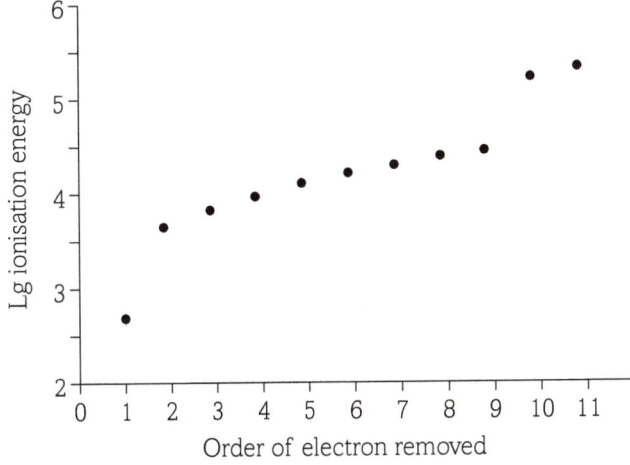

**fig B** Graph showing the trend in successive ionisation energies for sodium.

## Why do successive ionisation energies increase in magnitude?

In order to answer this question, we need to appreciate what ionisation energy represents.

The electron lost during ionisation is so far removed from the influence of the nucleus that it no longer experiences an attractive force from the nucleus. It is said to be at an infinite distance from the nucleus. The energy of the electron has to be increased to a particular value for it to be removed. For a given atom, the energy value that the electron has when it reaches this position will always be the same, regardless of where in the atom the electron has come from.

If an electron already has a high energy, then the energy it needs to gain in order to be removed will not be very large. If, however, the electron is in an orbital of a low-energy quantum shell, for example the 1s orbital, then it will need to gain considerably more energy to be removed. The difference in energy between the electron when it has been removed and the energy it has when it is in its original orbital in the quantum shell is known as the ionisation energy. This can be represented by the equation:

Ionisation energy (IE) = energy of electron when removed
− energy of electron when in the orbital

The ionisation energy for a particular electron in a given atom depends solely on the energy it has when it is in its orbital within the atom.

### WORKED EXAMPLE

Look back at the table of successive ionisation energies of sodium. The first electron to be removed has the highest energy of any of the electrons in the sodium atom. It is in the third quantum shell in a 3s orbital. So the amount of energy it has to gain in order to be removed is the lowest for any of the electrons. The first ionisation energy is therefore the lowest of all the ionisation energies.

There is a large jump from the first to the second ionisation energy. This is because the second electron to be removed is in a quantum shell of considerably lower energy, the second quantum shell. There is a steady rise in ionisation energy from the second to the ninth electron. This indicates that the eight electrons all exist within the same quantum shell. As each successive electron is removed from this shell, the electron–electron repulsion within the shell decreases. This results in a decrease in the energies of the remaining electrons, and therefore a steady increase in ionisation energy from the second to the ninth electron.

The large jump from the ninth to the tenth electron indicates another significant change in energy of the electron. This corresponds to an electron being removed from the first quantum shell: the 1s orbital. The tenth ionisation energy is larger than the ninth, since as one electron is removed from the 1s orbital, the remaining electron now experiences zero repulsion, so its energy decreases.

### Learning tip

Remember that electron–electron repulsion:
- exists between two electrons that are in the same orbital
- exists between electrons in different orbitals within a given quantum shell
- is sometimes most significant between electrons in adjacent quantum shells
- is sometimes called shielding or screening.

## Did you know?

The terms 'shielding' and 'screening' are in some ways an unfortunate choice for the effect that electron–electron repulsion has on the energies of the electrons in their respective orbitals. The terms seem to imply that, for example, the electrons in an inner quantum shell set up a barrier to the attractive force of the nucleus to those electrons in an outer quantum shell. This is not necessarily the case. The effect of electron–electron repulsion is to raise the energy of the electrons involved above the value they would have if there was no repulsion between them. This then has an effect on the amount of energy required to remove the electron from the atom or ion. That is, it has an effect on the magnitude of the ionisation energy.

## What determines the energy of an electron?

You might think the answer to this question is very straightforward. Surely it is solely determined by the orbital in which the electron has been placed? As you will often find in chemistry, there are a number of different factors that affect the answer!

## Hydrogen and helium

First, consider the atoms of hydrogen (H) and helium (He).

The electronic configuration of each is:

    H    $1s^1$
    He   $1s^2$

The outer electrons of both atoms are in a 1s orbital, so you might think that they will have the same energy. However, the 1s orbital of helium contains two electrons. This increases the electron–electron repulsion within the orbital: each electron is said to shield the other from the effect of the nuclear charge. The effect of this factor alone is to increase the energy of the electrons.

However, the nuclear charge of helium is double that of hydrogen, since it contains two protons as opposed to hydrogen's one. The effect of this increased nuclear charge is to decrease the energy of the 1s electrons since they are attracted more strongly.

In this case, the effect of the increased nuclear charge is greater than that of the increased shielding. So the first ionisation energy of helium ($2370\,kJ\,mol^{-1}$) is larger than that of hydrogen ($1310\,kJ\,mol^{-1}$).

## Helium and lithium

Now consider helium (He) and lithium (Li).

The electronic configuration of each is:

    He   $1s^2$
    Li    $1s^2 2s^1$

The nuclear charge of lithium is +3 (three protons), which is larger than that of helium (+2).

However, the outer electron is in the second quantum shell in a 2s orbital. The second quantum shell is at a higher energy level than the first quantum shell. Added to this, the electron in the 2s orbital experiences repulsion from the two inner 1s electrons, i.e. it experiences shielding from the nuclear charge.

The latter two effects are more significant than the increased nuclear charge. So the first ionisation energy of lithium ($519\,kJ\,mol^{-1}$) is smaller than that of helium.

## Summary

The factors that affect the energy an electron has are:
- the orbital in which the electron exists
- the nuclear charge of the atom (i.e. the number of protons in the nucleus)
- the repulsion (shielding) experienced by the electron from all the other electrons present.

## Trends in ionisation energies

The two major trends in ionisation energies in the Periodic Table are:
- across a period
- down a group.

### Across a period

To illustrate this trend consider Period 2; and the elements lithium (Li) to neon (Ne).

As we move from Li to Ne across Period 2, the nuclear charge increases as the number of protons increase. This, on its own, would lead to an increased attraction between the nucleus and the electron, and therefore a decrease in the energy of the outermost electron, and an increase in first ionisation energy.

To counteract this, one more electron is added to the same quantum shell on each occasion and this increases the electron–electron repulsion within the quantum shell. This, on its own, would cause an increase in energy of the outermost electron, and would lead to a decrease in first ionisation energy.

The increase in nuclear charge is more significant than the increase in electron–electron repulsion. So there is a general increase in first ionisation energy across Period 2.

The same can be said of the Period 3 elements, sodium (Na) to argon (Ar), excluding the d-block elements, scandium (Sc) to zinc (Zn).

## Learning tip

This trend is not perfect and there are two exceptions in Periods 2 and 3: in Period 2, beryllium and boron (Be and B); and in Period 3, magnesium and aluminium (Mg and Al).

### Down a group

To illustrate this trend consider the Group 1 elements, lithium (Li) to caesium (Cs).

As we descend through Group 1 from Li to Cs, the nuclear charge increases as the number of protons increases. This would lead to an increased attraction between the nucleus and the electron, and therefore a decrease in energy of the outer electron. This, in turn, would lead to an increase in the first ionisation energy.

However, one new quantum shell is added on each occasion. This increases the energy of the outermost electron for two reasons. Firstly, the third quantum shell has a higher energy value than the second; the fourth quantum shell is higher in energy than the third.

This continues down the group. Secondly, as each new quantum shell is added, the outer electron experiences increased repulsion (i.e. increased shielding) from the inner electrons.

On this occasion, the combined effect of adding an extra shell and increasing the shielding is more significant than the increase in nuclear charge. So the first ionisation energy decreases down Group 1 from Li to Cs.

This trend is repeated in:
- Group 2: beryllium to barium (Be to Ba)
- Group 5: nitrogen to bismuth (N to Bi)
- Group 6: oxygen to polonium (O to Po)
- Group 7: fluorine to astatine (F to At)
- Group 8: neon to radon (Ne to Rn).

In Group 4, lead (Pb) is an anomaly because it has a first ionisation energy that is higher than that of tin (Sn), the element immediately above it.

There is no general trend in first ionisation energy in Group 3 (boron to thallium, or B to Tl).

The explanations for the anomalies in Groups 3 and 4 are beyond the scope of this book, but they illustrate that, in chemistry, you can rarely apply a simple pattern or trend to all situations.

## Questions

1. State the three factors that determine the magnitude of the first ionisation energy of an element.

2. Write equations to represent:
   (a) the first ionisation energy of sodium
   (b) the second ionisation energy of calcium
   (c) the fourth ionisation energy of carbon.

3. Draw a sketch graph for the logarithm to base 10 for the successive ionisation energies of phosphorus.

4. The table below shows the first four ionisation energies, in kJ mol$^{-1}$, of five elements: **A, B, C, D** and **E**.

| Element | First ionisation energy | Second ionisation energy | Third ionisation energy | Fourth ionisation energy |
|---|---|---|---|---|
| A | 496 | 4563 | 6913 | 9544 |
| B | 738 | 1451 | 7733 | 10 541 |
| C | 578 | 1817 | 2745 | 11 578 |
| D | 900 | 1757 | 14 849 | 21 007 |
| E | 631 | 1235 | 2389 | 7089 |

   (a) Which two elements are in the same group of the Periodic Table? Explain your answer.
   (b) In which group of the Periodic Table is element **C** likely to occur? Explain your answer.
   (c) Which element requires the least amount of energy to form a 2+ ion. Explain your answer.

5. The first four ionisation energies, in kJ mol$^{-1}$, of calcium are 590, 1145, 4912 and 6474.
   (a) Explain why the second ionisation energy of calcium is larger than the first.
   (b) Explain why the third ionisation energy is much larger than the second.

### Key definitions

The **first ionisation energy** of an element is the energy required to remove an electron from each atom in one mole of atoms in the gaseous state.

The **second ionisation energy** of an element is the energy required to remove an electron from each singly charged positive ion in one mole of positive ions in the gaseous state.

# 1.2 1 The Periodic Table

**By the end of this section, you should be able to...**

- understand that electronic configuration determines the chemical properties of an element
- know that elements can be classified as s-, p- and d-block elements

## Groups, periods and blocks

### Groups

The vertical columns in the Periodic Table are called **groups**. All the elements in a main group (i.e. Group 1 to 8 inclusive) contain the same outer electronic configuration. In the table below, n represents the quantum shell. For lithium to neon (Li to Ne), n = 2; for sodium to argon (Na to Ar), n = 3, etc.

| | Group 1 | Group 2 | Group 3 | Group 4 | Group 5 | Group 6 | Group 7 | Group 8 |
|---|---|---|---|---|---|---|---|---|
| outer electronic configuration | $ns^1$ | $ns^2$ | $ns^2np^1$ | $ns^2np^2$ | $ns^2np^3$ | $ns^2np^4$ | $ns^2np^5$ | $ns^2np^6$ |

**table A** Outer electronic configuration of Groups 1–8 in the Periodic Table.

The elements within a group have similar chemical properties because they have the same outer electronic configuration.

### Periods

The horizontal rows in the Periodic Table are called **periods**. All elements in a period have the same number of quantum shells containing electrons. For example, all elements in Period 2, lithium to neon (Li to Ne), have electrons in both the first and second quantum shells.

### Blocks

The Periodic Table is also divided into **blocks**.

**Did you know?**

We cannot define the d-block in the same way as the s- and p-blocks, since the electrons in the d orbitals are of lower energy then those in the s orbital.

| s-block | | | d-block | | | | | | | | | p-block | | | | | |
|---|---|---|---|---|---|---|---|---|---|---|---|---|---|---|---|---|---|
| 1 | 2 | | | | | | | | | | | 3 | 4 | 5 | 6 | 7 | 8 |
| Li | Be | | | | | | | | | | | B | C | N | O | F | Ne |
| Na | Mg | | | | | | | | | | | Al | Si | P | S | Cl | Ar |
| K | Ca | Sc | Ti | V | Cr | Mn | Fe | Co | Ni | Cu | Zn | Ga | Ge | As | Se | Br | Kr |
| Rb | Sr | Y | Zr | Nb | Mo | Tc | Ru | Rh | Pd | Ag | Cd | In | Sn | Sb | Te | I | Xe |

**fig A** The s-, d- and p-blocks in the Periodic Table.

The s-block consists of the elements in Groups 1 and 2. An s-block element has its highest energy electron in an s orbital.

The p-block consists of the elements in Groups 3 to 8 inclusive. A p-block element has its highest energy electron in a p-orbital.

The d-block consists of the elements scandium to zinc (Sc to Zn) in Period 4 and yttrium to cadmium (Y to Cd) in Period 5. The number of electrons in the d orbitals gradually increases from left to right across the table. Hence, a d-block can be defined as one in which the d subshell is being filled.

## Questions

1. What is meant by the terms
   (a) s-block element and
   (b) p-block element?

2. Explain why the chemical properties of the elements in Group 1 are very similar.

3. The outer electronic configuration of carbon is $2s^2 2p^2$.
   Deduce the outer electronic configuration of arsenic (As). Explain how you arrived at your answer.

### Key definitions

The vertical columns in the Periodic Table are called **groups**.
The horizontal rows in the Periodic Table are called **periods**.
The Periodic Table is divided into **blocks**.

# 1.2 2 Periodicity

By the end of this section, you should be able to...
- understand periodicity in terms of a repeating pattern across different periods
- understand reasons for the trends in the melting and boiling temperatures and in the first ionisation energies of the elements from Periods 2 and 3 of the Periodic Table
- know how to illustrate periodicity using data (electronic configurations, atomic radii, melting and boiling temperatures, and first ionisation energies)

## What is periodicity?

The elements in a period exhibit **periodicity**.

We can best illustrate periodicity by looking at the elements in Periods 2 and 3.

We have already seen one example of periodicity in Periods 2 and 3 – the regular repeating pattern of electronic configurations from $ns^1$ through to $ns^2np^6$.

Other examples are the trends in atomic radii, melting and boiling temperatures, and first ionisation energies.

## Atomic radii

The atomic radius of an element is a measure of the size of its atoms. It is the distance from the centre of the nucleus to the boundary of the electron cloud. Since the atom does not have a well-defined boundary, we can find the atomic radius by determining the distance between the two nuclei and dividing it by two.

radius = $d/2$

**fig A** Measuring the covalent and van der Waals radii.

The diagram on the left shows two bonded atoms. The atoms are closer together than they are in the right-hand diagram (where they are just touching). On the left-hand side of the diagram, we are measuring the covalent radius.

The radius being measured in the diagram on the right is called the van der Waals radius. This is the only radius that can be determined for neon and argon, because they do not bond with other elements.

There is a third radius that is used for metals. It is called the metallic radius.

### Did you know?

Different types of radii give different measurements for the same element. The van der Waals radius is always larger. So always compare like with like when you are looking at the trends in atomic radii.

The diagram below shows the trend in covalent radii across the second period (lithium to fluorine, or Li to F) and third period (sodium to chlorine, or Na to Cl) measured in nanometres ($10^{-9}$ m).

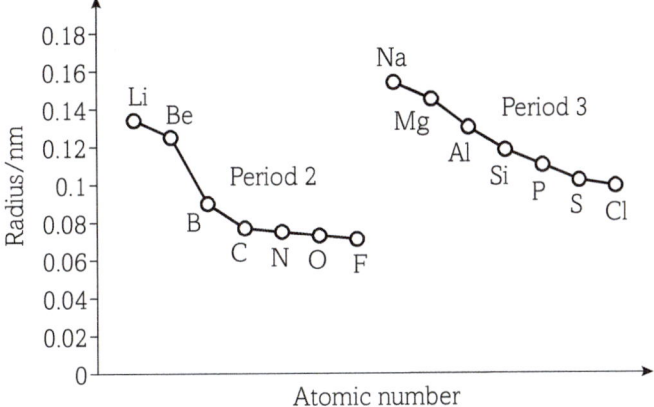

**fig B** Trend in covalent radii across the second and third periods.

You can see that the radius decreases across each period. This is because as the number of protons in the nucleus increases, so does the nuclear charge. This results in an increase in the attractive force between the nucleus and the outer electrons. This increase in attractive force offsets the increase in electron–electron repulsion as the number of electrons in the outer quantum shell increases.

## Melting and boiling temperatures

The tables below illustrate the changes in melting and boiling temperatures for the Period 2 and 3 elements.

**Period 2**

|  | Li | Be | B | C (diamond) | N | O | F |
|---|---|---|---|---|---|---|---|
| melting temperature/°C | 181 | 1278 | 2300 | 3550 | −210 | −218 | −220 |
| boiling temperature/°C | 1342 | 2970 | 3927 | 4827 | −196 | −183 | −188 |
| type of bonding | metallic | metallic | covalent | covalent | covalent | covalent | covalent |
| structure | giant lattice | giant lattice | giant lattice | giant lattice | simple molecular | simple molecular | simple molecular |

**Period 3**

|  | Na | Mg | Al | Si | P | S | Cl |
|---|---|---|---|---|---|---|---|
| melting temperature/°C | 98 | 649 | 660 | 1410 | 44 | 113 | −101 |
| boiling temperature/°C | 883 | 1107 | 2467 | 2355 | 280 | 445 | −35 |
| type of bonding | metallic | metallic | metallic | covalent | covalent | covalent | covalent |
| structure | giant lattice | giant lattice | giant lattice | giant lattice | simple molecular | simple molecular | simple molecular |

**table A** Changes in melting and boiling temperatures for Period 2 and 3 elements.

You may have noticed that the elements with giant lattice structures have high melting and boiling temperatures, and those with simple molecular structures have low melting and boiling temperatures. We will explain why this is later in the book.

## First ionisation energies

The diagram below is a plot of first ionisation energy for the first three Periods.

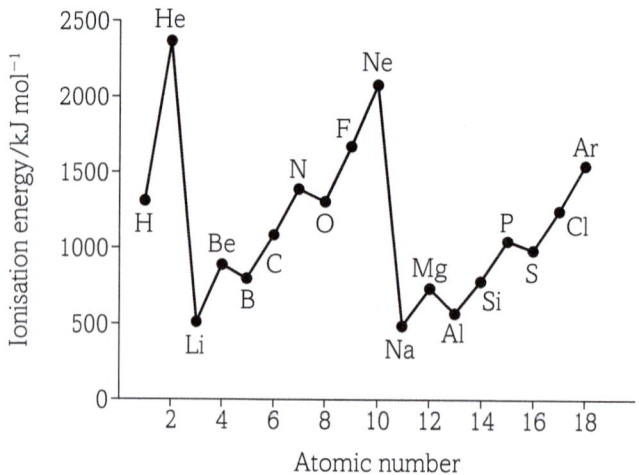

**fig C** First ionisation energy for Periods 1–3.

### Hydrogen and helium

The electronic configurations of hydrogen and helium are $1s^1$ and $1s^2$, respectively.

We can explain the increase in first ionisation energy from hydrogen to helium by the increase in nuclear charge from 1 to 2 as an extra proton is added. This increase in nuclear charge more than offsets the increase in electron–electron repulsion in the 1s orbital as a second electron is added.

### Anomalies

We have already explained the general increase in first ionisation energy across the period from lithium to neon (Li to Ne) and from sodium to argon (Na to Ar), in terms of the increasing nuclear charge.

However, you will notice that there are two anomalies in each case. The first ionisation of the Group 3 element is less than that of the Group 2 element, and the first ionisation energy of the Group 6 element is less than that of the Group 5 element.

First, consider the case of beryllium (Be) and boron (B).

The electronic configurations are:

Be: $1s^2\ 2s^2$
B: $1s^2\ 2s^2\ 2p^1$

Although the nuclear charge of the boron atom is greater than that of the beryllium atom, the outer electron has more energy, since it is in a 2p orbital as opposed to the 2s orbital for beryllium. So the energy required to remove this electron is less than that required to remove a 2s electron from a beryllium atom. In addition, the 2p electron in boron experiences greater electron–electron repulsion (i.e. greater shielding) because there are two inner electron sub-shells as opposed to only one in the beryllium atom.

A similar argument applies to magnesium and aluminium (Mg and Al), except that in this case it is the 3s and 3p electrons that are involved.

### Nitrogen and oxygen

Now consider nitrogen (N) and oxygen (O).

The electronic configurations are:

N: $1s^2\ 2s^2\ 2p_x^1 2p_y^1 2p_z^1$
O: $1s^2\ 2s^2\ 2p_x^2 2p_y^1 2p_z^1$

The first electron removed from the oxygen atom is one of the two paired electrons in the $2p_x$ orbital.

---

**Learning tip**

It is possible you may come across a suggestion that the first ionisation energy of boron is less than that of beryllium, since by losing an electron the boron atom will acquire a full 2s orbital and so become more stable. In fact, because energy has to be supplied to the boron atom in order for an electron to be removed, the ion formed is energetically less stable than the atom since it has a higher energy value.

The presence of two electrons in a single orbital increases the electron–electron repulsion in this orbital. Less energy is therefore required to remove one of these electrons than is required to remove a 2p electron from a nitrogen atom, despite the larger nuclear charge of the oxygen atom.

### Learning tip

You may see an explanation stating that the first ionisation energy of oxygen is less than that of nitrogen, since by losing an electron the oxygen atom will acquire a stable half-full 2p subshell. However, as has been mentioned earlier, there is no 'magical' stability associated with a half-full subshell. Also, as with the boron atom, energy has to be supplied to the oxygen atom in order for an electron to be removed, so the ion formed is energetically less stable than the atom.

## Questions

1. An element has a very high melting and boiling temperature.
   (a) What type of structure is it likely to have?
   (b) Which physical property would help you determine whether the bonding was metallic or covalent? Explain your answer.

2. Explain why the first ionisation energy of helium is higher than that of hydrogen.

3. Explain why the first ionisation energy of lithium is much lower than that of helium, even though the lithium has greater nuclear charge.

4. Would you expect the first ionisation energy of gallium (Ga) to be higher or lower than that of calcium (Ca)? Explain your answer.

5. Why does neon have the highest first ionisation energy of all the elements in Period 2?

### Key definition

**Periodicity** is a regularly repeating pattern of atomic, physical and chemical properties with increasing atomic number.

# THINKING BIGGER

## ELEMENTAL FINGERPRINTS

How can we possibly know what stars in distant galaxies are made from? The following extract is taken from *From Stars to Stalagmites: How Everything Connects* by Paul Braterman (2012).

## ELEMENTAL "FINGERPRINTS"

In 1825, the French philosopher Auguste Comte said that there are some things we will never know, among these the chemical composition of the stars. This was an unfortunate example. We do not need actual material from the stars in order to analyse them. What we do need, of course, is information, and this they send us plentifully, in the form of light. To reveal this information, we must separate the light into its different wavelengths or colours.

Many years earlier Isaac Newton had passed sunlight through a glass prism, and found that this gave him all the colours of the rainbow. In 1802 William Wollaston in England … repeated the experiment, using an up-to-date high quality glass prism, and discovered that the continuous spectrum was interrupted by narrow dark lines. (Rather unfairly these are now known as Frauenhofer lines, after the German physicist Joseph von Frauenhofer, who confirmed and extended Wollaston's observations). These lines later proved to match exactly the light given out by different elements when heated or in electric discharges. A now familiar example is provided by the element sodium, whose yellow emission is used in street lamps. The lines can be used as a kind of fingerprint for each element. We can take this fingerprint using electrical discharges here on Earth and compare it with the lines found in the spectrum of the Sun. In this way we can get a good chemical analysis of the surface layer of the Sun or of any other star whose light we can collect and astronomers now extend the same process to the most distant galaxies.

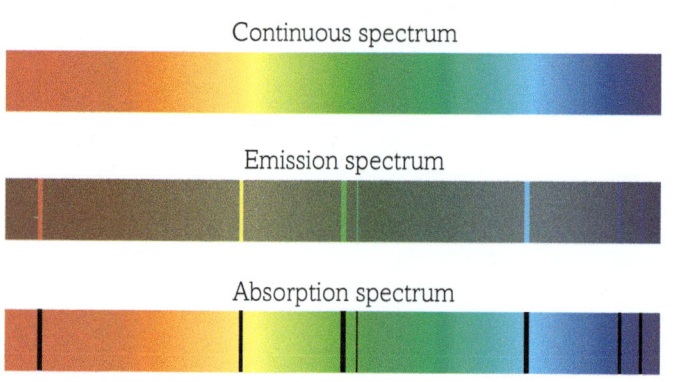

fig A

### Learning tip
An emission spectrum shows the frequencies of electromagnetic radiation that are emitted by a substance. An absorption spectrum shows the frequencies of electromagnetic radiation that are absorbed by the substance. This means that the absorption spectrum is essentially the 'negative' of the emission spectrum.

### Did you know?
Many images of galaxies taken by the Hubble telescope use the line emission spectra of elements such as hydrogen, sulfur and oxygen to build up a colour image of the galaxy. The different elements are assigned red, green and blue frequencies to build up a full colour image.

Where else will I encounter these themes?

Let us start by considering the nature of the writing in the article.

1. **a.** Do you think this article is aimed at scientists, the general public, or people who are not scientists but have an interest in science? Look back through the extract and find examples to support your answer.
   **b.** Explain how the book's tagline *How Everything Connects* is supported by the extract.

> As well as considering the complexity of the scientific ideas, think about the level of detail in which they are discussed. Notice how many different branches of natural science are mentioned in the extract. Are all of the scientists mentioned chemists or did they work in different scientific disciplines?

Now we will look at the chemistry in, or connected to, this article. Don't worry if you are not ready to give answers to these questions yet. You may like to return to the questions once you have covered other topics later in the book. The timeline at the bottom of the page highlights where else in the book you will meet the concepts covered, helping you to see how different areas of Chemistry link together.

2. **a.** Describe the structure of the sodium atom in terms of protons, neutrons and electrons.
   **b.** Give the electronic structure of a sodium atom using s, p, d notation.
   **c.** What is giving rise to (i) the absorption and (ii) the emission spectrum of sodium?
   **d.** Can you suggest why sodium vapour is used in preference to potassium vapour in street lamps?
3. Would you expect heavier isotopes of sodium to give a different 'fingerprint'? Explain your answer.
4. The element helium was first discovered in the outer layer of the Sun. Suggest why it was discovered there before it was discovered here on Earth. Think back to work you have done on identifying alkali metals using flame tests.

> **Command word**
> If the word explain is used in a question, your answer should clearly describe the thing you are explaining. You'll need to use your reasoning and maybe examples to support your point.

## Activity

Today, the elemental fingerprints in the light collected from distant galaxies have provided evidence for an expanding universe. The elemental fingerprints are subject to a phenomenon called 'red shift'.

Explain in 200–300 words how red shift has provided evidence for an expanding universe. Try to structure your explanation so that the concepts can be understood by an audience of 16-year-old GCSE students.

**fig B** A picture taken by the Hubble Telescope showing over 10 000 galaxies!

- From *From Stars to Stalagmites. How Everything Connects* by Paul Braterman

# 1 Exam-style questions

1  A sample of helium from a rock was found to contain two isotopes with the following composition by mass: $^3$He, 0.992%; $^4$He, 99.008%.
   (a) State what is meant by the term **isotopes**. [1]
   (b) State the difference in the atomic structures of $^3$He and $^4$He. [1]
   (c) (i) Which isotope is used as the basis for relative atomic mass measurements? [1]
       (ii) Calculate the relative atomic mass of helium in the rock sample. [2]
   (d) Helium has the largest first ionisation energy of all the elements.
       (i) State what is meant by the term first ionisation energy. [3]
       (ii) Write an equation, including state symbols, to represent the first ionisation energy of helium. [2]
       (iii) Explain why the first ionisation energy of helium is larger than that of hydrogen. [2]
   [Total: 12]

2  The five ionisation energies of boron are:
   801   2427   3660   25 026   32 828
   (a) State and justify the group in the Periodic Table in which boron is placed. [2]
   (b) Which of the following represents the **second** ionisation energy of boron?
       A  B(g) → B$^{2+}$(g) + 2e$^-$  $\Delta H^\ominus$ = + 2427 kJ mol$^{-1}$
       B  B$^+$(g) → B$^{2+}$(g) + e$^-$  $\Delta H^\ominus$ = + 2427 kJ mol$^{-1}$
       C  B(g) → B$^{2+}$(g) + 2e$^-$  $\Delta H^\ominus$ = − 2427 kJ mol$^{-1}$
       D  B$^+$(g) → B$^{2+}$(g) + e$^-$  $\Delta H^\ominus$ = − 2427 kJ mol$^{-1}$ [1]
   (c) Explain why the second ionisation energy of boron is larger than the first. [2]
   (d) Give the electronic configuration of a boron atom. [1]
   (e) Is boron classified as an s-, a p- or a d-block element? Justify your answer. [2]
   (f) Explain why the first ionisation energy of boron is less than that of beryllium, even though a boron atom has a greater nuclear charge. [2]
   [Total: 10]

3  (a) State what is meant by the term **periodicity**. [1]
   (b) Explain the **general** trend in first ionisation energy of the elements Na to Ar in Period 3 of the Periodic Table. [3]
   (c) Explain the trend in atomic radii of the elements sodium to chlorine in Period 3 of the Periodic Table. [4]
   (d) The table shows the melting temperatures of the elements sodium to chlorine in Period 3 of the Periodic Table.

| Symbol of element | Na | Mg | Al | Si | P | S | Cl |
|---|---|---|---|---|---|---|---|
| melting temperature/°C | 98 | 649 | 660 | 1410 | 44 | 113 | −101 |
| bonding | | | | | | | |
| structure | | | | | | | |

   Complete the table using the following guidelines.
   (i) Complete the 'bonding' row using **only** the words metallic or covalent.
   (ii) Complete the 'structure' row using **only** the words simple molecular or giant lattice.
   (iii) Explain why the melting temperature of phosphorus is different from that of silicon. [5]
   [Total: 13]

4  (a) Complete the table to show the properties of the three major sub-atomic particles. [3]

| Particle | Relative charge | Relative mass |
|---|---|---|
| proton | | |
| neutron | | |
| electron | | |

   (b) The particles in each pair differ **only** in the number of protons **or** the number of neutrons **or** the number of electrons that they contain. State the difference in each pair.
       (i) $^{16}$O and $^{17}$O
       (ii) $^{24}$Mg and $^{24}$Mg$^{2+}$
       (iii) $^{39}$K$^+$ and $^{40}$Ca$^{2+}$ [3]
   [Total: 6]

5 (a) State what is meant by the term **orbital**. [2]
  (b) Draw the shapes of:
      (i) an s-orbital
      (ii) a p-orbital. [2]
  (c) How many electrons can occupy
      (i) an s-orbital
      (ii) a p-subshell? [2]
  (d) Using s and p notation, give the electronic configurations of each of the following atoms and ions:
      (i) Na
      (ii) $O^{2-}$
      (iii) $Mg^{2+}$
      (iv) Cl [4]
  (e) Which of these ions contains the fewest number of electrons?
      **A** $NH_4^+$
      **B** $P^{3-}$
      **C** $S^{2-}$
      **D** $Cl^-$ [1]
      [Total: 11]

6 Chlorine has two isotopes, $^{35}Cl$ and $^{37}Cl$.
  The diagram shows part of the mass spectrum of a sample of chlorine gas:

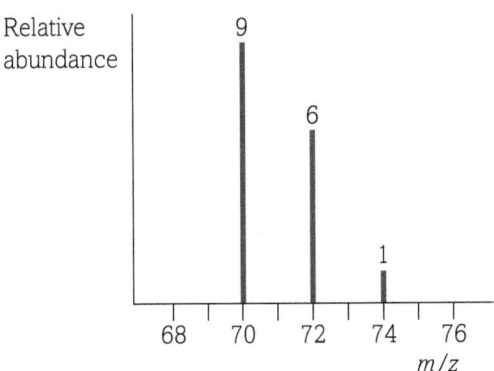

  (a) Give the formula of the species responsible for each peak at m/z of 70, 72 and 74. [3]
  (b) Give the m/z ratio for the two other peaks in the mass spectrum of chlorine. Give the formula of the species responsible for each peak and state the relative abundance of these two species. [5]
      [Total: 8]

7 The table gives the first four ionisation energies of the elements sodium, magnesium and aluminium.

| Element | Ionisation energy/kJ mol$^{-1}$ | | | |
| --- | --- | --- | --- | --- |
|  | 1st | 2nd | 3rd | 4th |
| sodium | 496 | 4563 | 6913 | 9544 |
| magnesium | 738 | 1451 | 7733 | 10 541 |
| aluminium | 578 | 1817 | 2745 | 11 578 |

Explain why:
  (a) the first ionisation energy of sodium is lower than that of the first ionisation energy of magnesium. [2]
  (b) the first ionisation energy of magnesium is higher than the first ionisation energy of aluminium. [2]
  (c) the second ionisation energy of magnesium is lower than the second ionisation energy of aluminium. [2]
  (d) The fourth ionisation energy of aluminium is higher than its third ionisation energy. [2]
      [Total: 8]

8 Lithium is an element in the s-block of the Periodic Table. Naturally occurring lithium contains a mixture of two isotopes, $^6Li$ and $^7Li$.
  (a) Complete the table to show the atomic structure of each of the two isotopes.

| Isotope | Number of protons | Number of neutrons | Number of electrons |
| --- | --- | --- | --- |
| $^6Li$ | | | |
| $^7Li$ | | | |

  (b) A sample of lithium was found to contain 7.59% of $^6Li$ and 92.41% of $^7Li$.
      (i) Give the name of the instrument that can be used to determine this information. [1]
      (ii) The $^7Li$ isotope has a relative isotopic mass of 7.016 and the $^6Li$ isotope has a relative isotopic mass of 6.015. State what is meant by the term **relative isotopic mass**. [2]
      (iii) Use the relative isotopic masses to calculate the relative atomic mass of lithium. Give your answer to three significant figures. [2]
  (c) (i) Give the electronic configuration of an atom of lithium. [1]
      (ii) State why lithium is described as an s-block element. [1]
      (iii) Would you expect the two isotopes of lithium to have the same chemical properties? Justify your answer. [1]
      [Total: 8]

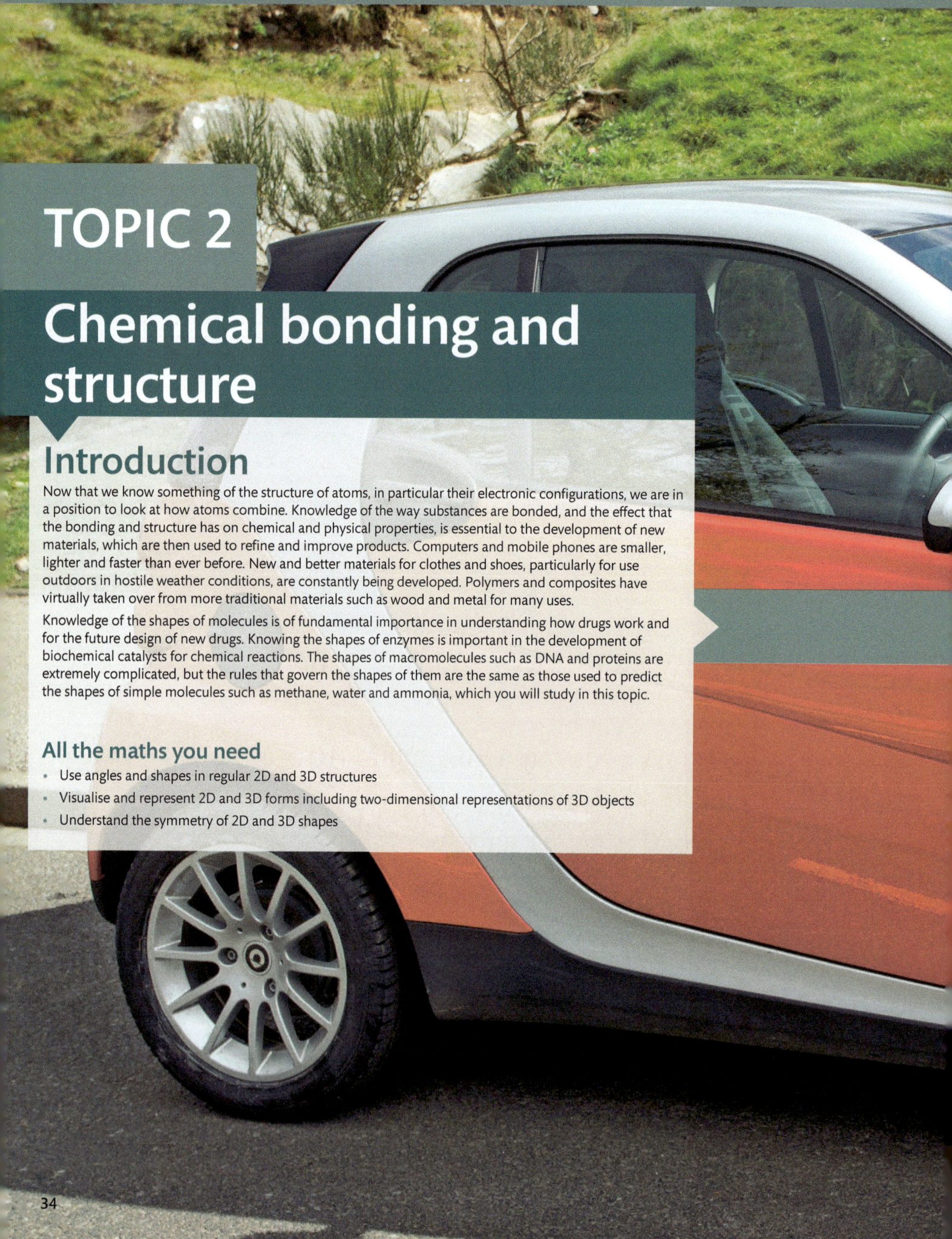

# TOPIC 2

# Chemical bonding and structure

## Introduction

Now that we know something of the structure of atoms, in particular their electronic configurations, we are in a position to look at how atoms combine. Knowledge of the way substances are bonded, and the effect that the bonding and structure has on chemical and physical properties, is essential to the development of new materials, which are then used to refine and improve products. Computers and mobile phones are smaller, lighter and faster than ever before. New and better materials for clothes and shoes, particularly for use outdoors in hostile weather conditions, are constantly being developed. Polymers and composites have virtually taken over from more traditional materials such as wood and metal for many uses.

Knowledge of the shapes of molecules is of fundamental importance in understanding how drugs work and for the future design of new drugs. Knowing the shapes of enzymes is important in the development of biochemical catalysts for chemical reactions. The shapes of macromolecules such as DNA and proteins are extremely complicated, but the rules that govern the shapes of them are the same as those used to predict the shapes of simple molecules such as methane, water and ammonia, which you will study in this topic.

### All the maths you need

- Use angles and shapes in regular 2D and 3D structures
- Visualise and represent 2D and 3D forms including two-dimensional representations of 3D objects
- Understand the symmetry of 2D and 3D shapes

## What have I studied before?
- The electronic configurations of the first 36 elements in the Periodic Table
- Metallic, ionic and covalent bonding
- Using dot-and-cross diagrams to represent ions and molecules
- The physical properties of metals, ionic compounds and covalent compounds, both simple molecular and giant structures

## What will I study later?
- The existence of isomerism in organic compounds
- Trends in the properties of Group 2 and Group 7 elements (A level)
- The mechanisms of some reactions of alkanes, alkenes, halogenoalkanes and alcohols (A level)
- The mechanisms of some reactions of carbonyls, carboxylic acids, arenes and organic nitrogen compounds (A level)
- The nature of the bonding in and the shapes of transition metal complexes (A level)
- The characteristic properties of transition metals, e.g. the ability to have more than one oxidation state and the ability to act as catalysts (A level)

## What will I study in this topic?
- The nature of metallic, ionic, covalent, polar covalent and dative covalent bonding
- The nature of intermolecular interactions, including hydrogen bonding
- The shapes of discrete (simple) molecules
- Electronegativity and polarity of molecules
- An explanation of the physical properties of substances based on their bonding and structure

# 2.1 / 1 Metallic bonding

By the end of this section, you should be able to...

- know that metallic bonding is the strong electrostatic attraction between the nuclei of metal cations and delocalised electrons

## The nature of metallic bonding

Metals typically have the following physical properties:

- high melting temperatures
- good electrical conductivity
- good thermal conductivity
- malleability
- ductility.

Any theory of the way that the atoms in a metal are bonded together must explain the above properties.

Metals typically have one, two or three electrons in the outer shell of their atoms and have low ionisation energies. The electrical conductivity of a metal generally increases as the number of outer-shell electrons increases.

Since electrical conductivity depends on the presence of mobile carriers of electric charge, we can build a picture of a metal as consisting of an array of atoms with at least some of their outer-shell electrons removed and free to move throughout the structure. These **delocalised electrons** are largely responsible for the characteristic properties of metals.

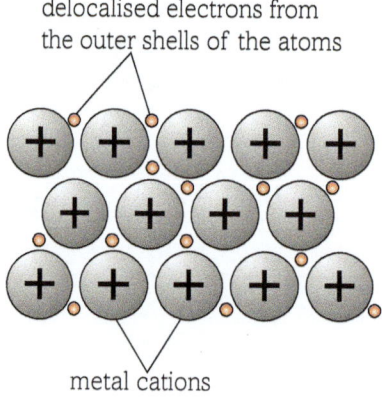

**fig A** Structure of a metal.

The electrons are said to be delocalised since they are free to move throughout the structure and are not confined, i.e. localised, between any pair of cations.

There are electrostatic forces of attraction between the nuclei of the cations and the delocalised electrons. This is known as **metallic bonding**.

## Explaining the physical properties of metals
### Melting temperature

In order to melt a metal it is necessary to overcome the forces of attraction between the nuclei of the cations and the delocalised electrons to such an extent that the cations are free to move around the structure. Metals have a giant lattice structure where many of these forces must be overcome. The energy required to do this is usually very large, so the melting temperatures are typically high.

The number of delocalised electrons per cation plays a part in determining the melting temperature of a metal.

- Group 1 metals have low melting temperatures.
- Group 2 metals have higher melting temperatures.
- Metals in the d-block typically have high melting temperatures since they have more delocalised electrons per ion.

Another factor that affects the melting temperature of a metal is the size of the cation. The smaller the cation, the closer the delocalised electrons to the nucleus of the cation. This results, in general, in an increase in the forces of attraction between the nuclei and the delocalised electrons, and so an increase in melting temperature.

## Electrical conductivity

When a potential difference is applied across the ends of a metal, the delocalised electrons will be attracted to, and move towards, the positive terminal of the cell. This movement of electrical charge constitutes an electric current.

## Thermal conductivity

Two factors contribute to the ability of metals to transfer heat energy.

1. The free-moving delocalised electrons pass kinetic energy along the metal.
2. The cations are closely packed and pass kinetic energy from one cation to another.

## Malleability and ductility

Metals can be hammered or pressed into different shapes (malleability). They can also be drawn into a wire (ductility). Both of these properties depend on the ability of the delocalised electrons and the cations to move throughout the structure of the metal.

When a stress is applied to a metal, the layers of cations may slide over one another.

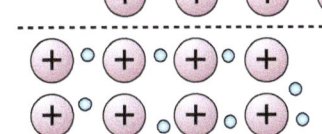

**fig B** Effect of stress on a metal.

However, since the delocalised electrons are free moving, they move with the cations and prevent strong forces of repulsion forming between the cations in one layer and the cations in another layer.

### Did you know?

The above argument based on size of cation is a slight oversimplification, since the number of protons in the nucleus, and the effect to which the delocalised electrons are shielded from the nucleus, also contribute to the force of attraction. However, these extra complications are beyond the scope of A level and are best ignored.

It is also tempting to think that the magnitude of the charge on the cation will also affect the melting temperature of a metal. This is an argument that is best avoided, since the attraction is not between the cations and the delocalised electrons, but between the nuclei of the cations and the delocalised electrons.

## Questions

1. Explain what is meant by the term 'metallic bonding'.
2. Suggest why the melting temperatures and electrical conductivities of sodium, magnesium and aluminium are in the order Na < Mg < Al.

### Key definitions

**Delocalised electrons** are electrons that are not associated with any single atom or any single covalent bond.

**Metallic bonding** is the electrostatic force of attraction between the nuclei of metal cations and delocalised electrons.

# 2 Ionic bonding

## 2.1

By the end of this section, you should be able to...
- know that ionic bonding is the strong electrostatic attraction between oppositely charged ions
- understand the effects that ionic radius and ionic charge have on the strength of ionic bonding
- understand reasons for the trends in ionic radii down a group and for a set of isoelectronic ions, e.g. $N^{3-}$ to $Al^{3+}$
- understand that the physical properties of ionic compounds and the migration of ions provide evidence for the existence of ions
- understand the formation of ions in terms of electron loss or gain
- draw electronic configuration diagrams of cations and anions using the dot-and-cross convention

## The nature of ionic bonding

Ionic bonding is confined to solid materials consisting of a regular array of oppositely charged ions extending throughout a giant lattice network.

The most familiar ionic compound is sodium chloride, NaCl. It consists of a regular array of sodium ions, $Na^+$, and chloride ions, $Cl^-$, as shown in the diagrams below.

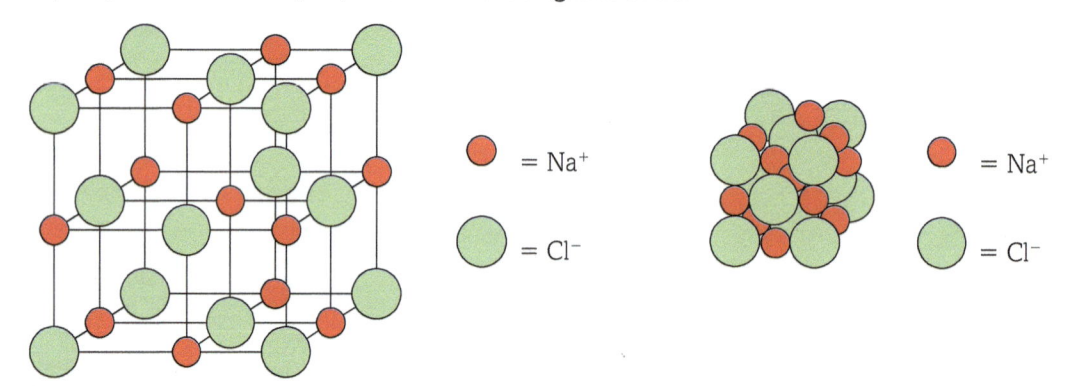

**fig A** Structure of sodium chloride.

The diagram on the left is an 'exploded' version of the structure, which is often drawn for the sake of clarity. In practice, the ions are touching one another, as shown in the diagram on the right.

In an ionic solid, there are strong electrostatic interactions between the ions. The ions are arranged in such a way that the electrostatic attractions between the oppositely charged ions are greater than the electrostatic repulsions between ions with the same charge. The electrostatic interaction between ions is not directional: all that matters is the distance between two ions, not their orientation with respect to one another. (Compare this with covalent bonding in **Section 2.1.3**.)

When ions are present, the electrostatic interaction between them tends to be dominant. However, it is possible for there to be significant covalent interactions between ions such that pure ionic bonding is best regarded as an idealised bonding situation. This concept will be developed further in **Book 2**.

## The strength of ionic bonding

You can determine the strength of ionic bonding by calculating the amount of energy required in one mole of solid to separate the ions to infinity (i.e. in the gas phase). When they are at an infinite distance from on another, the ions can no longer interact.

### Learning tip

It is best to avoid saying that there is an 'ionic bond' between two ions.

This is because in an ionic solid each ion interacts with many other ions, of both the same and opposite charge to itself. The energy that binds the structure does not come from single interactions between ions (so-called ionic bonds), but from the interactions between all of the ions in the lattice. This energy is called the 'lattice energy' and will be discussed further in **Book 2**.

The table below shows the energy required to separate to infinity the ions in one mole of various alkali metal halides.

increasing size of anion →

| | Amount of energy required to separate the ions to infinity/kJ mol$^{-1}$ | | | |
|---|---|---|---|---|
| | F$^-$ | Cl$^-$ | Br$^-$ | I$^-$ |
| Li$^+$ | 1031 | 848 | 803 | 759 |
| Na$^+$ | 918 | 780 | 742 | 705 |
| K$^+$ | 817 | 711 | 679 | 651 |
| Rb$^+$ | 783 | 685 | 656 | 628 |

↓ increasing size of cation

**table A** Energy required to break up a lattice of an ionic compound.

For ions of the same charge, the smaller the ions the more energy is required to overcome the electrostatic interactions between the ions and to separate them.

So, one factor that affects the strength of ionic bonding is the size of the ions, which in turn determines how closely packed they are in the lattice.

The equivalent energy for magnesium fluoride, Mg$^{2+}$(F$^-$)$_2$, is 2957 kJ mol$^{-1}$. The radius of the Mg$^{2+}$ ion (0.072 nm) is very similar to that of the Li$^+$ ion (0.074 nm). The increased charge of the Mg$^{2+}$ ion over the Li$^+$ ion results in a significant increase in the strength of the ionic bonding.

When both cation and anion are doubly charged, the energy required to separate the ions is even larger. For magnesium oxide, Mg$^{2+}$O$^{2-}$, the value is 3791 kJ mol$^{-1}$.

There is no simple mathematical relationship to describe the effects that ionic radius and ionic charge have on the strength of ionic bonding. The situation is complicated by the way in which the ions pack together to form the lattice, and by the extent to which there are covalent interactions between the ions. In general, however, the smaller the ions and the larger the charge on the ions, the stronger the ionic bonding.

## Trends in ionic radii

Ionic radii are difficult to measure with any degree of certainty, and vary according to the environment of the ion. For example, it matters how many oppositely charged ions are touching it (i.e. its co-ordination number) and also the nature of the ions.

There are several different ways of measuring ionic radii in use and they all produce slightly different values. So, if you are going to make reliable comparisons using ionic radii, they must come from the same source.

Remember that there are quite large uncertainties when using ionic radii. Trying to explain things in fine detail is made difficult by those uncertainties.

| Group 1 | | |
|---|---|---|
| Ion | Electronic configuration | Ionic radius/nm |
| Li$^+$ | 2 | 0.076 |
| Na$^+$ | 2.8 | 0.102 |
| K$^+$ | 2.8.8 | 0.138 |
| Rb$^+$ | 2.8.18.8 | 0.152 |

| Group 7 | | |
|---|---|---|
| Ion | Electronic configuration | Ionic radius/nm |
| F$^-$ | 2.8 | 0.133 |
| Cl$^-$ | 2.8.8 | 0.181 |
| Br$^-$ | 2.8.18.8 | 0.196 |
| I$^-$ | 2.8.18.18.8 | 0.220 |

**table B** Trends in ionic radii in Groups 1 and 7.

As you go down each group, the ions have more electron shells; therefore, the ions get larger.

| Period 2 | $N^{3-}$ | $O^{2-}$ | $F^-$ |
|---|---|---|---|
| Number of protons | 7 | 8 | 9 |
| Electronic configuration | 2.8 | 2.8 | 2.8 |
| Ionic radius/nm | 0.146 | 0.140 | 0.133 |

| Period 3 | $Na^+$ | $Mg^{2+}$ | $Al^{3+}$ |
|---|---|---|---|
| Number of protons | 11 | 12 | 13 |
| Electronic configuration | 2.8 | 2.8 | 2.8 |
| Ionic radius/nm | 0.102 | 0.072 | 0.054 |

**table C** Trends in ionic radii across a period.

### Did you know?
When you compare chemical values, you should make sure all the data you are comparing come from the same source. For example, the values in tables B and C have all been taken from the Database of Ionic Radii from Imperial College London. However, Chemistry Data Book (J G Stark and H G Wallace) gives a value of 0.171 for the ionic radius of the nitride ion, $N^{3-}$. In either case, the value is larger than that for the oxide ion, $O^{2-}$, as expected.

All six of the ions listed in the above tables are isoelectronic, i.e. they have the same number of electrons and so the same electronic configuration.

The ionic radius decreases as the number of protons increases.

As the positive charge of the nucleus increases, the electrons are attracted more strongly and are therefore pulled closer to the nucleus.

## Physical properties of ionic compounds
Ionic compounds typically have the following physical properties:
- high melting temperatures
- brittleness
- poor electrical conductivity when solid but good when molten
- often soluble in water.

### High melting temperatures
Ionic solids consist of a giant lattice network of oppositely charged ions (see **Section 2.3.1**). There are many ions in the lattice and the combined electrostatic forces of attraction among all of the ions is large.

So, a large amount of energy is required to overcome the forces of attraction sufficiently for the ions to break free from the lattice and be able to slide past one another.

### Brittleness
If a stress is applied to a crystal of an ionic solid, then the layers of ions may slide over one another.

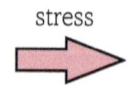

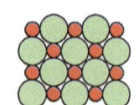

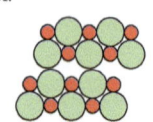

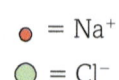

**fig B** Effect of stress on an ionic crystal.

Ions of the same charge are now side by side and repel one another. The crystals break apart.

### Electrical conductivity
Solid ionic compounds do not, in general, conduct electricity. This is because there are no delocalised electrons and the ions are also not free to move under the influence of an applied potential difference.

However, molten ionic compounds will conduct since the ions are now mobile and will migrate to the electrodes of opposite sign when a potential difference is applied. If direct current is used, the compound will undergo electrolysis as the ions are discharged at the electrodes.

As nearly always in chemistry, there are exceptions to this. For example, solid lithium nitride ($Li_3N$) will conduct electricity and is used in batteries for this reason.

Aqueous solutions of ionic compounds also conduct electricity and undergo electrolysis, since the lattice breaks down into separate ions when the compound dissolves.

### Solubility
Many ionic compounds are soluble in water. We will explain this solubility more fully in **Book 2**, including the part played by entropy changes.

At the moment, you just need to appreciate that the energy required to break apart the lattice structure and separate the ions can, in some instances, be supplied by the hydration of the separated ions produced. Both positive and negative ions are attracted to water molecules because of the polarity that water molecules possess (see **Section 2.1.4** for an explanation of polarity).

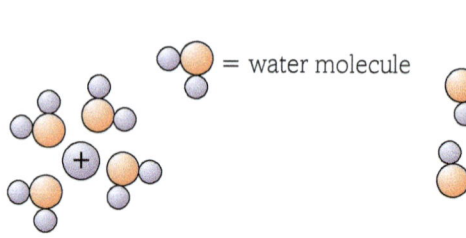

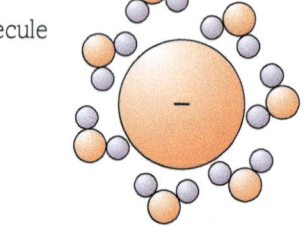

The oxygen end of the water molecules are attracted to positive ions.

The hydrogen end of the water molecules are attracted to negative ions.

**fig C** Hydration of ions.

### The formation of cations and anions
Some ionic compounds can be formed by the direct combination of two elements.

### Formation of sodium and chloride ions
For example, sodium chloride can be formed by burning sodium in chlorine:

$$2Na(s) + Cl_2(g) \rightarrow 2NaCl(s)$$

The reaction that occurs can be represented by two ionic half-equations:

$$2Na \rightarrow 2Na^+ + 2e^- \text{ and } Cl_2 + 2e^- \rightarrow 2Cl^-$$

Each sodium atom has lost one electron to become a positive sodium ion. The chlorine molecule has gained two electrons to become two chloride ions.

The electronic changes involved can be represented by dot-and-cross diagrams.

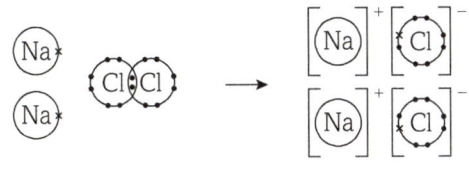

**fig D** Dot-and-cross diagram showing the formation of sodium and chloride ions.

## Formation of magnesium and oxide ions

Here is the equation for the formation of magnesium oxide:

$$2Mg(s) + O_2(g) \rightarrow 2MgO(s)$$

A dot-and-cross diagram for the reaction between magnesium and oxygen is:

**fig E** Dot-and-cross diagram showing the formation of magnesium and oxide ions.

### Learning tip
It is important when drawing a dot-and-cross diagram to represent the electrons of one atom using a cross and the other using a dot. It is also necessary to show only the outer electrons.

## Evidence for the existence of ions

The most convincing evidence for the existence of ions is the ability of an ionic compound to conduct electricity and undergo electrolysis when either molten or in aqueous solution.

For example, when a direct electric current is passed through molten sodium chloride, sodium is formed at the negative electrode and chlorine is formed at the positive electrode.

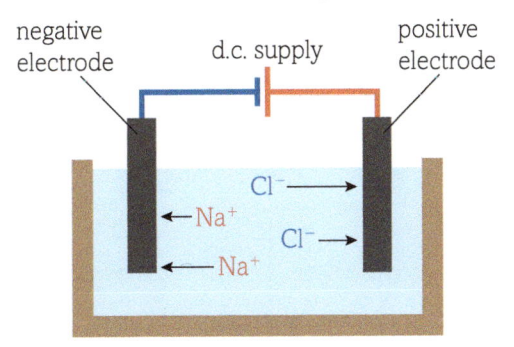

**fig F** Electrolysis of molten sodium chloride.

The explanation for this phenomenon is that:
- the positive sodium ions migrate towards the negative electrode where they gain electrons and become sodium atoms
- the negative chloride ions migrate towards the positive electrode where they lose electrons and become chlorine molecules.

At the negative electrode: $2Na^+ + 2e^- \rightarrow 2Na$

At the positive electrode: $2Cl^- \rightarrow Cl_2 + 2e^-$

Overall equation: $2NaCl \rightarrow 2Na + Cl_2$

The movement of ions can be demonstrated by passing a direct current through copper(II) chromate(VI) solution. Aqueous copper(II) ions, $Cu^{2+}(aq)$, are blue and aqueous chromate(VI) ions, $CrO_4^{2-}(aq)$, are yellow.

The $Cu^{2+}(aq)$ ions migrate towards the negative electrode and the solution around this electrode turns blue. The $CrO_4^{2-}(aq)$ ions migrate towards the positive terminal and the solution around this electrode turns yellow.

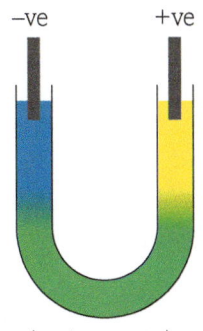

**fig G** The effect of passing an electric current through aqueous copper(II) chromate.

# Questions

1. Explain what is meant by the term 'ionic bonding'.

2. Calcium reacts with fluorine to form the ionic compound calcium fluoride:

    $Ca(s) + F_2(g) \rightarrow CaF_2(s)$

    Use a dot-and-cross diagram to show the electronic changes that occur in this reaction.

3. (a) Suggest why the strength of ionic bonding is greater in sodium fluoride than in potassium fluoride.

    (b) Suggest why the strength of ionic bonding in calcium oxide is approximately four times larger than that in potassium fluoride.

4. Explain the trend in the following ionic radii:
    (a) $Ca^{2+} > Mg^{2+} > Be^{2+}$
    (b) $P^{3-} > S^{2-} > Cl^-$

# 2.1 3 Covalent bonding

By the end of this section, you should be able to...

- know that a covalent bond is formed by the overlap of two atomic orbitals each containing a single electron
- know that a covalent bond is the strong electrostatic attraction between the nuclei of two atoms and the bonding pair of electrons
- understand the relationship between bond length and bond strength for covalent bonds

## Formation of covalent bonds

A covalent bond is formed between two atoms when an atomic orbital containing a single electron from one atom overlaps with an atomic orbital, also containing a single electron, of another atom.

The atomic orbitals involved can be any of those found in the atoms, but we shall confine our discussion to those involving only s- and p-orbitals.

The figure shows three ways in which these orbitals may overlap.

### Did you know?

There is a fourth way that overlap can occur. An s- and p-orbital can overlap end on.

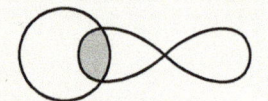

**fig B** Overlap of an s- and a p-orbital to form a sigma bond.

This overlap, however, can only result from atoms of two different elements. This invariably leads to the formation of a variant of a covalent bond, known as a 'polar' covalent bond (see Section 2.1.4).

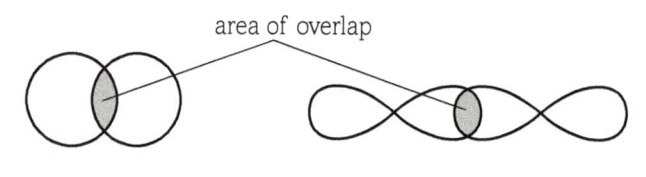

end on overlap of two s-orbitals (sigma bond)

end on overlap of two p-orbitals (sigma bond)

sideways overlap of two p-orbitals (pi bond)

**fig A** Formation of sigma bonds by an end-on overlap of atomic orbitals and a pi bond by sideways overlap of p orbitals.

An end-on overlap leads to the formation of sigma ($\sigma$) bonds. This leads to the formation of a single covalent bond between the two atoms.

A sideways overlap of two p-orbitals leads to the formation of a pi ($\pi$) bond. A feature of a $\pi$ bond is that it cannot form until a $\sigma$ bond has been formed. So, $\pi$ bonds only exist between atoms that are joined by double or triple bonds.

The different types of orbital overlap are shown in the following examples.

### Example 1 – Hydrogen

A hydrogen atom has an electronic configuration of $1s^1$.

When two hydrogen atoms bond together to form a hydrogen molecule, the two s orbitals overlap to form a new molecular orbital. The two electrons then exist in this new orbital; the highest electron density is between the two nuclei.

diagram showing orbital overlap

space filling model of a hydrogen molecule

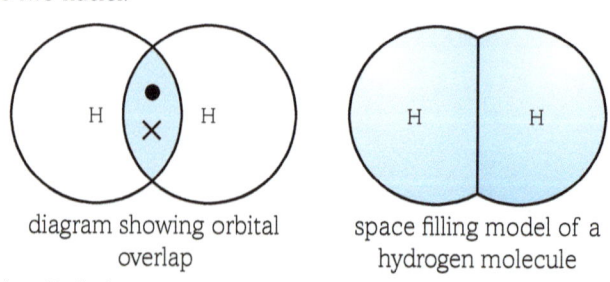

**fig C** Formation of the $\sigma$ bond in hydrogen.

## Giant structures 2.1

## Example 2 – Chlorine

A chlorine atom has an electronic configuration of $1s^2\ 2s^2\ 2p^6\ 3s^2\ 3p_x^2\ 3p_y^2\ 3p_z^1$.

When two chlorine atoms bond together, the two p orbitals (each containing a single electron) overlap.

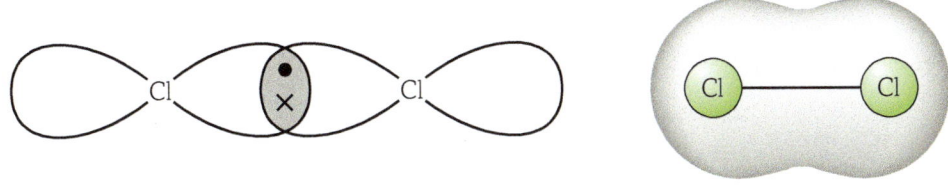

diagram showing the
orbital overlap

space filling model of a
chlorine molecule

**fig D** Formation of the $\sigma$ bond in chlorine.

### Did you know?

This is a simplistic view of the bonding in chlorine.

A more robust theory describes the bonding as the overlap between two $sp^3$ hybrid orbitals. This theory is beyond the scope of this book but we would encourage you to carry out your own research if you are interested. Look up 'orbital hybridisation'.

## Example 3 – $\pi$ bond formation

Once a $\sigma$ bond has been formed, it is possible, in certain circumstances, for a $\pi$ bond to form.

The $\pi$ bond results in a high electron density both above and below the molecule, as shown in the figure.

sideways overlap of p-orbitals

electron destiny above and
below the molecule

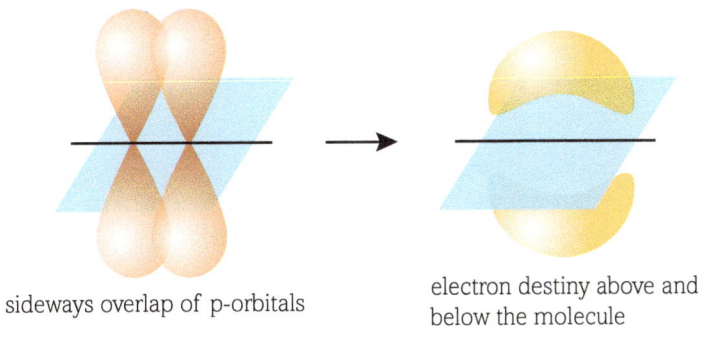

**fig E** Formation of a $\pi$ bond.

This is what happens in the ethene molecule. One of the bonds between the carbon atoms is a $\sigma$ bond; the other is a $\pi$ bond.

The $\pi$ bond in ethene is weaker than the $\sigma$ bond. This is the reason for the increased reactivity of alkenes compared with alkanes, and why alkenes can readily undergo addition reactions. (See **Sections 6.2.5** and **6.2.6** for more information.)

The triple bond in the nitrogen molecule (N≡N) is made up of one $\sigma$ bond and two $\pi$ bonds.

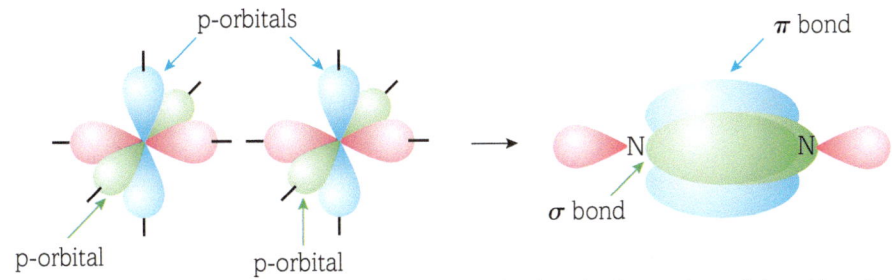

p-orbitals

$\pi$ bond

$\sigma$ bond

p-orbital    p-orbital

The p-orbitals marked are those that are involved in the formation of the pi bonds.

**fig F** Formation of the two $\pi$ bonds in nitrogen.

## Bond length and bond strength

The **bond length** is the distance between nuclei of the two atoms that are covalently bonded together.

The strength of a covalent bond is measured in terms of the amount of energy required to break one mole of the bond in the gaseous state (see **Section 8.2.1**).

The tables below show the relationship between the bond length and bond strength of a selection of covalent bonds.

| Bond | Bond length/nm | Bond strength/kJ mol$^{-1}$ |
|---|---|---|
| Cl–Cl | 0.199 | 243 |
| Br–Br | 0.228 | 193 |
| I–I | 0.267 | 151 |

| Bond | Bond length/nm | Bond strength/kJ mol$^{-1}$ |
|---|---|---|
| C–C | 0.154 | 347 |
| C=C | 0.134 | 612 |
| C≡C | 0.120 | 838 |

| Bond | Bond length/nm | Bond strength/kJ mol$^{-1}$ |
|---|---|---|
| N–N | 0.145 | 158 |
| N=N | 0.120 | 410 |
| N≡N | 0.110 | 945 |

| Bond | Bond length/nm | Bond strength/kJ mol$^{-1}$ |
|---|---|---|
| O–O | 0.148 | 144 |
| O=O | 0.121 | 498 |

**table A** Relationship between bond length and bond strength for a range of covalent bonds.

The general relationship between bond length and bond strength, for bonds that are of a similar nature, is the shorter the bond, the greater the bond strength. This is a result of an increase in electrostatic attraction between the two nuclei and the electrons in the overlapping atomic orbitals.

### Learning tip

When making a comparison between bond length and bond strength, it is important to compare 'like with like'.

For example, the strength of the C–C bond (348 kJ mol$^{-1}$) is greater than that of the N–N bond (163 kJ mol$^{-1}$) despite being longer (0.154 nm compared with 0.146 nm). In a molecule such as hydrazine ($H_2N–NH_2$), each nitrogen atom has a non-bonding (lone) pair of electrons and these repel one another, weakening the bond. In a molecule such as ethane ($H_3C–CH_3$), the carbon atoms do not have any lone pairs.

### Key definition

**Bond length** is the distance between nuclei of the two atoms that are covalently bonded together.

## Questions

1. Suggest a reason for the following trend in bond strengths:

   C–C > Si–Si > Ge–Ge

2. The F–F bond in fluorine is much shorter (0.142 nm) than the Cl–Cl (0.199 nm) bond in chlorine, and yet it is much weaker (158 kJ mol$^{-1}$ compared with 242 kJ mol$^{-1}$). Suggest a reason for this.

3. Suggest a reason why the sigma ($\sigma$) bond between the two carbon atoms in the ethene molecule is stronger that the pi ($\pi$) bond.

# 2.1 4 Electronegativity and bond polarity

**By the end of this section, you should be able to...**

- know that electronegativity is the ability of an atom to attract a bonding pair of electrons
- know that ionic and covalent bonding are the extremes of a continuum of bonding type and that electronegativity differences lead to bond polarity
- understand what is meant by a polar covalent bond

## What is electronegativity?

**Electronegativity** is the ability of an atom to attract a bonding pair of electrons.

The electronegativity of elements, in general:

- decreases down a group of the Periodic Table
- increases from left to right across a period.

This is demonstrated in the following section of the Periodic Table.

| | | | | | | | | | | | | | | | | | |
|---|---|---|---|---|---|---|---|---|---|---|---|---|---|---|---|---|---|
| | | | | | | | H 2.1 | | | | | | | | | | He |
| Li 1.0 | Be 1.5 | | | | | | | | | | | B 2.0 | C 2.5 | N 3.0 | O 3.5 | F 4.0 | Ne |
| Na 0.9 | Mg 1.2 | | | | | | | | | | | Al 1.5 | Si 1.8 | P 2.1 | S 2.5 | Cl 3.0 | Ar |
| K 0.8 | Ca 1.0 | Sc 1.3 | Ti 1.5 | V 1.6 | Cr 1.6 | Mn 1.5 | Fe 1.8 | Co 1.8 | Ni 1.8 | Cu 1.9 | Zn 1.6 | Ga 1.6 | Ge 1.8 | As 2.0 | Se 2.4 | Br 2.8 | Kr |
| Rb 0.8 | Sr 1.0 | Y 1.2 | Zr 1.4 | Nb 1.6 | Mo 1.8 | Tc 1.9 | Ru 2.2 | Rh 2.2 | Pd 2.2 | Ag 1.9 | Cd 1.7 | In 1.7 | Sn 1.8 | Sb 1.9 | Te 2.1 | I 2.5 | Xe |
| Cs 0.7 | Ba 0.9 | La 1.1 | Hf 1.3 | Ta 1.5 | W 1.7 | Re 1.9 | Os 2.2 | Ir 2.2 | Pt 2.2 | Au 2.4 | Hg 1.9 | Tl 1.8 | Pb 1.8 | Bi 1.9 | Po 2.0 | At 2.2 | Rn |

**fig A** Table of electronegativities.

## Distribution of electron density

If two atoms of the same element are bonded together by the overlap of atomic orbitals, the distribution of electron density between the two nuclei will be symmetrical. This is because the ability of each atom to attract the bonding pair of electrons is identical.

The diagram below is an electron density map for chlorine ($Cl_2$):

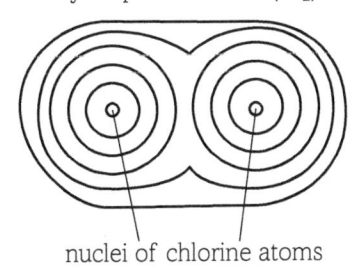

**fig B** Electron density map of a chlorine molecule.

The diagram looks like a contour map. The contour lines correspond to electron density and can be thought of as the likelihood that a bonding electron will fall within that contour at a given instant in time. For a normal covalent bond, the contours fall symmetrically about the nuclei.

## Polar covalent bonds

However, if the two atoms bonded together are from elements that have different electronegativities, then the distribution of electron density will not be symmetrical about the two nuclei. This is shown by the following electron density map for the hydrogen chloride (HCl) molecule.

nucleus of hydrogen atom (electronegativity 2.1)

nucleus of chlorine atom (electronegativity 3.0)

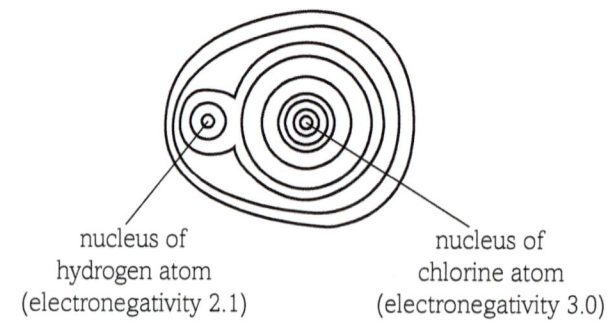

**fig C** Electron density map of a hydrogen chloride molecule.

The contour lines are more closely spaced near to the chlorine atom, which is the atom with the higher electronegativity.

Since the electron density is higher around the chlorine atom, that end of the molecule has acquired a slightly negative charge. This is represented by the symbol $\delta-$. The other end of the molecule carries a slightly positive charge, represented by the symbol $\delta+$.

$$H^{\delta+}\text{—}Cl^{\delta-}$$

A bond like this is called a **polar covalent bond** or sometimes just a 'polar bond'.

Another way of representing a polar covalent bond is to use an arrow to show the direction of electron drift.

$$H\rightarrow Cl$$

Other examples of polar covalent bonds are:

$$C^{\delta+}\rightarrow Cl^{\delta-} \quad H^{\delta+}\rightarrow O^{\delta-} \quad H^{\delta+}\rightarrow N^{\delta-} \quad H^{\delta+}\rightarrow C^{\delta-} \quad C^{\delta+}\rightarrow Cl^{\delta-}$$

> **Learning tip**
>
> Covalent bonds that are non-polar are sometimes called 'normal' covalent bonds or 'pure' covalent bonds. This is to distinguish them from polar covalent bonds.
>
> Both types of bond are formed by the overlap of atomic orbitals.
>
> In both cases the overlapping area contains two electrons per bond formed.

## Continuum of bonding type

Polar covalent bonds can be thought of as being between two ideals on bonding types. These ideals are:

- pure (100%) covalent
- pure (100%) ionic.

Consider a polar covalent bond as a covalent bond that has some degree of ionic character.

## Questions

1. Suggest why the electronegativity of fluorine is greater than that of chlorine, despite the fact that the nucleus of a chlorine atom contains more protons.

2. Ionic bonding and covalent bonding are two extremes of chemical bonding. Many compounds have bonding that is intermediate in character.

   (a) Giving an example in each case, explain what is meant by the terms:

   (i) ionic bonding, and  (ii) covalent bonding.

   (b) Select a compound that has bonding of an intermediate character and explain why it has this type of bonding.

3. Place the following bonds in order of decreasing polarity (i.e. place the most polar first).

   C–Br   C–Cl   C–F   C–I

   Explain how you arrived at your answer.

### Key definitions

**Electronegativity** is the ability of an atom to attract a bonding pair of electrons in a covalent bond.

A **polar covalent bond** is a type of covalent bond between two atoms where the bonding electrons are unequally distributed. Because of this, one atom carries a slight negative charge and the other a slight positive charge.

### Did you know?

Although we have suggested that complete electron transfer occurs in the formation of ionic compounds, in practice there is always some degree of covalency. We will deal with the reasons behind this in **Book 2**.

# 2.2　1　Bonding in discrete molecules

**By the end of this section, you should be able to...**

- understand what is meant by the term discrete (simple) molecule
- draw dot-and-cross diagrams to show electrons in discrete molecules, including those with multiple bonds
- draw displayed formulae to represent the bonding in discrete molecules

### Did you know?

Molecules as components of matter are common in organic substances (and therefore biochemistry). They also make up most of the oceans and atmosphere.

However, ionic crystals (salts) and giant covalent crystals (network solids) are often composed of repeating unit cells that extend either in a plane (such as in graphene) or three-dimensionally (such as in diamond or sodium chloride). These substances are not composed of molecules. Solid metals are also not composed of molecules.

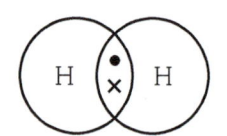

**fig A** Dot-and-cross diagram for hydrogen with overlapping circles.

## Discrete molecules

A **discrete (simple) molecule** is an electrically neutral group of two or more atoms held together by chemical bonds.

## Dot-and-cross diagrams

Covalent and polar covalent bonding in discrete molecules can be shown by dot-and-cross diagrams.

**Fig A** shows the example of hydrogen, $H_2$.

Further examples of dot-and-cross diagrams are shown in the table below.

| Substance | Dot-and-cross diagram | |
|---|---|---|
| Water, $H_2O$ | H:Ö:H | |
| Ammonia, $NH_3$ | H:N̈:H  H | |
| Methane, $CH_4$ | H:C̈:H  H  H | |

**table A** Dot-and-cross diagrams for water, ammonia and methane.

## Discrete molecules 2.2

## What about the octet rule?

You might read that in order to form a stable compound, the outer shell of each atom must have the same number of electrons as the outer shell of a noble gas. In most cases this will be eight electrons. This has led to a rule that is often referred to as the 'octet rule'.

This is not always true, as you can see from the examples below. In each case, the outer shell of the central atom of the molecule does not contain eight electrons.

| Substance | Dot-and-cross diagram | Number of electrons around central atom |
|---|---|---|
| Beryllium chloride, $BeCl_2$ | :Cl×Be×Cl: | 4 |
| Boron trichloride, $BCl_3$ | :Cl×B×Cl: with :Cl: above | 6 |
| Phosphorus(V) chloride, $PCl_5$ | (dot-and-cross diagram) | 10 |
| Sulfur hexafluoride, $SF_6$ | (dot-and-cross diagram) | 12 |

**table B** Examples disproving the octet rule.

### Dot-and-cross diagrams of molecules containing multiple bonds

Table C shows the dot-and-cross diagrams for three molecules that contain a double or triple bond.

### Displayed formulae (full structural formulae)

A **displayed (full structural) formula** shows each bonding pair as a line drawn between the two atoms involved.

Table C gives some examples of displayed formulae together with the dot-and-cross diagram.

| Substance | Dot-and-cross diagram | Displayed formula |
|---|---|---|
| Water, $H_2O$ | H:O:H | H−O−H |
| Ammonia, $NH_3$ | H:N:H with H above | H−N−H with H above |
| Oxygen, $O_2$ | :O::O: | O=O |
| Nitrogen, $N_2$ | :N:::N: | N≡N |
| Carbon dioxide, $CO_2$ | :O::C::O: | O=C=O |

**table C** Examples of displayed formulae with the corresponding dot-and-cross diagram.

### Learning tip

Although it is essential to show all of the non-bonding (lone) pairs of electrons in a dot-and-cross diagram, it is not necessary to show them in a displayed formula.

### Key definitions

A **discrete (simple) molecule** is an electrically neutral group of two or more atoms held together by chemical bonds.

A **displayed (full structural) formula** shows each bonding pair as a line drawn between the two atoms involved.

## Questions

1. Draw a dot-and-cross diagram for each of the following molecules:
   (a) $H_2S$  (b) $PH_3$  (c) $PF_3$  (d) $SCl_2$  (e) $AsF_5$  (f) HCN  (g) $SO_2$

2. Draw the displayed formula for each of the molecules in Question 1.

## 2.2 2 Dative covalent bonds

By the end of this section, you should be able to...

- know that in a dative covalent bond both electrons in the bond are supplied by only one of the atoms involved in forming the bond
- draw dot-and-cross diagrams for some molecules and ions that contain dative covalent bonds

### Dative covalent bond formation

A dative covalent bond is formed when an empty orbital of one atom overlaps with an orbital containing a non-bonding pair (lone pair) of electrons of another atom.

It is often represented by an arrow starting from the atom providing the pair of electrons and going towards the atom with the empty orbital.

### Example 1 – the hydroxonium ion, $H_3O^+$

The dot-and-cross diagram and the displayed formula of a hydroxonium ion are:

$$\left[ \begin{array}{c} H \overset{\times\times}{\underset{\times\times}{:\!O\!:}} H \\ H \end{array} \right]^+ \quad \left[ \begin{array}{c} H-O-H \\ \downarrow \\ H \end{array} \right]^+$$

**fig A** Dot-and-cross diagram and displayed formula of the hydroxonium ion.

The empty 1s orbital of the $H^+$ ion overlaps with the orbital of the oxygen atom that contains the lone pair of electrons.

### Example 2 – the ammonium ion, $NH_4^+$

The dot-and-cross diagram and the displayed formula of an ammonium ion are:

$$\left[ \begin{array}{c} H \\ H \overset{\times\bullet}{\underset{\times\times}{:\!N\!:}} H \\ H \end{array} \right]^+ \quad \left[ \begin{array}{c} H \\ | \\ H-N-H \\ \downarrow \\ H \end{array} \right]^+$$

**fig B** Dot-and-cross diagram and displayed formula for the ammonium ion.

The empty 1s orbital of the $H^+$ ion overlaps with the orbital of the nitrogen atom that contains the lone pair of electrons.

### Example 3 – aluminium chloride, $Al_2Cl_6$

The aluminium atom in the $AlCl_3$ molecule has only six electrons in its outer shell and so has an empty orbital.

**fig C** Dot-and-cross diagram for aluminium chloride.

In the gas phase, just above its boiling temperature, aluminium chloride exists as $Al_2Cl_6$ molecules.

Two $AlCl_3$ molecules bond together. One of the atomic orbitals of a chlorine atom of one $AlCl_3$ molecule that contains a lone pair overlaps with the empty orbital of the aluminium atom of a second $AlCl_3$ molecule. The same happens between the chlorine atom of the second molecule and the aluminium atom of the first molecule.

One chlorine atom from each molecule acts as a bridge connecting the two molecules with dative covalent bonds.

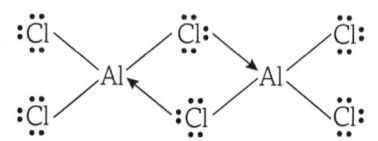

**fig D** Displayed formula for aluminium chloride dimer.

## Questions

1. (a) Draw a dot-and-cross diagram for a molecule of $NH_3$ and a molecule of $BF_3$.
   (b) Draw a dot-and-cross diagram for a molecule of $NH_3 \cdot BF_3$.

2. Draw a dot-and-cross diagram and displayed formula for the $AlCl_4^-$ ion and identify the dative covalent bond.

3. One way of describing the bonding in a molecule of carbon monoxide (CO) is to state that it contains two covalent bonds and one dative bond. Using this description, draw a dot-and-cross diagram and displayed formula for a molecule of carbon monoxide.

## 2.2 3 Shapes of molecules and ions

By the end of this section, you should be able to...

- understand that the shape of a simple molecule or ion is determined by the repulsion between the electron pairs that surround the central atom
- understand reasons for the shapes of, and bond angles in, simple molecules and ions with up to six outer pairs of electrons (with any combination of bonding pairs and lone pairs)
- predict the shapes of, and bond angles in, simple molecules and ions with up to six outer pairs of electrons using electron-pair repulsion theory

### Electron pair repulsion theory

The electron pair repulsion (or EPR) theory states that:

- the shape of a molecule or ion is caused by repulsion between the pairs of electrons, both bond pairs and lone (non-bonding) pairs, that surround the central atom
- the electron pairs arrange themselves around the central atom so that the repulsion between them is at a minimum
- lone pair–lone pair repulsion > lone pair–bond pair repulsion > bond pair–bond pair repulsion.

**Learning tip**

This theory is sometimes also called the valence shell electron pair repulsion theory, abbreviated to VSEPR.

The first two rules are used to obtain the basic shape of the molecule or ion. The third rule is used to estimate values for the bond angles.

### The shapes of molecules and ions

To obtain the shape of a molecule or ion it is first necessary to obtain the number of bond pairs and lone pairs of electrons around the central atom.

This is best done by drawing a dot-and-cross diagram. You can then apply the guidelines listed in the **table A**.

### Molecules with multiple bonds

In order to determine the shape of a molecule containing one or more multiple bonds, treat each multiple bond as if it contained only one pair of electrons.

### Example 1 – carbon dioxide, $CO_2$

The displayed formula for carbon dioxide is $O=C=O$. There are no lone pairs on the carbon atom.

If each double bond is treated as an electron pair, then the molecule is linear, like $BeCl_2$.

| Number of bond pairs | Number of lone pairs | Shape | Example |
|---|---|---|---|
| 2 | 0 | linear | Cl—Be—Cl |
| 3 | 0 | trigonal planar | $BCl_3$ |
| 4 | 0 | tetrahedral | $CH_4$ |
| 5 | 0 | trigonal bipyramidal | $PCl_5$ |
| 6 | 0 | hexagonal | $SF_6$ |
| 3 | 1 | trigonal pyramidal | $NH_3$ |
| 2 | 2 | V-shaped | $H_2O$ |

**table A** Shapes of molecules.

### Example 2 – sulfur trioxide, $SO_3$

The displayed formula of sulfur trioxide is:

$$O=S=O$$
$$\|$$
$$O$$

There are no lone pairs on the sulfur atom.

Treating each double bond as an electron pair produces a trigonal planar shape for this molecule.

## The bond angles in molecules and ions

The table below shows the bond angles of a range of molecules and ions.

| | |
|---|---|
| 1 Linear, e.g. $BeCl_2$<br><br>The bond angle is 180°. 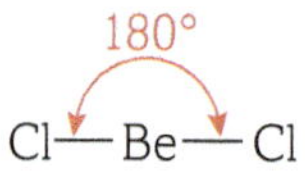 | 2 Trigonal planar, e.g. $BCl_3$<br><br>The bond angle is 120°. 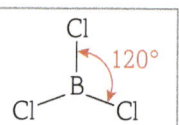 |
| 3 Tetrahedral, e.g. $CH_4$<br><br>The bond angle is 109.5°. 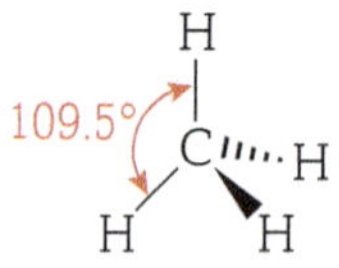 | 4 Trigonal pyramidal, e.g. $NH_3$<br><br>The bond angle is 107°. 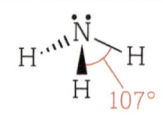<br><br>Lone pair–bond pair repulsion is greater than bond pair–bond pair repulsion, so the angle is slightly less than 109.5°. |
| 5 V-shaped, e.g. $H_2O$<br><br>The bond angle is 104.5°. 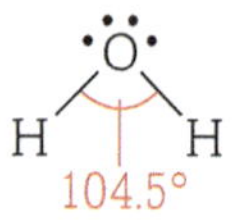<br><br>Lone pair–lone pair repulsion is greater than lone pair–bond pair repulsion, so the bond angle is even further depressed from 109.5°, and is slightly less than the 107° in $NH_3$. | 6 Trigonal bipyramidal, e.g. $PCl_5$<br><br>There are two bond angles: 90° and 120°. 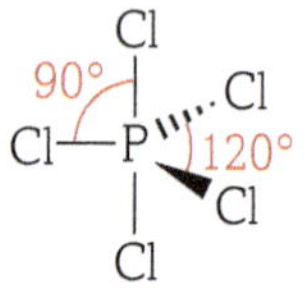 |
| 7 Hexagonal, e.g. $SF_6$<br><br>There are two bond angles: 90° and 180°. 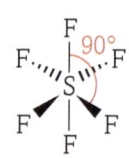<br><br>The angle between the bonds of two fluorine atoms opposite one another is 180°. | 8 Tetrahedral, e.g. $NH_4^+$<br><br>As with $CH_4$, the bond angles are 109.5°. 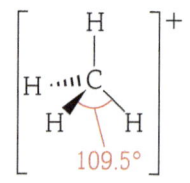<br><br>Note the change from 107° in ammonia to 109.5° in the ammonium ion. |

**table B** The bond angles of a range of molecules and ions.

# Questions

1. (a) Draw a diagram to show the shape of each of the following molecules:
   (i) $H_2S$  (ii) $PH_3$  (iii) $PF_3$  (iv) $SCl_2$  (v) $AsF_5$  (vi) HCN  (vii) $SO_2$
   (b) Give the name of each shape.

2. Solid phosphorus pentachloride has the formula $[PCl_4]^+[PCl_6]^-$.
   (a) Draw a diagram to show the shape of each ion.
   (b) State the bond angles present in each ion.

3. Two possible ways of arranging the bonding pairs and lone pairs of electrons in a molecule of $XeF_4$ are:

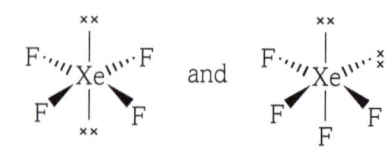

   Suggest which of these two arrangements is the more likely and justify your answer.

# 2.2 4 Non-polar and polar molecules

**By the end of this section, you should be able to...**

- understand the difference between non-polar and polar molecules and be able to predict whether or not a given molecule is likely to be polar

## Shape and polarity

The drift of bonded electrons towards the more electronegative element (see **Section 2.1.4**) results in a separation of charge. This separation of charge is called a **dipole**.

Each of the bonds in a molecule has its own dipole associated with it. The overall dipole of a molecule depends on its shape. Depending on the relative angles between the bonds, the individual dipoles can either reinforce one another or cancel out each other.

- If the cancellation is complete, the resulting molecule will have no overall dipole and is said to be 'non-polar'.
- If the dipoles reinforce one another, the molecule will possess an overall dipole and is said to be 'polar'.

## Diatomic molecules

Hydrogen and chlorine are examples of diatomic molecules that are non-polar. The two atoms in each molecule are the same and so have the same electronegativity. The distribution of electron density of the bonding electrons in either molecule is totally symmetrical (see **Section 2.1.4**). The bond in each is therefore non-polar, making the molecules non-polar.

However, the bond in the hydrogen chloride molecule is polar since the electronegativity of chlorine (3.0) is greater than that of hydrogen (2.1).

$$H^{\delta+} \rightarrow Cl^{\delta-}$$

Since this is the only polar bond in the molecule, the molecule itself is polar.

The following symbol is used to represent a dipole: $\rightarrow$

So, the dipole in the hydrogen chloride molecule is shown as:

$$H^{\delta+} - Cl^{\delta-}$$
$$\rightarrow$$

## Polyatomic molecules

1. Linear molecules

Example: carbon dioxide, $CO_2$

Both of the bonds in the carbon dioxide molecule are polar, but the dipoles cancel out one another.

$$O^{\delta-}=C^{\delta+}=O^{\delta-}$$

**fig A** Dipoles in carbon dioxide.

The carbon dioxide molecule is therefore non-polar.

2. Trigonal planar molecules

Example: boron chloride, $BCl_3$

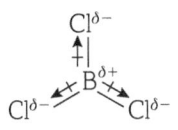

**fig B** Dipoles in boron trichloride.

All three B—Cl bonds are polar, but because the molecule is symmetrical the dipoles cancel out one another. The molecule is non-polar.

3. Tetrahedral molecules

Example 1: tetrachloromethane, $CCl_4$

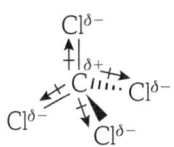

**fig C** Dipoles in tetrachloromethane.

All four C—Cl bonds are polar, but because the molecule is symmetrical the dipoles cancel out one another. The molecule is non-polar.

Example 2: trichloromethane, $CHCl_3$

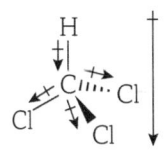

**fig D** Dipoles in trichloromethane.

All four bonds are polar but, although the molecule is symmetrical, the dipoles reinforce one another and so the molecule is polar.

## 4. Trigonal pyramidal molecules

Example: ammonia, $NH_3$

All three N—H bonds are polar and the dipoles reinforce one another. The molecule is polar.

**fig E** Dipoles in ammonia.

## 5. V-shaped molecules

Example: water, $H_2O$

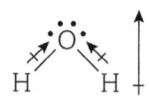

**fig F** Dipoles in water.

Both O—H bonds are polar and the dipoles reinforce one another. The molecule is polar.

### Additional reading

**Dipole moments**

The polarity of the molecule is measured by its **dipole moment**.

- For a diatomic molecule such as hydrogen chloride, this is defined as the difference in charge (i.e. the difference in magnitude between δ+ and δ−) multiplied by the distance of separation between the charges.
- For a polyatomic molecule, it is more complicated because the polarities of each bond have to be taken into account, as well as any lone pairs on the central atom.

The table below gives the dipole moments of a number of molecules.

| Molecule | Dipole moment/D |
|---|---|
| $H_2$ | 0 |
| $Cl_2$ | 0 |
| HCl | 1.05 |
| $CO_2$ | 0 |
| $BCl_3$ | 0 |
| $CCl_4$ | 0 |
| $CHCl_3$ | 1.02 |
| $NH_3$ | 1.48 |
| $H_2O$ | 1.84 |

**table A** Dipole moments of some molecules.

The unit of dipole moment is the Debye, symbol D. You do not need to understand this unit; just focus on the magnitude of the numbers. The larger the number, the more polar the molecule.

# Questions

1. A bond between two atoms in a molecule may possess a 'dipole'.
   (a) Explain how this dipole arises.
   (b) Some bonds that you are likely to meet in organic chemistry are listed. Which of these bonds are likely to possess a dipole? In each case indicate which atom is δ+ and which is δ−.

   C–Cl   O–H   C–C   C–O   C=C   C–N   N–H

2. State whether each of the following molecules are non-polar or polar. In each case, explain your reasoning.
   (a) $H_2S$
   (b) $CH_4$
   (c) $SO_2$
   (d) $SO_3$
   (e) $AlBr_3$
   (f) $PBr_3$

3. There are two stereoisomers of dichloroethene. The structure and shape of each molecule is:

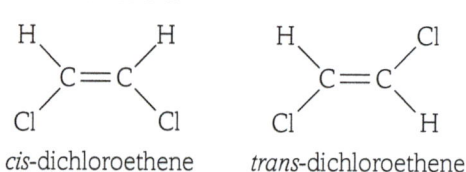

   cis-dichloroethene        trans-dichloroethene

   Suggest why the *cis* isomer is polar, while the *trans* isomer is non-polar.

### Key definition

A **dipole** is said to exist when two charges of equal magnitude but opposite signs are separated by a small distance.

## 2.2 5 Intermolecular interactions

By the end of this section, you should be able to...

- understand the nature of intermolecular forces resulting from the following interactions:
  (i) London forces
  (ii) permanent dipoles
  (iii) hydrogen bonds
- understand the interactions in molecules, such as $H_2O$, liquid $NH_3$ and liquid HF, which give rise to hydrogen bonding
- predict the presence of hydrogen bonding in molecules analogous to those mentioned above

## Background to non-bonded intermolecular interactions

A number of interactions between molecules are considerably weaker than typical covalent and polar covalent bonds. These interactions are usually described as:

- 'non-bonded interactions'
- 'intermolecular' because they occur between molecules.

The most important non-bonded interaction gives rise to forces known as 'London forces', since their existence was first postulated by Fritz London, a German physicist, in 1930. London forces are also sometimes referred to as 'dispersion forces'. Despite being weaker than covalent and polar covalent bonds, London forces play an important part in determining the physical and chemical properties of many molecules.

Other intermolecular interactions arise from the permanent dipoles that exist in some molecules.

## London forces

It is easiest to describe this interaction by considering two non-polar molecules of nitrogen, labelled **A** and **B**.

molecule **A**     molecule **B**

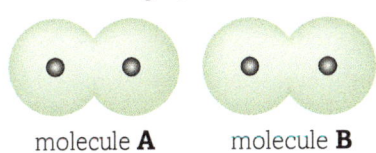

**fig A** Electron density in nitrogen.

Each molecule is non-polar because, on average, the electron density is symmetrically distributed throughout the molecule (see **Section 2.2.4**).

However, electron density fluctuates over time. If at any instant the electron density becomes unsymmetrical in molecule **A**, a dipole will be generated, as shown in the diagram below.

instantaneous dipole in molecule **A**

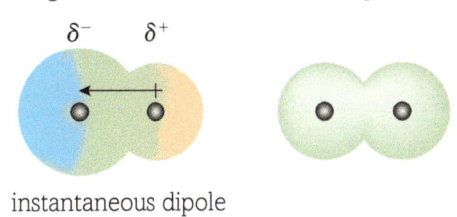

**fig B** An instantaneous dipole in nitrogen.

The electron density on the left of the molecule has increased, giving that end of the molecule a partial negative charge ($\delta-$), whereas the electron density on the right has decreased, giving that end of the molecule a partial positive charge ($\delta+$). So, an instantaneous dipole is created in the molecule **A**.

The $\delta+$ end of molecule **A** is closer to molecule **B**, so the electron density of molecule **B** is pulled to the left, generating a partial negative charge on the left-hand end of the molecule, and a partial positive charge on the right-hand end. The result is the creation of an induced dipole in molecule **B**.

induced dipole in molecule **B**

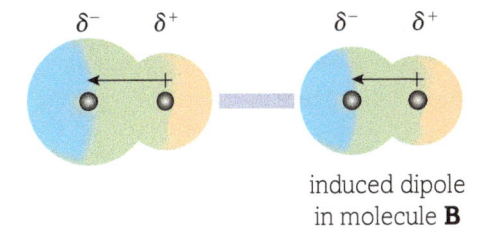

**fig C** An induced dipole in nitrogen.

Since it was the dipole of **A** that led to the induction of the dipole in **B**, the two dipoles are necessarily arranged so that they will interact favourably with one another. It is this favourable interaction that is responsible for the London force of attraction between the two molecules.

The important point to realise is that the induced dipole will *always* be aligned in such a way that the interaction with the instantaneous dipole is favourable. The fluctuations that lead to the generation of an instantaneous dipole, and the subsequent induction of a dipole in a nearby molecule, are very rapid processes compared with the rates at which the molecules are moving owing to the kinetic energy and rotational energy they have. So, as the molecules move around, they will continue to attract each other regardless of the orientation of the two molecules.

A feature of London forces is that the attractive force increases with increasing number of electrons in the molecule. This is easily demonstrated by observing the boiling temperatures of the noble gases, all of which exist as monatomic molecules (i.e. single atom molecules). The London force is the only force of attraction between the molecules. The stronger the force of attraction, the more energy is required to separate the molecules, so the boiling temperature increases.

| Gas | helium | neon | argon | krypton | xenon | radon |
|---|---|---|---|---|---|---|
| Boiling temperature /K | 4.3 | 27.1 | 87.4 | 121 | 165 | 211 |

**table A** Boiling temperatures of the noble gases.

The more electrons there are in a molecule, the greater the fluctuation in electron density and the larger the instantaneous and induced dipoles created. There is a similar trend in the boiling temperatures of the halogens from $F_2$ to $I_2$ (see **Section 4.2.1**).

A second feature of London forces is that they depend on the shape and size of the molecules. The more points of contact between the molecules, the greater the overall London force. We will look at this in more detail in **Section 2.2.6**.

A third feature is that London forces are always present between molecules, regardless of whether they have a permanent dipole or whether or not they hydrogen bond with each other (see later in this section).

## Permanent dipoles

If the molecules possess permanent dipoles, they will also interact with one another. If the dipoles are aligned correctly, then there will be a favourable interaction and the two molecules will attract one another.

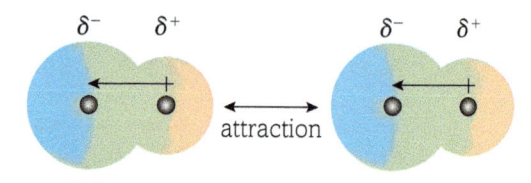

**fig D** Attraction between permanent dipoles.

The problem here is that the random movement of the molecules is such that the dipoles are not always aligned to produce a favourable interaction.

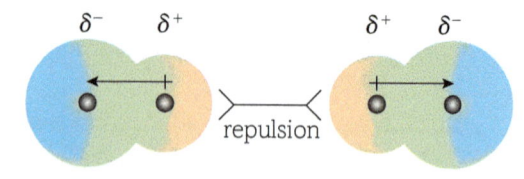

**fig E** Repulsion between permanent dipoles.

As a result, when averaged out, the interaction between permanent dipoles is usually much less than the interaction between instantaneous-induced dipoles. The London force is usually the most significant interaction between molecules.

### Learning tip

At first glance it seems counterintuitive that the interaction between permanent dipoles should be weaker than that between instantaneous and induced dipoles.

The key point is that induced dipoles are always aligned so that their interaction is favourable. This is not true for permanent dipoles.

It is possible for a molecule with a permanent dipole to induce a dipole in a nearby molecule. The two types of interaction, permanent dipole–permanent dipole and permanent dipole–induced dipole, are usually put together at A level under the heading permanent dipole–dipole forces.

## Summary of non-bonded intermolecular interactions

The table is a useful reminder of the origins of non-bonded intermolecular interactions.

| Name of interaction | Origin |
|---|---|
| London force | instantaneous dipole–induced dipole interaction |
| Permanent dipoles | permanent dipole–dipole interaction |

**table B** Summary of non-bonded intermolecular interactions.

### Learning tip

It is important to realise that London forces exist between all types of molecules, whether non-polar or polar.

So London forces, as well as permanent dipole–dipole interactions, exist between the polar molecules of hydrogen chloride.

### Did you know?

The term 'van der Waals force' is often used in conjunction with intermolecular interactions.

The van der Waals force is the sum of all the intermolecular interactions between the molecules. The term includes London forces and permanent dipole–dipole interactions.

## The hydrogen bond

There is one other intermolecular interaction that is, in some cases, very important. It is called a **hydrogen bond**.

The key to understanding the nature of the hydrogen bond is appreciating that the atom bonded to hydrogen has to be more electronegative than hydrogen, and that there must be some evidence of bond formation between the hydrogen and another atom, either within the same molecule ('intramolecular hydrogen bonding') or a different molecule ('intermolecular hydrogen bonding').

We are only looking at *inter*molecular hydrogen bonding, although *intra*molecular hydrogen bonding plays a significant role in many biochemical molecules such as DNA and proteins.

Because there has to be evidence of some bond formation (the nature of the evidence, largely from various types of spectroscopy, is beyond the scope of this book), there is an argument that hydrogen bonds could be considered to be bonded intermolecular interactions. However, most academic texts still prefer to use the term non-bonded intermolecular interaction to describe them.

Hydrogen bonding is most significant when hydrogen is bonded to very small, highly electronegative atoms such as oxygen, nitrogen and fluorine, although it is not confined to these atoms.

## Hydrogen bonding through oxygen

All compounds containing an –O—H group form intermolecular hydrogen bonds.

The most important example is water.

The hydrogen bond forms between the oxygen atom of one water molecule and the hydrogen atom of a second water molecule.

The interaction is not just that of an extreme dipole–dipole interaction; there is some partial bond formation utilising a lone pair of electrons on the oxygen atom. Since the oxygen atom has two lone pairs, it can form hydrogen bonds with two other water molecules.

Another feature of a hydrogen bond that indicates partial bond formation is that, like covalent bonds, hydrogen bonds are directional in nature. The bond angle between the three atoms involved is often 180°, or close to it, but this is not always so.

Alcohols (see **Section 6.4.1**) also form intermolecular hydrogen bonds.

The diagram shows the formation of a hydrogen bond between two ethanol molecules.

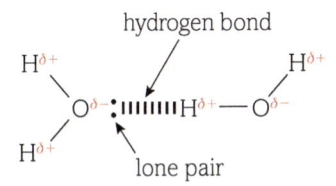

**fig F** Hydrogen bonding between water molecules.

**fig G** Hydrogen bonding between ethanol molecules.

### Did you know?

Hydrogen bonds are often quoted as being the strongest of all intermolecular interactions. This is true for some molecules such as water, but it is not true for the majority of molecules containing hydrogen bonds.

For example, while the hydrogen bond is significant in the case of the short chain alcohols, e.g. methanol ($CH_3OH$) and ethanol ($CH_3CH_2OH$), it becomes less significant for longer chain alcohols such as pentan-1-ol ($CH_3CH_2CH_2CH_2CH_2OH$) (see **Section 2.2.6**). The same trend is exhibited by the amines (see the section on amines in **Book 2**).

So, it is not correct to state that the hydrogen bond is always the most significant intermolecular interaction in any given group of analogous compounds. In fact, we will see that some hydrogen bonds are very weak.

## Hydrogen bonding through nitrogen

### Example: ammonia, $NH_3$

All compounds containing an –N—H group can form intermolecular hydrogen bonds. An example is the organic group of compounds known as primary amines, which have the general formula $RNH_2$.

## Hydrogen bonding through fluorine

The only fluorine compound with intermolecular hydrogen bonding is hydrogen fluoride.

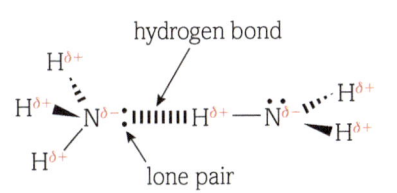

**fig H** Hydrogen bonding between ammonia molecules.

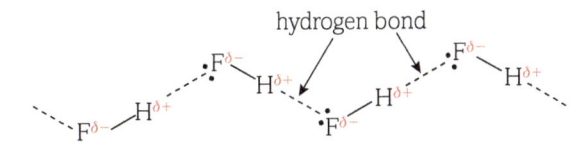

**fig I** Hydrogen bonding between hydrogen fluoride molecules.

## Additional reading

### Strong and weak hydrogen bonds

It was once firmly believed that hydrogen bonds were formed only when hydrogen was bonded to oxygen, nitrogen or fluorine. Richard Nelmes of Edinburgh University (now retired) surprised the world of chemistry when he discovered that solid hydrogen sulfide had an extended system of hydrogen bonding similar to that in ice.

This is how Gautam Desiraju, of the University of Hyderabad, defined the weak hydrogen bond.

*"The weak hydrogen bond is an interaction X–H....Y in which a hydrogen atom forms a bond between X and Y, of which one or even both are of moderate to low electronegativity"* (1999).

Since sulfur has a low electronegativity, the hydrogen bond between $H_2S$ molecules is considered to be weak. In fact, it has a magnitude of 7 kJ mol$^{-1}$, compared with 22 kJ mol$^{-1}$ in ice.

Some strengths of hydrogen bonds are shown in the table below. For comparison, the strengths of the 'full' bonds in some of these species are also shown.

**table C** Strengths of some hydrogen bonds.

You will notice that, in most cases, the hydrogen bonds are very much weaker than the full bonds. You can also see that the strengths of the hydrogen bonds increase with the electronegativity of the element to which the hydrogen is attached. So, hydrogen bonding increases as follows:

$H_2S < NH_3 < H_2O < HF$

Also note that the hydrogen bond to an *ion* is much stronger than that to a neutral molecule.

The strongest hydrogen bond shown in the table is that between HF and F$^-$. In this species the two bond lengths and bond strengths are identical. The resulting ion, [F–H–F]$^-$, is so stable that it is obtainable as the solid sodium salt, sodium hydrogendifluoride, NaHF$_2$.

## Questions

1. Molecules of ethanoic acid dimerise through hydrogen bonding when dissolved in certain organic solvents. The structure and shape of an ethanoic acid molecule is:

    Draw a diagram to show how hydrogen bonds are formed between two molecules of ethanoic acid.

2. Explain how it is possible for a hydrogen bond to form between a molecule of propanone ($CH_3COCH_3$) and trichloromethane ($CHCl_3$). The structure and shapes of the molecules are:

3. The boiling temperature of ethanol ($CH_3CH_2OH$, 78.5 °C) is considerably higher than that of methoxymethane ($CH_3OCH_3$, −24.8 °C). The structure and shape of each molecule is:

    ethanol     methoxymethane

    In terms of the intermolecular interactions involved, explain the difference in boiling temperatures.

### Key definition

The **hydrogen bond** is an intermolecular interaction (in which there is some evidence of bond formation) between a hydrogen atom of a molecule (or molecular fragment) bonded to an atom which is more electronegative than hydrogen and another atom in the same or a different molecule.

## 2.2 | 6 | Intermolecular interactions and physical properties

**By the end of this section, you should be able to...**

- understand in terms of intermolecular interactions and the physical properties shown by materials, including:
    (i) the trends in boiling temperatures of alkanes with increasing chain length
    (ii) the effect of branching in the carbon chain on the boiling temperatures of alkanes
    (iii) the relatively low volatility (higher boiling temperatures) of alcohols compared with alkanes with a similar number of electrons
    (iv) the trends in boiling temperatures of the hydrogen halides HF to HI
- understand the following anomalous properties of water resulting from hydrogen bonding:
    (i) its relatively high melting and boiling temperatures
    (ii) the density of ice compared with that of water
- understand the reasons for the choice of solvents, including:
    (i) water, to dissolve some ionic compounds, in terms of the hydration of the ions
    (ii) water, to dissolve simple alcohols, in terms of hydrogen bonding
    (iii) water, as a poor solvent for some compounds, in terms of inability to form strong hydrogen bonds
    (iv) non-aqueous solvents, for compounds which have similar intermolecular interactions to those in the solvent

## Boiling temperatures of alkanes and alcohols

### Unbranched alkanes

The alkanes are a homologous series of hydrocarbons with the general formula $C_nH_{2n+2}$ (see **Section 6.2.5**).

The graph shows the relationship between the boiling temperature and the relative molecular mass for the first 10 unbranched alkanes (i.e. $CH_4$ to $C_{10}H_{22}$ inclusive).

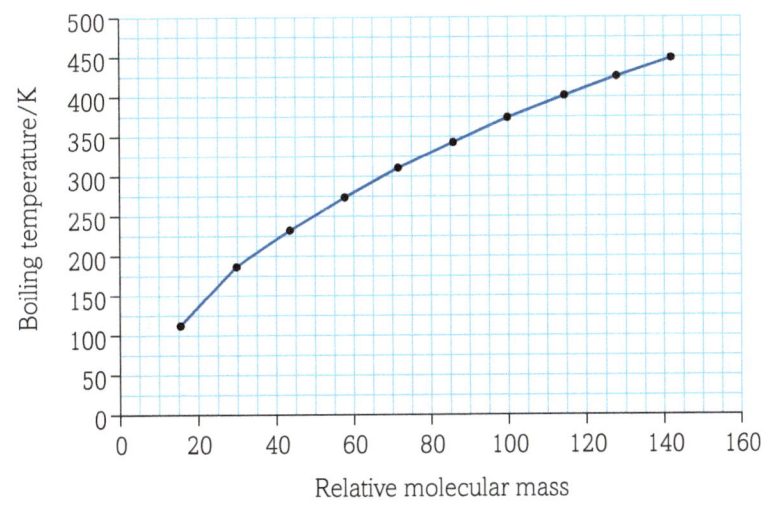

**fig A** The relationship between boiling temperature and relative molecular mass for the first 10 alkanes.

The only significant intermolecular interaction between alkane molecules is the London force.

There are two reasons for the increase in boiling temperature with increasing molecular mass.
1. As molecular mass increases, the number of electrons per molecule increases and so the instantaneous and induced dipoles increase (see **Section 2.2.5**).
2. As the length of the carbon chain increases, the number of points of contact between adjacent molecules increases. Instantaneous dipole-induced dipole forces exist at each point of contact between the molecules, so the more points of contact the greater the overall intermolecular (London) force of attraction.

You can see the relationship between chain length and points of contact using the skeletal formulae of the alkanes.

propane    butane    pentane

**fig B** Skeletal formulae of propane, butane and pentane.

Because of their shapes, the molecules of the alkanes fit together very well and pack very closely. There are points of contact all the way along the chain. The longer the chain, the more points of contact there are.

### Branched alkanes

Branched chain alkanes have lower boiling temperatures than their unbranched isomers.

| Name of alkane | Structural formula | Boiling temperature/K |
|---|---|---|
| Pentane | $H-C(H_2)-C(H_2)-C(H_2)-C(H_2)-C(H_2)-H$ (pentane chain) | 309 |
| 2-methylbutane | 2-methylbutane structure | 301 |
| 2,2-dimethylpropane | 2,2-dimethylpropane structure | 283 |

The more branching in the molecule, the fewer points of contact between adjacent molecules; i.e. they do not pack together as well. This leads to a decrease in the overall intermolecular force of attraction between molecules and a decrease in boiling temperature.

### Alcohols

Alcohols are a homologous series of compounds with the general formula $C_nH_{2n+1}OH$ (see **Section 6.4.1**).

They contain an –O—H group and can therefore form intermolecular hydrogen bonds in addition to London forces. This additional bonding has an effect on the boiling temperature of alcohols when compared with the equivalent alkane, as is shown in the following table.

| Formula of alcohol | Number of electrons in molecule | Boiling temperature/K | Formula of alkane | Number of electrons in molecule | Boiling temperature/K |
|---|---|---|---|---|---|
| $CH_3OH$ | 18 | 338 | $CH_3CH_3$ | 18 | 184 |
| $CH_3CH_2OH$ | 26 | 352 | $CH_3CH_2CH_3$ | 26 | 231 |
| $CH_3CH_2CH_2OH$ | 34 | 370 | $CH_3CH_2CH_2CH_3$ | 34 | 267 |
| $CH_3CH_2CH_2CH_2OH$ | 42 | 390 | $CH_3CH_2CH_2CH_2CH_3$ | 36 | 303 |

**table A** Boiling temperatures of alcohols and alkanes.

Let us consider the case of methanol ($CH_3OH$) and ethane ($CH_3CH_3$). Both molecules have a similar chain length and also have the same number of electrons. If the intermolecular interactions in each were London forces, then their boiling points would be almost identical. However, the boiling point of the alcohol is higher than that of the alkane. This increase can be attributed to the hydrogen bonding that exists between methanol molecules, which does not exist between molecules of ethane. The additional force of attraction increases the energy required to separate the molecules.

You will notice the same trend in the other compounds in the table.

It is sometimes stated that the predominant bonding in alcohols is hydrogen bonding. We have already mentioned that this is not always the case (**Section 2.2.5**). The following table provides evidence that for the first few members of the alcohol series hydrogen bonding is predominant, but that London forces eventually predominate as the chain length increases.

| Alcohol | Enthalpy change of vaporisation/kJ mol$^{-1}$ |
|---|---|
| $CH_3CH_2OH$ | 38.6 |
| $CH_3CH_2CH_2OH$ | 47.5 |
| $CH_3CH_2CH_2CH_2OH$ | 52.4 |
| $CH_3CH_2CH_2CH_2CH_2OH$ | 57.0 |
| $CH_3CH_2CH_2CH_2CH_2CH_2OH$ | 61.6 |

| Alkane | Enthalpy change of vaporisation/kJ mol$^{-1}$ |
|---|---|
| $CH_3CH_2CH_3$ | 15.7 |
| $CH_3CH_2CH_2CH_3$ | 21.0 |
| $CH_3CH_2CH_2CH_2CH_3$ | 26.4 |
| $CH_3CH_2CH_2CH_2CH_2CH_3$ | 31.6 |
| $CH_3CH_2CH_2CH_2CH_2CH_2CH_3$ | 36.6 |

**table B** Enthalpy changes of vaporisation of alcohols and alkanes.

The enthalpy change of vaporisation is a measure of the amount of energy that is required to *completely* separate the molecules of a liquid and convert it into a gas at the same temperature. It is, therefore, a direct measure of the strength of the intermolecular interactions. The greater the enthalpy change of vaporisation, the greater the forces of attraction between the molecules.

In the case of ethanol, the total energy required to separate one mole of molecules is 38.6 kJ.

Of this, approximately 15.7 kJ mol$^{-1}$ can be attributed to London forces. So, the hydrogen bonding is the predominant bonding, providing approximately 59% of the total.

You can calculate the percentage contribution for the other alcohols in a similar way. The following table shows the results of these calculations.

| Alcohol | Approximate percentage contribution of hydrogen bonding |
|---|---|
| $CH_3CH_2CH_2OH$ | 56 |
| $CH_3CH_2CH_2CH_2OH$ | 50 |
| $CH_3CH_2CH_2CH_2CH_2OH$ | 45 |
| $CH_3CH_2CH_2CH_2CH_2CH_2OH$ | 41 |

**table C** Percentage contribution of hydrogen bonding in alcohols.

The calculations demonstrate the danger of making generalised statements such as, 'The predominant bonding in alcohols is hydrogen bonding'.

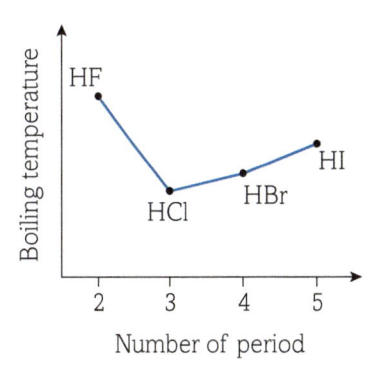

# Boiling temperatures of the hydrogen halides

The graph shows the boiling temperatures of the hydrogen halides HF to HI.

**fig C** Boiling temperatures of the hydrogen halides.

The steady increase in boiling temperature from HCl to HI is the result of the increasing number of electrons per molecule, which in turn results in an increase in London forces.

> **Did you know?**
>
> There are in fact three intermolecular interactions within each compound. They are:
> 1. London forces
> 2. permanent dipole–dipole interactions
> 3. hydrogen bonding.
>
> However, with HCl, HBr and HI, the electronegativity of the halogen is low enough for the permanent dipole–dipole interactions and the hydrogen bonding to be relatively weak. The predominant interaction in each compound is the London force.
>
> The boiling temperature of HF is significantly higher than that of the other three hydrogen halides, despite having fewer electrons per molecule. The London force is weaker in HF, but the hydrogen bonding is significantly greater because of the high electronegativity of fluorine.

## Anomalous properties of water

Water has a number of anomalous properties. The following two are particularly important:

1. It has a relatively high melting and boiling temperature for a molecule with so few electrons.
2. The density of ice at 0 °C is less than that of water at 0 °C.

### Melting and boiling temperatures

The hydrogen bonds between water molecules are relatively strong (see **Section 2.2.5**). As a result, the overall intermolecular forces of attraction in water are greater than would be expected from the number of electrons (10) in the molecule.

Water, therefore, has an abnormally high melting temperature (0 °C, 273 K) and boiling temperature (100 °C, 373 K) at 100 kPa pressure.

It is interesting to compare the boiling temperatures of water, ammonia and hydrogen fluoride.

| | Boiling temperature | Number of electrons per molecule | Strength of hydrogen bonding/ kJ mol$^{-1}$ |
|---|---|---|---|
| $H_2O$ | 373 K (100 °C) | 10 | 22 |
| $NH_3$ | 240 K (−33 °C) | 10 | 17 |
| HF | 189 K (−84 °C) | 10 | 29 |

**table D** Strength of hydrogen bonding in water, ammonia and hydrogen fluoride.

The number of electrons per molecule is identical, so we would expect the London forces to be similar between each set of molecules. The differences are a result of different extents of hydrogen bonding.

The hydrogen bond strength for HF is greater than that for $H_2O$, and yet its boiling temperature is lower. This is because of two factors.

1. HF forms an average of one hydrogen bond per molecule, whereas water molecules form an average of two hydrogen bonds per molecule so the hydrogen bonding is much more extensive in water.
2. Not all of the hydrogen bonds in HF are broken on vaporisation, since HF is substantially polymerised, even in the gas phase.

Ammonia falls in between the two. It has the weakest hydrogen bonding and also only forms an average of one hydrogen bond per molecule. It is highly likely, therefore, that all, or nearly all, of the hydrogen bonds between the molecules are broken on vaporisation, which would explain why its boiling temperature is higher than that of HF.

### Density of ice

As we mentioned above, water has another unusual property. The density of the solid (ice) is less than the density of the liquid at 0 °C.

The molecules in ice are arranged in rings of six, held together by hydrogen bonds.

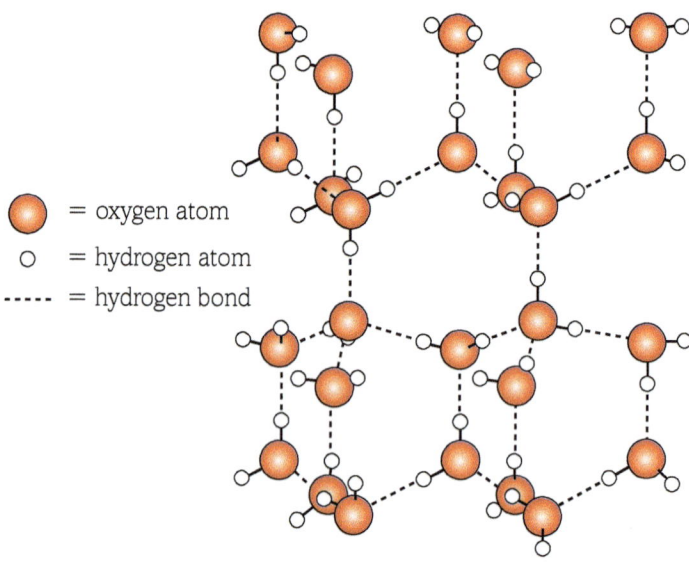

**fig D** Hydrogen bonding in ice.

The structure creates large areas of open space inside the rings. When ice melts, the ring structure is destroyed and the average distance between the molecules decreases, causing an increase in density.

### Choosing suitable solvents

In order for a substance to dissolve, the following two conditions must be met.

- The solute particles must be separated from each other and then become surrounded by solvent particles.
- The forces of attraction between the solute and solvent particles must be strong enough to overcome the solvent–solvent forces and the solute–solute forces.

## Dissolving ionic solids

Many ionic solids dissolve in water. The energy required to separate the ions in the solid is either completely, or partially, supplied by the hydration of the ions.

The diagram shows the process of dissolving for sodium chloride.

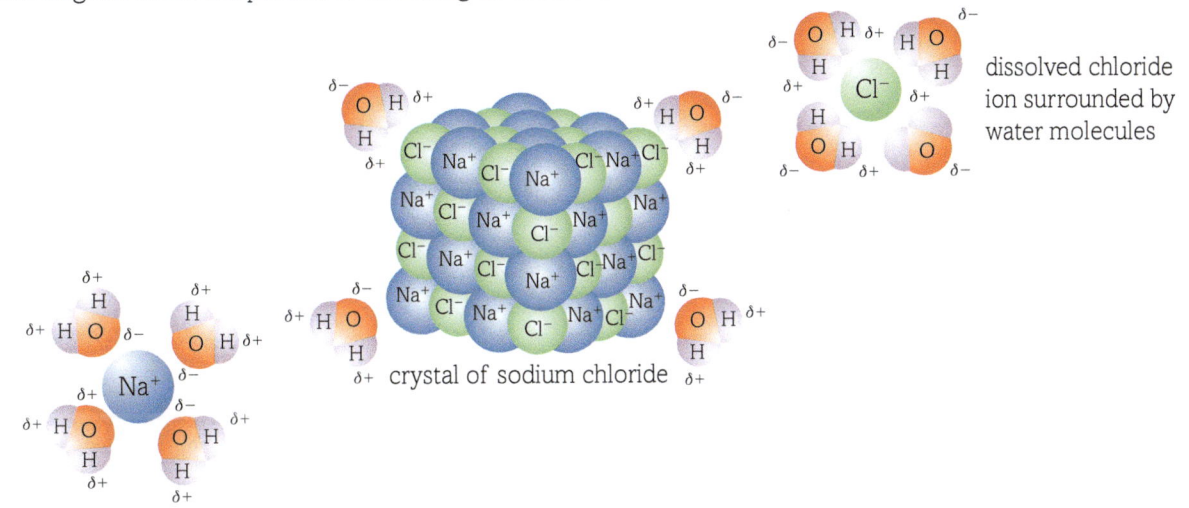

**fig E** Dissolving of sodium chloride.

The δ− end of the water molecules attract the sodium ions sufficiently to remove them from the lattice. The sodium ions then become surrounded by water molecules, as shown. The interaction between the sodium ions and the water molecules is called an ion–dipole interaction.

The δ+ end of the water molecules attract the chloride ions. Once in solution the chloride ions become surrounded by water molecules, as shown. The chloride ions are hydrogen bonded to the water molecules.

The above processes are known as 'hydration' and the energy released is known as the 'hydration energy'.

A more detailed explanation of the dissolving of ionic solids in water will be discussed in **Book 2**.

> ### Did you know?
> This explanation is a simplification of the processes that occur. A more detailed explanation of the dissolving of ionic solids in water will be discussed in **Book 2**.

## Compounds that can form hydrogen bonds with water

Alcohols contain an –O—H group and can therefore form hydrogen bonds with water.

The diagram shows a molecule of ethanol forming a hydrogen bond to water:

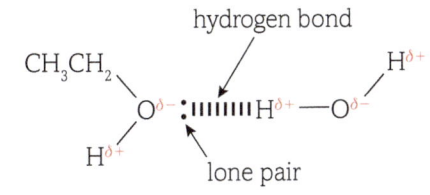

**fig F** Hydrogen bonding between ethanol and water.

Ethanol and water mix in all proportions. The hydrogen bonding between the ethanol and water molecules is similar in strength to the hydrogen bonding in pure ethanol and in pure water.

The solubility of alcohols in water decreases with increasing hydrocarbon chain length as London forces predominate between the alcohol molecules.

### Compounds that cannot form hydrogen bonds with water

Non-polar molecules such as the alkanes do not dissolve in water. The attraction between the alkane molecules and water molecules is not sufficiently strong to disrupt the hydrogen bonded system between the water molecules.

Many polar molecules also have limited solubility in water. This is because they either do not form hydrogen bonds with water, or the hydrogen bonds they form are weak compared with the hydrogen bonds in water.

Ethoxyethane, $CH_3CH_2OCH_2CH_3$, is polar (dipole moment = 1.15 D) and yet is almost totally immiscible with water. The forces of attraction between ethoxyethane and water molecules are not large enough to replace the relatively strong hydrogen bonding between the water molecules.

Halogenoalkanes (see **Sections 6.3.1–6.3.3**) are also not very soluble in water for similar reasons. They are much more soluble in ethanol, and this is why some reactions of halogenoalkanes are carried out in a medium of aqueous ethanol.

### Non-aqueous solvents

The 'rule of thumb' is that like dissolves like. So, if you are searching for a solvent for a non-polar substance, or for a substance that has a substantial non-polar part to its molecule, then liquids that contain similar molecules are often the answer.

For example, alkanes are soluble in one another. Crude oil is a complex mixture of alkanes dissolved in each other.

Non-polar bromine dissolves readily in non-polar hexane ($C_6H_{14}$), and this solution is commonly used to test for unsaturation in molecules. It is more convenient to use than bromine water, since the molecules being tested will be also soluble in hexane, whereas they are likely to be insoluble in water.

## Questions

1 The hydrides of Group 4 exist as tetrahedral molecules.
   Explain the trend in their boiling temperatures:
   $CH_4$ (112 K)   $SiH_4$ (161 K)   $GeH_4$ (185 K)   $SnH_4$ (221 K)

2 Suggest why water has a relatively high surface tension for a molecule of such low molecular mass.

3 A propanone molecule ($CH_3COCH_3$) has 32 electrons. A butane molecule ($CH_3CH_2CH_2CH_3$) has 34 electrons. Explain why propanone has a much higher boiling temperature than butane.

4 Suggest why magnesium chloride is soluble in water despite the fact that the energy required to break up the lattice is 2494 kJ mol$^{-1}$.

# 2.3 1 Solid lattices

### By the end of this section, you should be able to...

- know that giant lattices are present in:
  (i) solid metals (giant metallic lattices)
  (ii) ionic solids (giant ionic lattices)
  (iii) covalently bonded solids, such as diamond, graphite and silicon(IV) oxide (giant covalent lattices)
- know the different structures formed by carbon atoms, including graphite, diamond and graphene
- know that the structure of solid iodine and ice is discrete (simple) molecular

## Introduction to solid lattices

The four types of solid structures we shall deal with in this section are:

- giant metallic lattices
- giant ionic lattices
- giant covalent lattices
- discrete (simple) molecular lattices.

### Did you know?

There is one other important type of solid structure described as 'polymeric'. These solids are composed of macromolecules such as natural polymers and man-made polymers. We will look at these in **Section 6.2.8**.

## Metallic lattices

Metallic lattices are composed of a regular arrangement of positive metal ions surrounded by delocalised electrons.

Substances that have a giant metallic lattice typically have the following properties:

- high melting and boiling temperatures
- good electrical conductivity
- good thermal conductivity
- malleability
- ductility.

We explained these properties in **Section 2.1.1**.

## Giant ionic lattices

Giant ionic lattices are composed of a regular arrangement of positive and negative ions.

Substances that consist of giant ionic lattices typically have the following properties:

- fairly high melting temperatures
- brittleness
- poor electrical conductivity when solid but good when molten
- often soluble in water.

We explained these properties in **Section 2.1.2**.

## Giant covalent lattices

Giant covalent lattices are sometimes called network covalent lattices. They consist of a giant network of atoms linked to each other by covalent bonds.

Four of the most common giant covalent substances are:

- diamond
- graphite
- graphene
- silicon(IV) oxide (not discussed below).

### Diamond

In diamond each carbon atom forms four sigma ($\sigma$) bonds to four other carbon atoms, in a giant three-dimensional tetrahedral arrangement. All bond angles are 109.5°.

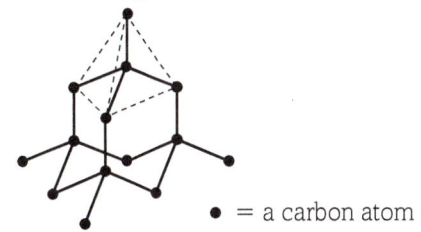

**fig A** Structure of diamond.

Diamond is extremely hard because of the very strong C—C bonding throughout the structure. It also has a very high melting temperature because a great number of strong C—C bonds have to be broken in order to melt it. This requires a large amount of heat energy.

### Did you know?

You may have heard diamond referred to as 'ice'. A diamond first acquired this nickname because of its properties rather than its appearance. Owing to the wave nature of electrons, the electron wave is able to travel more smoothly through a perfect crystal without bouncing, in much the same way as light travels through a clear crystal. In addition, thermal conduction is enhanced by the network of strong covalent bonds. The use of a heat probe is one of the major tests in determining if a diamond is authentic. This is because real diamonds are able to conduct heat and are actually one of the best conductors of heat. Even as a diamond heats up, it is cool to the touch, just like ice. This is why the points of diamond cutting tools do not become overheated.

## Graphite

Graphite has a layered structure.

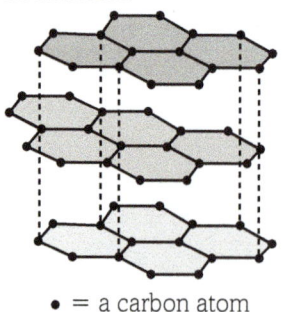

• = a carbon atom

**fig B** Structure of graphite.

Each carbon atom is bonded to three others by sigma bonds, forming interlocking hexagonal rings. The fourth electron on each carbon atom is in a p-orbital. The carbon atoms are close enough for the p-orbitals to overlap with one another to produce a cloud of delocalised electrons above and below the plane of the rings (we will compare this with the structure of benzene in **Book 2**).

Graphite can be used as a solid lubricant since the layers slide easily over one another. Although there are only weak London forces of attraction between the layers, this does not account for its lubricating properties. The ability of graphite to act as a lubricant decreases five-fold at high altitude, and by a factor of eight in a vacuum. Its lubricating properties are as a result of adsorbed gases on the surface of the carbon atoms. Graphite's inability to act as an effective lubricant in a vacuum is the reason why it is not used as a lubricant in spacecraft. Either molybdenum disulfide ($MoS_2$) or hexagonal boron nitride (BN) can be used instead.

Graphite is a fairly good conductor of electricity. The delocalised electrons between the layers are free to move under the influence of an applied potential difference. An interesting feature of graphite is that it can only conduct electricity parallel to its layers. The delocalised electrons are not free to move from one layer to the next. Compare this with a metal, which is able to conduct electricity in all directions throughout the structure.

Graphite has a high melting temperature for the same reason as diamond.

## Graphene

Graphene is pure carbon in the form of a very thin sheet, one atom thick. The carbon atoms are bonded in exactly the same way as in graphite and it can, therefore, be described as a one-atom thick layer of graphite.

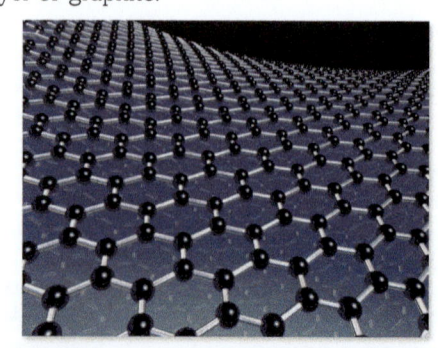

**fig C** Structure of graphene.

### Did you know?

Graphene is the thinnest material on Earth; one million times thinner than a human hair.

It is 200 times stronger than steel.

It is an excellent thermal conductor, better even than diamond.

It absorbs light.

It can be considered the basic unit of most other forms of carbon.

Graphite consists of layers of graphene on top of one another joined by London forces.

A sheet of graphene can be rolled into a ball to produce fullerene molecules.

A sheet of graphene can be rolled into a cylinder to produce a carbon nanotube.

Graphene can self-repair holes in its sheets when exposed to molecules containing carbon, such as hydrocarbons. When bombarded with pure carbon atoms, the atoms perfectly align into hexagons, completely filling the holes.

## Molecular lattices

Two common solid molecular lattices are iodine and ice. Ice has already been discussed (**Section 2.2.6**), so we will concentrate on solid iodine.

Iodine exists as diatomic molecules, $I_2$. In solid iodine these molecules are arranged in a regular pattern, which explains its crystalline nature.

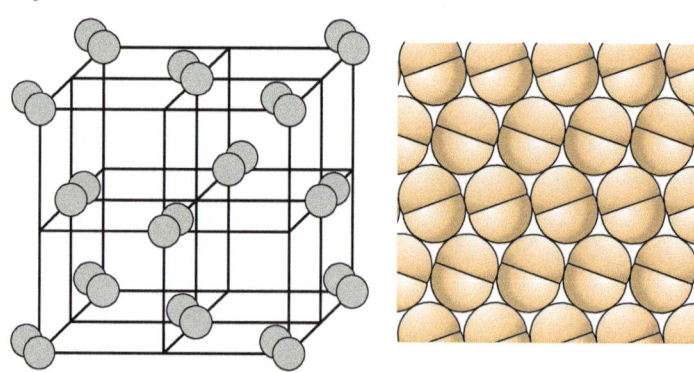

**fig D** Structure of iodine.

The diagram on the left shows the arrangement of the iodine molecules. The structure is described as 'face-centred cubic'. In practice, the iodine molecules will be touching one another (right-hand diagram); they have been drawn apart for the sake of clarity. The molecules of iodine are held together by London forces.

**fig E** Large crystals of iodine.

Other molecular solids include sulfur ($S_8$), white phosphorus ($P_4$), Buckminster fullerene ($C_{60}$), dry ice (solid carbon dioxide, $CO_2$), sucrose ($C_{12}H_{22}O_{11}$) and solid alkanes (e.g. paraffin wax).

**fig F** Sucrose.

**fig G** Buckminster fullerene.

**fig H** Dry ice.

## Physical properties of molecular solids

Molecular solids will, in general, have low melting and boiling temperatures. In order to melt a molecular solid it is not necessary to break the covalent bonds within the molecule (the intramolecular bonds); it is only necessary to overcome the intermolecular forces of attraction.

Since intermolecular forces of attraction tend to be much weaker than covalent bonds, little energy is required to either break down the lattice structure of the solid and cause it to melt, or to separate the molecules and cause the liquid to boil and vaporise.

As mentioned before, London forces tend to increase with both an increase in the number of electrons per molecule and also with increasing length of molecule. So, a macromolecular solid such as poly(ethene) will have a much higher melting temperature than its monomer, ethene.

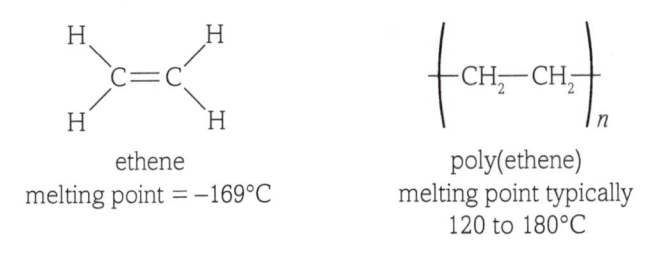

ethene
melting point = −169°C

poly(ethene)
melting point typically 120 to 180°C

**fig I** Ethene and poly(ethene).

# Questions

1. Magnesium oxide ($Mg^{2+}O^{2-}$) has the same structure as sodium chloride ($Na^+Cl^-$).
   (a) Draw a diagram to show the arrangement of the ions in a crystal of magnesium oxide.
   (b) Explain why the melting temperature of magnesium oxide is higher than that of sodium chloride.

2. Explain the following observations:
   (a) Magnesium and magnesium fluoride both have giant lattice structures containing ions. Solid magnesium conducts electricity, but solid magnesium fluoride does not.
   (b) Silicon and phosphorus are both described as covalent substances, but silicon has a much higher melting temperature than phosphorus.

3. Hexagonal boron nitride has a structure similar to that of graphite.

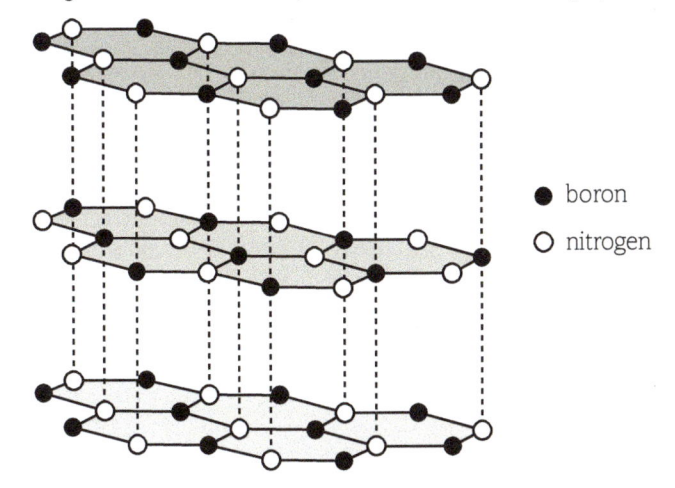

● boron
○ nitrogen

Suggest why solid boron nitride can act as a lubricant in a vacuum, whereas graphite cannot.

# 2.3 Structure and properties

**By the end of this section, you should be able to...**

- predict the type of structure and bonding present in a substance from numerical data and/or other information
- predict the physical properties of a substance, including melting and boiling temperature, electrical conductivity and solubility in water, in terms of:
  (i) the types of particle present (atoms, molecules, ions, electrons)
  (ii) the structure of the substance
  (iii) the type of bonding and the presence of intermolecular forces, where relevant

## Types of bonding and structure

The table below shows the types of bonding and structure that exist in elements and compounds.

| Bonding | Structure | Examples |
|---|---|---|
| Metallic | giant lattice | Mg, Al, Cu, Zn |
| Ionic | giant lattice | NaCl, MgO, CsF |
| Covalent (including polar covalent) | giant lattice | C (diamond), C (graphite), Si, $SiO_2$, BN |
| | molecular | $H_2O$ (ice), $I_2$, $P_4$, $S_8$, $C_{60}$, $C_{12}H_{22}O_{11}$ (sucrose) |
| | macromolecular | polymers (e.g. poly(ethene)), proteins, DNA |

**table A** Types of bonding and structure.

## Predicting physical properties

The physical properties of a substance are determined by the type of bonding and structure it has.

The following flow diagram enables you to determine the type of bonding and structure in a substance by considering some of its properties.

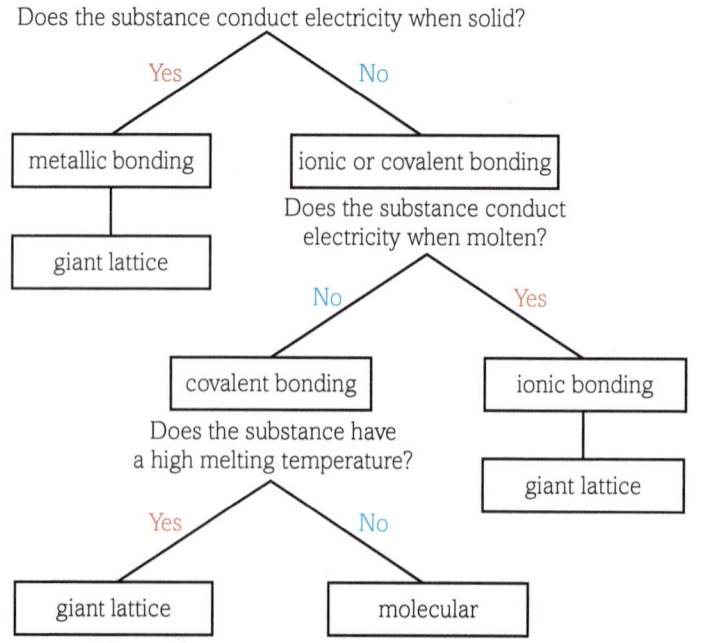

**fig A** Flow chart for determining bonding and structure.

There will, of course, always be exceptions. For example, graphite has a giant covalent structure and yet it is a relatively good conductor of electricity when solid.

Here is a summary of the major properties of each type of structure.

|  | Giant metallic | Giant ionic | Giant covalent | Molecular |
|---|---|---|---|---|
| Particles present | positive ions and delocalised electrons | positive and negative ions | atoms | molecules |
| Type of bonding | metallic | ionic | covalent | covalent |
| Are there any intermolecular forces of attraction? | no | no | no | yes |
| Melting and boiling temperatures | fairly high to high | fairly high to high | high to very high | generally low |
| Electrical conductivity | good when solid and when molten | non-conductor when solid; good when molten | non-conductor | non-conductor |
| Solubility in water | insoluble unless the metal reacts with water, e.g. sodium | generally soluble, but with notable exceptions, e.g. AgCl, AgBr, AgI and $BaSO_4$ | insoluble | generally insoluble, but may dissolve if hydrogen bonding is possible (e.g. sucrose), or if the substance reacts with water (e.g. $Cl_2$) |

**table B** Structure and properties of a substance.

# Questions

1. The table below gives some properties of four substances: **A**, **B**, **C** and **D**.

   Analyse the data and then decide what type of bonding and structure is likely to be present in each substance.

   |  | Melting temperature/°C | Ability to conduct electricity when solid | Ability to conduct electricity when molten | Solubility in water |
   |---|---|---|---|---|
   | A | 1083 | good | good | insoluble |
   | B | 119 | poor | poor | insoluble |
   | C | 2230 | poor | poor | insoluble |
   | D | 801 | poor | good | soluble |

2. Each of the substances listed below has at least one property that is unusual for its type of bonding and structure. In each case, state the most likely bonding and structure, and identify the unusual property or properties. Suggest a possible identity for each substance.

   (a) Substance **P**.
   
   Melting temperature 813 °C; good conductor of electricity when solid and when molten; reacts with water to form ammonia gas and an alkaline solution.
   
   (b) Substance **Q**.
   
   Melting temperature 1414 °C; semi-conductor of electricity when solid; insoluble in water.
   
   (c) Substance **R**.
   
   Melting temperature 98 °C; good conductor of electricity when solid; reacts with water to form hydrogen gas and an alkaline solution.

# THINKING BIGGER

## BORING BORON?

This article is taken from the news website of Brown University. Read the article and then answer the questions opposite.

## BORON 'BUCKYBALL' DISCOVERED

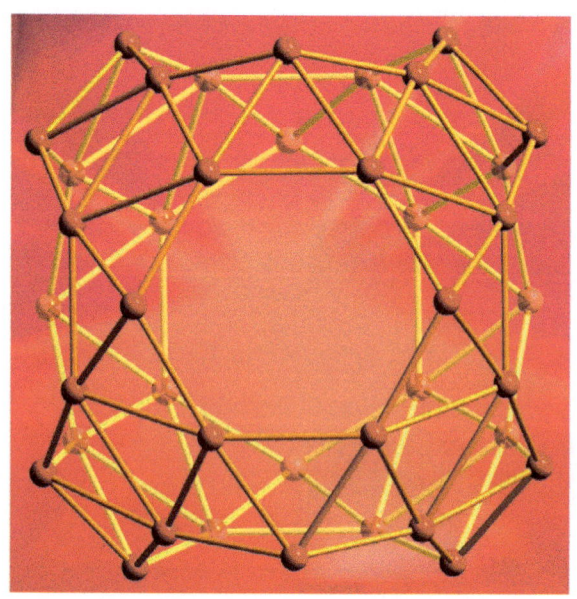

**fig A** Researchers have shown that clusters of 40 boron atoms form a molecular cage similar to the carbon buckyball. This is the first experimental evidence that such a boron cage structure exists.

The discovery 30 years ago of soccer-ball-shaped carbon molecules called buckyballs helped to spur an explosion of nanotechnology research. Now, there appears to be a new ball on the pitch.

Researchers from Brown University, Shanxi University and Tsinghua University in China have shown that a cluster of 40 boron atoms forms a hollow molecular cage similar to a carbon buckyball. It's the first experimental evidence that a boron cage structure – previously only a matter of speculation – does indeed exist.

'This is the first time that a boron cage has been observed experimentally,' said Lai-Sheng Wang, a professor of chemistry at Brown who led the team that made the discovery. 'As a chemist, finding new molecules and structures is always exciting. The fact that boron has the capacity to form this kind of structure is very interesting.'

Wang and his colleagues describe the molecule, which they've dubbed borospherene, in the journal *Nature Chemistry*.

Carbon buckyballs are made of 60 carbon atoms arranged in pentagons and hexagons to form a sphere – like a soccer ball. Their discovery in 1985 was soon followed by discoveries of other hollow carbon structures including carbon nanotubes. Another famous carbon nanomaterial – a one-atom-thick sheet called graphene – followed shortly after.

After buckyballs, scientists wondered if other elements might form these odd hollow structures. One candidate was boron, carbon's neighbor on the periodic table. But because boron has one less electron than carbon, it can't form the same 60-atom structure found in the buckyball. The missing electrons would cause the cluster to collapse on itself. If a boron cage existed, it would have to have a different number of atoms.

Wang and his research group have been studying boron chemistry for years. In a paper published earlier this year, Wang and his colleagues showed that clusters of 36 boron atoms form one-atom-thick disks, which might be stitched together to form an analog to graphene, dubbed borophene. Wang's preliminary work suggested that there was also something special about boron clusters with 40 atoms. They seemed to be abnormally stable compared to other boron clusters. Figuring out what that 40-atom cluster actually looks like required a combination of experimental work and modeling using high-powered supercomputers.

Where else will I encounter these themes?

# Thinking Bigger 2

Let us start by considering the nature of the writing in the article.

1. Assess what aspects of the article suggest that the piece has been written with a broad audience in mind.

> **Command word**
> If you are asked to assess something you will need to give careful consideration to the factors or events and identify the most important. Your answer should include a conclusion.

Questions 2–4 are about the chemistry in, or connected to, this article. Some of the questions will build on ideas that you covered at GCSE and others will be covered later in the course. The timeline below tells you about other sections in this book that are relevant to this question: Don't worry if you are not ready to give answers to these questions at this stage. You may like to read through the activity at this initial stage and return to the questions once you have covered other topics later in the book.

2. a. Name the type of bond present in borospherene.
   b. The melting point of a different form of boron is 2300 °C. What does this suggest about the structure formed in this case?
3. An important reducing agent in chemistry is the compound sodium borohydride. Describe the bonding in $NaBH_4$ and use your knowledge of electron pair repulsion theory to predict and explain the shape of the $BH_4^-$ ion.
4. Assign an oxidation number to boron in each of the following compounds:
   a. $Na_2B_4O_7$
   b. $NaBH_4$
   c. $BF_3$

> **Chemical vocabulary**
> As you read the article, identify any unfamiliar words. Look these up to check you understand their meaning. You could look in the glossary or on the internet, but make sure that any sources you use are reliable. It is worth using several sources to make sure the meanings are the same.

## Activity

Boron is currently being considered for its potential medical uses in a number of areas. Carry out a web search and find out what you can about its potential uses. Aim to deliver a presentation to your peers (no more than five slides) summarising as many aspects of research as you can. Be prepared to defend your presentation in a 5 minute Q&A session.

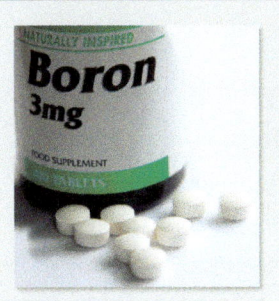

fig B Would you take a boron supplement?

> Be constructively critical about the nature of the source material on the web. What constitutes a reliable resource?

### Did you know?

Boron has two stable isotopes: $^{10}B$ and $^{11}B$, neither of which is radioactive. As a result, boron that is enriched with a higher proportion of $^{10}B$ can act as a neutron capture material and can be used in control rods in nuclear reactors. The uranium fission reaction which produces high energy neutrons in a chain reaction can be controlled by raising or lowering these control rods.

• From the news website of Brown University, http://news.brown.edu

# 2 Exam-style questions

1. Hydrogen bonding occurs in water and its presence has an effect on the properties of water.
   (a) State what is meant by the term **hydrogen bond**. [2]
   (b) Draw a diagram to show the formation of a hydrogen bond between two water molecules. [2]
   (c) State two ways in which the properties of water are affected by hydrogen bonding. [2]
   [Total: 6]

2. Lithium, magnesium and sodium all exhibit metallic bonding.
   (a) State what is meant by the term **metallic bonding**. [2]
   (b) State how a metal can conduct electricity. [1]
   (c) Explain why magnesium is a better conductor of electricity than sodium. [2]
   (d) Predict whether or not sodium is a better electrical conductor than lithium. Justify your answer. [2]
   [Total: 7]

3. The table shows some of the properties of four substances, **A, B, C** and **D**.
   State the type of bonding and structure that is likely to be present in each of the substances.
   In each case justify your answer. [8]

   | Substance | Solubility in water | Electrical conductivity | | Melting temperature/°C |
   |---|---|---|---|---|
   | | | of solid | in aqueous solution | |
   | A | insoluble | poor | — | 1610 |
   | B | soluble | poor | good | 801 |
   | C | insoluble | good | — | 1083 |
   | D | soluble | poor | fair | −78 |

   [Total: 8]

4. The electron pair repulsion theory (EPR) can be used to predict the shape of simple molecules and ions.
   (a) State the main assumptions of the EPR theory. [3]
   (b) Draw a dot-and-cross diagram for each of the flowing molecules/ions. [4]
       (i) $BF_3$  (ii) $NH_3$  (iii) $NH_4^+$  (iv) $SF_6$ [4]
   (c) Predict the shape and bond angles in each of the molecule/ions in part (b). [5]
   [Total: 16]

5. Nitrogen and carbon monoxide are both gases consisting of diatomic molecules. Both gases are colourless, odourless and tasteless, but unlike nitrogen, carbon monoxide is extremely toxic.
   (a) The dot-and-cross diagram, showing only the outer electrons, is:

   $$\overset{..}{\underset{..}{\times C \vdots O}}\!:$$

   Copy the diagram and add labels to identify:
   (i) a lone pair of electrons
   (ii) a covalent bonding pair of electrons and
   (iii) a dative covalent bonding pair of electrons. [3]
   (b) The nitrogen and carbon monoxide molecules are isoelectronic.
       (i) State what is meant by the term **isoelectronic**. [1]
       (ii) The HCN molecule is also isoelectronic with $N_2$ and CO.
       Draw a dot-and-cross diagram for HCN, showing only the outer shell electrons. [2]
       (iii) Explain why nitrogen is much less reactive than either carbon monoxide or hydrogen cyanide. [1]
   [Total: 7]

6. Ethanol ($CH_3CH_2OH$) and methoxymethane ($CH_3OCH_3$) are structural isomers.
   Ethanol is soluble in water whereas methoxymethane is not.
   Their boiling temperatures are: ethanol, 78.5 °C; methoxymethane, −24.8 °C.
   The structural formula and shape of each molecule is:

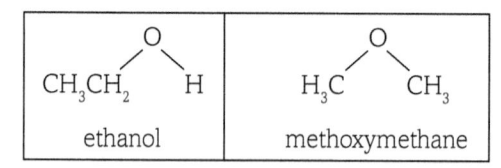

   | ethanol | methoxymethane |
   |---|---|
   | $CH_3CH_2$ — O — H | $H_3C$ — O — $CH_3$ |

   (a) Name types of intermolecular interactions (forces) that exist in
       (i) ethanol
       (ii) methoxymethane. [5]
   (b) Use your answers in part (a) to explain the difference in solubility and boiling temperatures of the two substances. [4]
   [Total: 9]

7 Sodium chloride is an ionic compound containing sodium ions ($Na^+$) and chloride ions ($Cl^-$).
It has a fairly high melting temperature, is soluble in water and is a poor conductor of electricity when solid, but good when molten.

(a) Draw dot-and-cross diagrams to show the changes in arrangement of electrons that take place when sodium (Na) reacts with chlorine ($Cl_2$) to form sodium chloride. Show only the outer electrons in each case. [4]

(b) Complete the diagram to show the structure of sodium chloride using the key provided. [2]

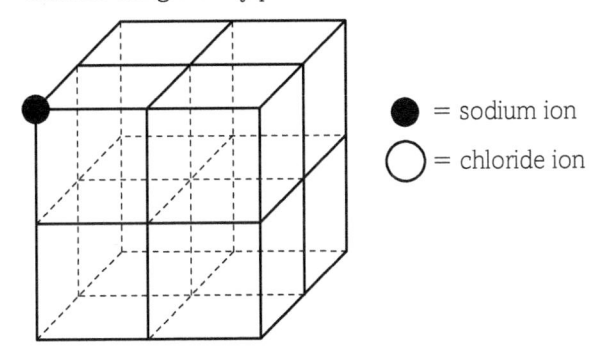

● = sodium ion
○ = chloride ion

(c) Explain why sodium chloride
  (i) has a fairly high melting temperature [3]
  (ii) is soluble in water [3]
  (iii) is a poor conductor of electricity when solid, but good when molten. [2]

**[Total: 14]**

8 The table shows the relative molecular masses ($M_r$) and boiling temperatures of water and the first eight unbranched members of the homologous series of alkanes, methane to octane.

| Formula of compound | $H_2O$ | $CH_4$ | $C_2H_6$ | $C_3H_8$ | $C_4H_{10}$ | $C_5H_{12}$ | $C_6H_{14}$ | $C_7H_{16}$ | $C_8H_{18}$ |
|---|---|---|---|---|---|---|---|---|---|
| $M_r$ | 18 | 16 | 30 | 44 | 58 | 72 | 86 | 100 | 114 |
| Boiling temperature/°C | 100 | −164 | −89 | −42 | −0.5 | 36 | 69 | 98 | 125 |

(a) Comment on the difference in boiling temperature between water and methane. [3]

(b) Explain the trend in boiling temperatures of the alkanes. [3]

(c) Butane has an isomer called methylpropane, $(CH_3)_3CH$. Explain whether the boiling temperature of methylpropane is the same as, lower or higher than that of butane. [3]

**[Total: 9]**

9 This question is about four simple molecular compounds: water, ammonia, methane and boron trichloride.

(a) Complete the table to show the number of bonding pairs and non-bonding (lone) pairs of electrons in one molecule of each compound. [4]

| Formula of molecule | $H_2O$ | $NH_3$ | $CH_4$ | $BCl_3$ |
|---|---|---|---|---|
| Number of bonding pairs | | | | |
| Number of non-bonding pairs | | | | |

(b) The ammonia molecule and the boron trichloride molecule both contain polar bonds.
  (i) State what is meant by a **polar bond**, and explain how the polarity arises. [3]
  (ii) Explain why the ammonia molecule is polar, but the boron trichloride molecule is not. [3]

**[Total: 10]**

# TOPIC 3
# Redox reactions

## Introduction

Redox reactions are amongst the most common and most important chemical reactions in everyday life. Here are three examples of redox reactions that occur throughout our daily lives.

One of the most common chemical reactions, **combustion**, is a classic redox reaction. For example, when petrol burns within the internal combustion engine of your car, carbon within fuel is oxidised to carbon dioxide and carbon monoxide. Meanwhile, oxygen within the air feeding your automobile engine is reduced when it combines with the hydrogen atoms of the fuel to form water. Similar reactions occur in most other fires, such as forest fires or the more controlled one when you light your barbecue.

**Glycolysis** is a process fundamental to all living creatures. It is what ultimately converts food into usable energy. Glycolysis is a multistep reaction that involves a series of electron transfers between molecules while releasing free energy that the organism can use for its various metabolic and physical functions. Glycolysis is an example of a redox reaction that does not involve the use of oxygen.

**Photosynthesis**, probably the single most important process on the planet, ultimately uses the energy of sunlight to grow plants and so provide food to all higher organisms. It is accomplished through a series of redox reactions in which energy from light is finally converted into carbohydrates on which animals and humans can feed. Oxygen is also a vital by-product of photosynthesis.

### All the maths you need
- Carry out calculations using numbers in ordinary form

## What have I studied before?
- Metal atoms, in general, lose electrons to form positive ions
- Non-metal atoms, in general, gain electrons to from negative ions
- Oxidation is the loss of electrons
- Reduction is the gain of electrons
- Oxidising agents gain electrons
- Reducing agents lose electrons
- Redox reactions

## What will I study later?
- Disproportionation reaction of the halogens
- Redox reactions of certain transition metals (A level)
- Redox titrations (A level)
- Standard electrode (redox) potentials and their uses to calculate the electromotive force of a cell and the feasibility of a reaction (A level)

## What will I study in this topic?
- How to calculate the oxidation number of an element in a compound or an ion
- How to use oxidation numbers to decide whether oxidation and/or reduction has taken place
- Disproportionation as a special type of redox reaction
- How to use oxidation numbers to write a balanced chemical equation for a redox reaction
- How to use ionic half-equations to write a full, balanced chemical equation for a redox reaction
- How to write chemical formulae given oxidation numbers
- How to name compounds using oxidation numbers in the form of Roman numerals

## 3.1  1  Electron loss and gain

**By the end of this section, you should be able to...**

- understand oxidation and reduction in terms of electron loss and gain
- know that oxidising agents gain electrons and that reducing agents lose electrons
- understand that a disproportionation reaction involves an element in a species being simultaneously oxidised and reduced

### Background to oxidation and reduction

You originally know 'oxidation' as the addition of oxygen and 'reduction' as the removal of oxygen.

The reaction between iron(III) oxide and carbon monoxide that takes place in the blast furnace provides an example of each.

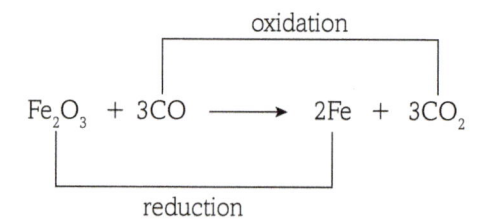

$$\underset{\text{reduction}}{\overset{\text{oxidation}}{Fe_2O_3 + 3CO \longrightarrow 2Fe + 3CO_2}}$$

**fig A** Example of oxidation and reduction during the reaction between iron(III) oxide and carbon monoxide.

We then expanded these definitions so that oxidation is described as the removal of hydrogen and reduction as the addition of hydrogen. The reaction between chlorine and hydrogen sulfide provides an example.

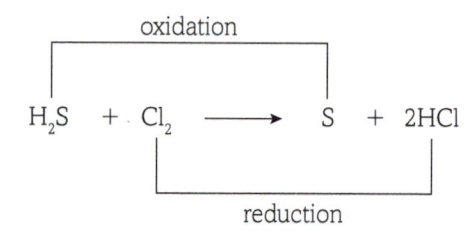

$$\underset{\text{reduction}}{\overset{\text{oxidation}}{H_2S + Cl_2 \longrightarrow S + 2HCl}}$$

**fig B** Example of oxidation and reduction during the reaction between chlorine and hydrogen sulfide.

A reaction that involves reduction and oxidation is called a **redox reaction**.

### Oxidation and reduction in terms of electron loss and gain

#### Electron transfer in redox reactions

The approach favoured by chemists today involves the transfer of electrons.

When magnesium burns in oxygen it forms magnesium oxide:

$2Mg + O_2 \rightarrow 2MgO$

The magnesium has been oxidised because it has gained oxygen. So the oxygen must have been reduced, but by what definition?

This reaction results in the formation of $Mg^{2+}$ and $O^{2-}$ ions. Each of the two magnesium atoms has lost two electrons:

$Mg \rightarrow Mg^{2+} + 2e^-$

The oxygen molecule has gained four electrons to become oxide ions:

$$O_2 + 4e^- \rightarrow 2O^{2-}$$

The magnesium has been *oxidised* because it has lost electrons.

Oxygen has been *reduced* because it has gained electrons.

## Definitions of oxidation and reduction

We now have new definitions of oxidation and reduction:

**Oxidation** is the *loss* of electrons
**Reduction** is the *gain* of electrons

These new definitions are easily remembered using the mnemonic **OIL RIG**.

**OXIDATION**

**IS**

**LOSS** (of electrons)

**REDUCTION**

**IS**

**GAIN** (of electrons)

The reduction of iron(III) oxide in the blast furnace can now be seen as the gain of three electrons by the iron(III) ion in the oxide:

$$Fe^{3+} + 3e^- \rightarrow Fe$$

But what about the carbon monoxide? It must have been reduced, but there is no easy way to see how it has gained electrons since it is a covalent compound and it forms another covalent compound, $CO_2$. We shall return to this problem later in the book.

## Oxidising and reducing agents

In the reaction between magnesium and oxygen to form magnesium oxide, the oxygen has oxidised the magnesium. Therefore, oxygen is an **oxidising agent**.

The magnesium has reduced the oxygen and is therefore a **reducing agent**.

## Disproportionation

Consider the following reaction, which occurs if a copper(I) compound is added to water:

$$2Cu^+(aq) \rightarrow Cu^{2+}(aq) + Cu(s)$$

One of the $Cu^+$ ions has lost an electron to become $Cu^{2+}$, whilst the other $Cu^+$ ion has gained an electron to become Cu. Both oxidation and reduction have occurred but the *same element*, copper, is involved in both changes.

This is an example of disproportionation. We will come back to this later in this topic.

# Questions

1. In the reaction:

    $$Zn(s) + CuSO_4(aq) \rightarrow ZnSO_4(aq) + Cu(s)$$

    (a) Which species has been oxidised and which species has been reduced?

    (b) Write an ionic half-equation to represent both the oxidation and the reduction reactions.

2. In each of the following reactions identify whether the underlined species has been oxidised, reduced or neither.

    (a) $\underline{Al}(s) + 1\frac{1}{2}Cl_2(g) \rightarrow AlCl_3(s)$
    (b) $4Na(s) + \underline{Ti}O_2(s) \rightarrow 2Na_2O(s) + Ti(s)$
    (c) $\underline{Ag}^+(aq) + Cl^-(aq) \rightarrow AgCl(s)$
    (d) $Cl_2(aq) + 2\underline{Br}^-(aq) \rightarrow 2Cl^-(aq) + Br_2(aq)$
    (e) $\underline{Cu}O(s) + 2H^+(aq) \rightarrow Cu^{2+}(aq) + H_2O(l)$
    (f) $\underline{Mn}O_2(s) + 4HCl(aq) \rightarrow MnCl_2(aq) + Cl_2(aq) + 2H_2O(l)$

3. In each of the following reactions identify the species that has been oxidised. In each case, justify your answer.

    (a) $Fe(s) + H_2SO_4(aq) \rightarrow FeSO_4(aq) + H_2(g)$
    (b) $Na(s) + \frac{1}{2}H_2(g) \rightarrow NaH(s)$
    (c) $CuO(s) + Cu(s) \rightarrow Cu_2O(s)$
    (d) $2Fe(OH)_2(s) + \frac{1}{2}O_2(aq) + H_2O(l) \rightarrow 2Fe(OH)_3(s)$
    (e) $2V^{3+}(aq) + Zn(s) \rightarrow 2V^{2+}(aq) + Zn^{2+}(aq)$

## Key definitions

A **redox reaction** is a reaction that involves both reduction and oxidation.

**Oxidation** is the loss of electrons.

**Reduction** is the gain of electrons.

An **oxidising agent** is a species (atom, molecule or ion) that oxidises another species by removing one or more electrons. When an oxidising agent reacts it gains electrons and is, therefore, reduced.

A **reducing agent** is a species that reduces another species by adding one or more electrons. When a reducing agent reacts it loses electrons and is, therefore, oxidised.

# 3.2   1   Calculating oxidation numbers

By the end of this section, you should be able to...
- understand what is meant by the term oxidation number
- calculate the oxidation number of an element in both a compound and an ion

## Oxidation number

So far we have restricted our discussion of oxidation and reduction to atoms and ions. For redox reactions involving these species it is easy to see which species are losing and which are gaining electrons. However, many compounds are covalent and for them a simple treatment involving ions is not appropriate.

To get around this difficulty, the concept of **oxidation number** has been developed.

For example, in MgO the oxidation number of magnesium is +2, since the charge on the magnesium ion is 2+. Similarly, the oxidation number of oxygen is −2.

In $SO_2$, the oxidation number of the sulfur is +4 because if the compound were fully ionic the sulfur ion would have a charge of 4+. The oxidation number of oxygen is once again −2.

## Rules for determining the oxidation number

Here are some rules to help you calculate the oxidation number.

- The oxidation number of an uncombined element is zero.
- The sum of the oxidation numbers of all the elements in a neutral compound is zero.
- The sum of the oxidation numbers of all the elements in an ion is equal to the charge on the ion.
- The more electronegative element in a substance is given a negative oxidation number.
- The oxidation number of fluorine is always −1.
- The oxidation number of hydrogen is +1, except when combined with a less electronegative element. Then it becomes −1.
- The oxidation number of oxygen is −2, except in peroxides where it is −1 and when combined with fluorine when it is positive.

The best way to get used to these rules is to put them into practice, which is what we shall now do.

> **Learning tip**
>
> Should we use the terminology 'oxidation number' or 'oxidation state'?
>
> To an extent it really does not matter. We say that an element has an oxidation number of ..., or is in the oxidation state of .... So it is really a matter of use of words, not of understanding.

### WORKED EXAMPLE 1

Deduce the oxidation number of chlorine in:
(a) NaCl
(b) NaClO
(c) $NaClO_3$

**Answers**

(a) The oxidation number of Na is +1. The two oxidation numbers must add up to zero, so the oxidation number of Cl must be −1.

(b) The oxidation number of Na is +1 and the oxidation number of O is −2. The oxidation numbers must add up to zero, so the oxidation number of Cl must be +1.

(c) The oxidation number of Na is +1 and the oxidation number of O is −2. The oxidation numbers must add up to zero, so the oxidation number of Cl must be +5.
$$+1 + (3 \times -2) + x = 0$$
$$\therefore \quad x = 0 - 1 + 6 = +5$$

## Oxidising agents and reducing agents 3.2

### WORKED EXAMPLE 2

Deduce the oxidation number of nitrogen in:

(a) $NH_3$
(b) $NO_2^-$
(c) $NO_3^-$

**Answers**

(a) N is more electronegative than H and must therefore have a negative oxidation number. The oxidation number of H is therefore +1. The oxidation numbers must add up to zero, so the oxidation number of N is −3.

$x + (3 \times 1) = 0$
∴ $x = -3$

(b) O is more electronegative than N, so the oxidation number of O is −2. The oxidation numbers must add up to −1 (the charge on the ion), so the oxidation number of N is +3.

$x + (2 \times -2) = -1$
∴ $x = +3$

(c) O is more electronegative than N, so the oxidation number of O is −2. The oxidation numbers must add up to −1 (the charge on the ion), so the oxidation number of N is +5.

$x + (3 \times -2) = -1$
∴ $x = +5$

# Question

1. Calculate the oxidation number of the underlined element in each of the following species.

$\underline{S}O_2$  $\underline{S}O_3$  $H_2\underline{S}$  $\underline{S}O_3^{2-}$  $\underline{S}O_4^{2-}$  $\underline{S}_2O_3^{2-}$

$\underline{Cr}O_4^{2-}$  $\underline{Cr}_2O_7^{2-}$  $\underline{Mn}O_4^{2-}$  $\underline{Mn}O_4^-$  $\underline{V}O^{2+}$  $\underline{V}O_2^+$

$\underline{Cl}O^-$  $\underline{Cl}O_2^-$  $\underline{Cl}O_3^-$  $\underline{Cl}O_4^-$  $[\underline{Cu}Cl_4]^{2-}$  $[\underline{V}(H_2O)_6]^{2+}$

$H_2\underline{O}_2$  $\underline{O}F_2$  $Na\underline{H}$  $Ba\underline{O}_2$

### Key definition

**Oxidation number** is the charge that an ion has or the charge that it would have if the species were fully ionic.

# 3.2   2   Recognising reactions using oxidation numbers

**By the end of this section, you should be able to...**

- know that the oxidation number is a useful concept in terms of classifying reactions as redox and as disproportionation

## Using oxidation numbers to classify reactions

Consider the following ionic half-equations:

$$Zn(s) \rightarrow Zn^{2+}(aq) + 2e^-$$
$$Fe^{2+}(aq) \rightarrow Fe^{3+}(aq) + e^-$$
$$2I^-(aq) \rightarrow I_2(aq) + 2e^-$$

In all three cases, electrons have been lost so the reactions are oxidations. But also notice that the oxidation number of the element has increased in each case.

- Zn has increased from 0 to +2.
- Fe has increased from +2 to +3.
- I has increased from −1 to 0.

This leads to another definition of **oxidation** and **reduction**.

Consider the following ionic half-equations:

$$Cl_2(aq) + 2e^- \rightarrow 2Cl^-(aq) \qquad \text{Cl changes from 0 to} -1$$
$$MnO_4^-(aq) + 8H^+(aq) + 5e^- \rightarrow Mn^{2+}(aq) + 4H_2O(l) \qquad \text{Mn changes from +7 to +2}$$
$$2H_2O(l) + 2e^- \rightarrow 2OH^-(aq) + H_2(g) \qquad \text{H changes from +1 to 0}$$

In each case, the oxidation number of one of the elements involved has decreased, so reduction has taken place.

This concept can now be applied to full equations.

### WORKED EXAMPLE

Use oxidation numbers to show whether the following reactions are an example of redox reactions.

(a)          $H_2S(g) + Cl_2(g) \rightarrow 2HCl(g) + S(s)$
Oxidation numbers:    +1 −2    0    +1 −1    0

**Answer**

The oxidation number of S has increased (−2 to 0), so it has been oxidised.
The oxidation number of Cl has decreased (0 to −1), so it has been reduced.
Since both oxidation and reduction have taken place, the reaction is classified as a redox reaction.

(b)          $NaOH(aq) + HCl(aq) \rightarrow NaCl(aq) + H_2O(l)$
Oxidation numbers:    +1 −2 +1    +1 −1    +1 −1    +1 −2

**Answer**

There is no change in oxidation number for any of the elements involved in the reaction, so this is not an example of a redox reaction.

(c)          $2NaOH(aq) + Cl_2(aq) \rightarrow NaCl(aq) + NaClO(aq) + H_2O(l)$
Oxidation numbers:    +1 −2 +1    0    +1 −1    +1 +1 −2    +1 −2

**Answer**

The oxidation number of Cl has both increased (0 in $Cl_2$ to +1 in NaClO) and decreased (0 in $Cl_2$ to +1 in NaCl), so this is an example of a redox reaction.

It is also an example of a **disproportionation** reaction.

## Question

1. Copy the following table. Use oxidation numbers to complete it. The first example has been done for you.

| Equation | Redox reaction (✓ or ✗) | Disproportionation (✓ or ✗) | Element oxidised | Element reduced |
|---|---|---|---|---|
| $Mg + 2HCl \rightarrow MgCl_2 + H_2$<br>   0     +1 -1       +2 -1      0 | ✓ | ✗ | Mg | H |
| $CaO + H_2O \rightarrow Ca(OH)_2$ | | | | |
| $2H_2O_2 \rightarrow 2H_2O + O_2$ | | | | |
| $2SO_2 + O_2 \rightarrow SO_3$ | | | | |
| $KOH + HNO_3 \rightarrow KNO_3 + H_2O$ | | | | |
| $Cl_2 + H_2O \rightarrow HCl + HClO$ | | | | |

### Key definitions

An element is **oxidised** when its oxidation number *increases*.

An element is **reduced** when its oxidation number *decreases*.

**Disproportionation** is the simultaneous oxidation and reduction of an element in a single reaction.

# 3.2 3 Use of oxidation numbers in nomenclature

By the end of this section, you should be able to...
- indicate the oxidation number of an element in a compound or ion, using a Roman numeral
- write formulae when given oxidation numbers

## Systematic names

When an element can have more than one oxidation state, the names of its compounds and its ions often include the oxidation number of the element. This is written as a Roman numeral in brackets and is often referred to as the 'systematic name'.

The table below shows some examples.

| Formula of compound or ion | Relevant oxidation number | Name of compound or ion |
|---|---|---|
| $FeCl_2$ | Fe +2 | iron(II) chloride |
| $FeCl_3$ | Fe +3 | iron(III) chloride |
| $KMnO_4$ | Mn +7 | potassium manganate(VII) |
| $K_2MnO_4$ | Mn +6 | potassium manganate(VI) |
| $CrO_4^{2-}$ | Cr +6 | chromate(VI) ion |
| $Cr_2O_7^{2-}$ | Cr +6 | dichromate(VI) ion |

**table A** Examples of how to indicate the oxidation number of an element in a compound or ion using the systematic name.

## When to use systematic names

We often use systematic names in chemistry so that we can be specific about the compounds and ions we are referring to. However, in the wider world the systematic names are often omitted.

For example:

- the systematic name for $Na_2SO_4$ is sodium sulfate(VI), but it is still called sodium sulfate
- $Na_2SO_3$ should be labelled sodium sulfate(IV), but sodium sulfite is still common
- $SO_2$ and $SO_3$ are more commonly referred to as sulfur dioxide and sulfur trioxide, rather than sulfur(IV) oxide and sulfur(VI) oxide, respectively.

You should use systematic names as often as possible, particularly during your studies.

## Writing formulae when you have the oxidation number

The other skill you need to develop is to work backwards from the oxidation number to work out the formula of the compound or ion concerned.

**WORKED EXAMPLE 1**

Deduce the formula for iron(II) sulfate.

**Answer**

The formula of the iron(II) ion is $Fe^{2+}$

The formula of the sulfate ion is $SO_4^{2-}$

So the two ions are present in a 1 : 1 ratio to produce a neutral compound, giving the formula $FeSO_4$

## WORKED EXAMPLE 2

Deduce the formula for iron(III) sulfate.

**Answer**

The formula of the iron(III) ion is $Fe^{3+}$

The formula of the sulfate ion is $SO_4^{2-}$

So the two ions are present in a 2:3 ratio to give a neutral compound, giving the formula $Fe_2(SO_4)_3$

## Questions

1. Give the systematic name for each of the following compounds:
   (a) $PCl_3$
   (b) $PCl_5$
   (c) $V_2O_5$
   (d) $NaClO$
   (e) $NaClO_3$

2. Give the systematic name for each of the following ions:
   (a) $NO_3^-$
   (b) $NO_2^-$
   (c) $ClO_4^-$
   (d) $VO^{2+}$
   (e) $VO_2^+$

3. Deduce the formula for each of the following compounds:
   (a) copper(I) oxide
   (b) copper(II) oxide
   (c) chromium(III) sulfate(VI)
   (d) lead(IV) iodide
   (e) cobalt(III) nitrate(V)

4. Why is it not necessary to refer to sodium chloride as sodium(I) chloride, or magnesium oxide as magnesium(II) oxide?

# 3.2 › 4 › Writing full equations from ionic half-equations

**By the end of this section, you should be able to...**

- write ionic half-equations and use them to construct full ionic equations

## Balancing by counting electrons

### Straightforward examples

When solid zinc is added to an aqueous solution of copper(II) sulfate, the following two changes take place:

$Zn(s) \rightarrow Zn^{2+}(aq) + 2e^-$ and $Cu^{2+}(aq) + 2e^- \rightarrow Cu(s)$

Both ionic half-equations involve two electrons, so to construct the full ionic equation for this reaction you simply add together the two half-equations so that the electrons cancel out:

$Zn(s) \rightarrow Zn^{2+}(aq) + 2e^-$
$Cu^{2+}(aq) + 2e^- \rightarrow Cu(s)$

---

$Zn(s) + Cu^{2+}(aq) \rightarrow Zn^{2+}(aq) + Cu(s)$

Now try an example where the electrons are not the same in the two ionic half-equations.

When chlorine gas is bubbled into an aqueous solution of iron(II) chloride, the iron(II) ions are oxidised to iron(III) ions and the chlorine molecules are reduced to chloride ions.

The two ionic half-equations are:

$Fe^{2+}(aq) \rightarrow Fe^{3+}(aq) + e^-$ and $Cl_2(g) + 2e^- \rightarrow 2Cl^-(aq)$

This time, one of the half-equations contains one electron, while the other contains two electrons. Before these can be added together to produce a full equation, the equation containing $Fe^{2+}$ must be multiplied by 2.

$2Fe^{2+}(aq) \rightarrow 2Fe^{3+}(aq) + 2e^-$
$Cl_2(g) + 2e^- \rightarrow 2Cl^-(aq)$

---

$2Fe^{2+}(aq) + Cl_2(g) \rightarrow 2Fe^{3+}(aq) + 2Cl^-(aq)$

### More complicated examples

And now for something a little more tricky!

When an acidified aqueous solution of potassium manganate(VII) is added to an aqueous solution of iron(II) sulfate, the following two changes occur:

$Fe^{2+}(aq) \rightarrow Fe^{3+}(aq) + e^-$ and $MnO_4^-(aq) + 8H^+(aq) + 5e^- \rightarrow Mn^{2+}(aq) + 4H_2O(l)$

In order to balance the electrons, the first half-equation must be multiplied by 5:

$5Fe^{2+}(aq) \rightarrow 5Fe^{3+}(aq) + 5e^-$
$MnO_4^-(aq) + 8H^+(aq) + 5e^- \rightarrow Mn^{2+}(aq) + 4H_2O(l)$

---

$5Fe^{2+}(aq) + MnO_4^-(aq) + 8H^+(aq) \rightarrow 5Fe^{3+}(aq) + Mn^{2+}(aq) + 4H_2O(l)$

Here's another challenging example.

The reaction between aqueous acidified potassium manganate(VII) and hydrogen peroxide involves two changes represented by the following ionic half-equations:

$MnO_4^-(aq) + 8H^+(aq) + 5e^- \rightarrow Mn^{2+}(aq) + 4H_2O(l)$ and

$H_2O_2(aq) \rightarrow 2H^+(aq) + O_2(g) + 2e^-$

The lowest common multiple of 2 and 5 is 10. So the first half-equation should be multiplied by 2 and the second by 5 before they are added together.

$2MnO_4^-(aq) + 16H^+(aq) + 10e^- \rightarrow 2Mn^{2+}(aq) + 8H_2O(l)$
$5H_2O_2(aq) \rightarrow 10H^+(aq) + 5O_2(g) + 10e^-$

$2MnO_4^-(aq) + 16H^+(aq) + 5H_2O_2(aq) \rightarrow$
$2Mn^{2+}(aq) + 8H_2O(l) + 10H^+(aq) + 5O_2(g)$

The electrons have now been cancelled, but we are left with an equation that has $H^+$ ions on both sides of the equation in unequal numbers. These have to be cancelled out so that they are present on only one side. To do this the $10H^+$ on the right-hand side is subtracted from the $16H^+$ on the left-hand side to give the final equation:

$2MnO_4^-(aq) + 6H^+(aq) + 5H_2O_2(aq) \rightarrow 2Mn^{2+}(aq) + 8H_2O(l) + 5O_2(g)$

## Balancing using oxidation numbers

### WORKED EXAMPLE 1

Use oxidation numbers to balance the following equation:

....$SO_2(g)$ + ....$H_2O(l)$ + ....$Ag^+(aq) \rightarrow$ .... $SO_4^{2-}(aq)$ + ....$H^+(aq)$ + ....$Ag(s)$

**Answer**

Identify the elements whose oxidation numbers have changed.

In this case:
- S changes from +4 to +6; this is a '2 electron' change
- Ag changes from +1 to 0; this is a '1 electron' change.

So the ratio of $SO_2$ to $Ag^+$ is 1:2. This gives:

$SO_2(g)$ + ....$H_2O(l)$ + $2Ag^+(aq) \rightarrow SO_4^{2-}(aq)$ + ....$H^+(aq)$ + $2Ag(s)$

We now need to balance the H and O atoms.

This gives:

$SO_2(g) + 2H_2O(l) + 2Ag^+(aq) \rightarrow SO_4^{2-}(aq) + 4H^+(aq) + 2Ag(s)$

Lastly, check the equation for balanced charges.
- The total charge on the left-hand side is 2+.
- The total charge on the right-hand side is also 2+ (−2 + +4).

The equation is now balanced.

### WORKED EXAMPLE 2

Use oxidation numbers to balance the following equation:

....$Fe^{2+}$ + ....$ClO_3^-$ + ....$H^+ \rightarrow$ .... $Fe^{3+}$ + ....$Cl^-$ + ....$H_2O$

**Answer**

Identify the elements whose oxidation numbers have changed.

In this case:

Fe changes from +2 to +3; this is a '1 electron' change

Cl changes from +5 to −1; this is a '6 electron' change.

So the ratio of Fe to $ClO_3^-$ is 6:1. This gives:

$6Fe^{2+} + ClO_3^- +$ ....$H^+ \rightarrow 6Fe^{3+} + Cl^- +$ ....$H_2O$

Once again, balance the H and O atoms.

This gives:

$6Fe^{2+} + ClO_3^- + 6H^+ \rightarrow 6Fe^{3+} + Cl^- + 3H_2O$

Check the charges.
- The total charge on the left-hand side is +17 ((6 × +2) + −1 + (6 × +1)).
- The total charge on the right-hand side is +17 ((6 × +3) + −1).

The equation is now balanced.

## Questions

1. Use each pair of ionic half-equations to construct a full ionic equation. Include state symbols.
   (a) $Zn(s) \rightarrow Zn^{2+}(aq) + 2e^-$ and $Fe^{3+}(aq) + e^- \rightarrow Fe^{2+}(aq)$
   (b) $\frac{1}{2}I_2(aq) + e^- \rightarrow I^-(aq)$ and $2S_2O_3^{2-}(aq) \rightarrow S_4O_6^{2-}(aq) + 2e^-$
   (c) $MnO_4^-(aq) + 8H^+(aq) + 5e^- \rightarrow Mn^{2+}(aq) + 4H_2O(l)$ and $Ce^{3+}(aq) \rightarrow Ce^{4+}(aq) + e^-$
   (d) $Cr_2O_7^{2-}(aq) + 14H^+(l) + 6e^- \rightarrow 2Cr^{3+}(aq) + 7H_2O(l)$ and $Fe^{2+}(aq) \rightarrow Fe^{3+}(aq) + e^-$
   (e) $FeO_4^{2-}(aq) + 8H^+(aq) + 3e^- \rightarrow Fe^{3+}(aq) + 4H_2O(l)$ and $C_2O_4^{2-}(aq) \rightarrow 2CO_2(g) + 2e^-$

2. Use oxidation numbers to balance each of the following equations.
   (a) ....$Cu(s)$ + ....$H^+(aq)$ + ....$NO_3^-(aq) \rightarrow$ ....$Cu^{2+}(aq)$ + ....$H_2O(l)$ + ....$NO(g)$
   (b) ....$Cu(s)$ + ....$H^+(aq)$ + ....$NO_3^-(aq) \rightarrow$ ....$Cu^{2+}(aq)$ + ....$H_2O(l)$ + ....$NO_2(g)$
   (c) ....$Cl_2(g)$ + ....$OH^-(aq) \rightarrow$ ....$Cl^-(aq)$ + ....$ClO_3^-(g)$ + ....$H_2O(l)$

# THINKING BIGGER

## NUTS ABOUT SELENIUM

Selenium is an essential dietary element in trace quantities, but it can be toxic in higher concentrations. The extract below describes one technique used to cut discharge of selenium salts into the San Francisco Bay.

## SOME OF OUR SELENIUM IS MISSING

In 1989, the multinational oil company Chevron discovered that it had cut its discharge of toxic selenium salts into the San Francisco Bay by almost three-quarters. The company should have been very pleased with itself. After all, the six oil companies in the area flush up to 3000 kilograms of selenium into the bay each year, and it seemed that by simply planting a 35-hectare wetland between its outfall and the bay, Chevron had found a practical answer to the problem.

But officials from the State Regional Water Quality Control Board were not so sanguine. They wanted to know where all the selenium was going and told Chevron to find out. The company searched in the sediment at the bottom of the wetland, it dug up the wetland plants, and it checked for a build-up of selenium compounds in the water. But despite all the effort, half of the selenium was still missing.

Alarm bells started to ring. Perhaps the local wildlife was eating the missing selenium. But there were no telltale signs – no dead or maimed animals. Then a team of biologists from the University of California at Berkeley made an educated guess. Norman Terry, Adel Zayed and their colleagues had been studying the way that some plants take toxic selenium salts from soil and water and turn them into the volatile gas dimethyl selenide. They suggested that Chevron's selenium could literally be vanishing into thin air.

**fig A** Wetlands are an important habitat for birds and other wildlife.

Where else will I encounter these themes?

1    2    3    YOU ARE HERE    4    5

Let us start by considering the nature of the writing in the article.

1. The article was written for *New Scientist* and was based on research presented in a scientific paper.

   a. Read the article and comment on the type of writing being used. Think about whether this is a scientist reporting the results of their experiments, a scientific review of a paper or a newspaper or magazine article for a specific audience.

   b. Is there any bias present in the report? What type of words would make you think that was the case?

   c. Criticise how the use of language has been adapted for the audience. Would the wording be different if this was aimed at an audience of 14–16 year olds? How would it be different if written by the press officer of the company Chevron?

> **Command word**
> Note that when the word criticise is given in this context it does not mean that you should be negative. Instead, your comments should show reasoned judgement – try to be constructive.

Now we will look at the chemistry in, or connected to, this article. Don't worry if you are not ready to give answers to these questions yet. You may like to return to the questions once you have covered other topics later in the book.

2. Calculate the number of moles of selenium present in 3000 kg of selenium.
3. Give the electronic configuration using s, p d notation of the Se atom and the $Se^{2-}$ ion.
4. Give the oxidation number of selenium in the following compounds:

   $Na_2SeO_3$  $Ag_2Se$  $H_2SeO_4$

5. Selenium is in Group 6 of the Periodic Table. Give some reasons why selenium chemistry is likely to be similar to sulfur chemistry.
6. Suggest a shape for the molecule $SeF_6$.
7. Give the formula for the molecule dimethyl selenide. Use your knowledge of Group 6 chemistry to suggest a shape for the molecule.
8. Dimethyl sulfide has a boiling temperature of 38 °C and dimethyl telluride has a boiling temperature of 82 °C. By drawing a suitable plot, suggest a boiling temperature for dimethyl selenide and explain your choice. What assumptions have you made?

> Consider the position of selenium relative to sulfur and tellurium in Group 6 when you are drawing your plot. Make sure you use suitable axes and scale in order to obtain a reasonable suggestion for the boiling point!

# Activity

In our world today there is an increasing need to communicate scientific ideas and concepts to people who consider themselves to be 'non-scientists'. Researchers may not always be the best people to communicate science (even their own!), but there are many occupations where this is an essential skill. Imagine you need to give a 5–10 minute presentation to the chief executive officer (CEO) of an oil company to argue the case for the use of genetically engineered plants to deal with selenium waste. How might you convince the CEO that the benefits are outweighed by any perceived concerns about genetic engineering?

> Refer back to the article's final paragraph. You might also like to do some research. If you do, make sure you carefully consider the nature of the source.

### Did you know?

Brazil nuts are a particularly rich source of selenium, but don't go overboard! Too much selenium in the diet can cause the body to break down and excrete excess selenium in the form of the gaseous hydrogen selenide ($H_2Se$). This can leave the unwitting consumer of too many brazil nuts breathing out an odour not unlike garlic.

**fig B** Brazil nuts.

- From an article in *New Scientist* magazine

# 3 Exam-style questions

1. Most chlorine based bleaches contain the ClO⁻ ion.
   (a) (i) State the oxidation number of the chlorine in the ClO⁻ ion. [1]
       (ii) Give the systematic name of the ClO⁻ ion. [1]
   (b) The ClO⁻ ion is produced when chlorine gas is reacted with water.
       Write an ionic equation for this reaction and explain why chlorine is said to undergo disproportionation. [3]
   (c) Chlorine forms another ion with the formula $ClO_3^-$. This ion exists in the compound $KClO_3$.
       $KClO_3$ decomposes on heating according to the equation
       $$4KClO_3(s) \rightarrow 3KClO_4(s) + KCl(s)$$
       Comment on the change in oxidation number of the chlorine in this reaction. [3]
       **[Total: 8]**

2. Bromine is present in sea water as aqueous bromide ions, Br⁻(aq). Bromine can be extracted from sea water by adding chlorine. The ionic equation for the reaction is
   $$Cl_2(aq) + 2Br^-(aq) \rightarrow 2Cl^-(aq) + Br_2(aq)$$
   (a) (i) State what is meant by oxidation and reduction in terms of electron transfer. [1]
       (ii) State which species is acting as an oxidising agent. Justify your answer. [2]
   (b) The chlorine required for extraction of bromine is manufactured from a concentrated solution of sodium chloride using electrolysis.
       The ionic half-equations for the reactions at the electrodes are
       At the anode:   $2Cl^-(aq) \rightarrow Cl_2(aq) + 2e^-$
       At the cathode: $2H_2O(l) + 2e^- \rightarrow 2OH^-(aq) + H_2(g)$
       Use these two half-equations to produce an overall ionic equation for the reaction taking place during the electrolysis. [2]
       **[Total: 5]**

3. Chlorine ($Cl_2$) and sodium hypochlorite (NaClO) can both be used to sterilise water.
   (a) State the oxidation number of chlorine in
       (i) $Cl_2$  and  (ii) NaClO [2]
       Chlorine dioxide ($ClO_2$) is used to prepare sodium chlorite ($NaClO_2$).
       $$2ClO_2 + Na_2O_2 \rightarrow 2NaClO_2 + O_2$$
   (b) Use oxidation numbers to explain if chlorine has been oxidised or reduced in this reaction. [3]
       Potassium chlorate ($KClO_3$) is used as the oxidiser in safety matches. When the match is struck the following reaction takes place
       $$KClO_3(s) \rightarrow KCl(s) + 1\tfrac{1}{2}O_2(g)$$
   (c) The change in oxidation number of the chlorine in this reaction is from
       A  −5 to +1
       B  −5 to −1
       C  +5 to −1
       D  +5 to +1 [1]
       **[Total: 6]**

4. Chlorine and iodine react differently with an aqueous solution containing thiosulfate ions.
   The ionic equations for the two reactions are:
   $$4Cl_2(aq) + S_2O_3^{2-}(aq) + 5H_2O(l)$$
   $$\rightarrow 2SO_4^{2-}(aq) + 10H^+(aq) + 8Cl^-(aq)$$
   $$I_2(aq) + 2S_2O_3^{2-}(aq) \rightarrow 2I^-(aq) + S_4O_6^{2-}(aq)$$
   (a) State the oxidation number of sulfur in each of the three ions $S_2O_3^{2-}$, $SO_4^{2-}$ and $S_4O_6^{2-}$. [3]
   (b) Using your answers from (a), comment of the relative oxidising power of chlorine and iodine. [2]
   (c) Chlorine and iodine both react with hot iron. The equations for the two reactions are:
       $$2Fe(s) + 3Cl_2 \rightarrow 2FeCl_3(s)$$
       $$Fe(s) + I_2(g) \rightarrow FeI_2(s)$$
       Explain whether these reactions support the comment you made in part (b). [3]
       **[Total: 8]**

5 The rusting of iron is a redox process that involves several stages.

**In stage 1**, the iron is converted into iron(II) ions. The ionic half-equations for the two reactions involved are:

$Fe(s) \rightarrow Fe^{2+}(aq) + 2e^-$

$O_2(aq) + 2H_2O(l) + 4e^- \rightarrow 4OH^-(aq)$

(a) Use these two half-equations to construct an overall ionic equation for the reaction. [2]

**In stage 2**, iron(II) ions and the hydroxide ions combine to form a precipitate of iron(II) hydroxide:

$Fe^{2+}(aq) + 2OH^-(aq) \rightarrow Fe(OH)_2(s)$

**In stage 3**, the iron(II) hydroxide is converted into iron(III) hydroxide by reaction with dissolved oxygen and water.

(b) Write an equation for this reaction. Include state symbols. [2]

(c) State whether or not the reaction in each of these three stages is a redox reaction. Justify your answers. [3]

[Total: 7]

6 This question is about the redox reactions of vanadium and manganese.

Vanadium(V) oxide is used as a catalyst in the contact process for the manufacture of sulfuric(VI) acid.

It increases the rate of the following reaction

$SO_2(g) + \frac{1}{2}O_2(g) \rightleftharpoons SO_3(g)$

The vanadium(V) oxide reacts with $SO_2(g)$ to form vanadium(IV) oxide and $SO_3$.

Vanadium(V) oxide is then regenerated by reaction with oxygen.

(a) Write equations for
 (i) the reaction of vanadium(V) oxide with $SO_2$ and
 (ii) the reaction of vanadium(IV) oxide with oxygen.
 State symbols are not required. [2]

An aqueous solution of ammonium vanadate(V) contains the vanadate(V) ion ($VO_2^+$).

This ion can be reduced to the vanadate(IV) ion ($VO^{2+}$) by reaction with sulfur(IV) oxide.

The two ionic half-equations are

$VO_2^+(aq) + 2H^+(aq) + e^- \rightleftharpoons VO^{2+}(aq) + H_2O(l)$

$SO_4^{2-}(aq) + 4H^+(aq) + 2e^- \rightleftharpoons SO_2(g) + 2H_2O(l)$

(b) Use the two ionic half-equations to construct an overall ionic equation for the reduction. [2]

(c) Iron(II) ions ($Fe^{2+}$) can be oxidised to iron(III) ions ($Fe^{3+}$) by manganate(VII) ions.

A dilute solution containing manganate(VII) ions ($MnO_4^-$) and an excess of dilute sulfuric acid is added to the solution containing iron(II) ions.

As the manganate(VII) ion solution is added it changes from purple to colourless.

 (i) The formula of the manganese species that is formed during this reaction is
 **A** $Mn^{2+}(aq)$
 **B** $Mn^{3+}(aq)$
 **C** $Mn^{4+}(aq)$
 **D** $MnO_4^{2-}(aq)$ [1]

 (ii) Use oxidation numbers to calculate the molar ratio of $MnO_4^-$ to $Fe^{2+}$ that would appear in the balanced chemical equation for this reaction. [3]

[Total: 8]

# TOPIC 4
# Inorganic chemistry and the Periodic Table

## Introduction

The Periodic Table of elements is a familiar feature in any study of chemistry. It helps us to make sense of the variety of chemical elements and their different reactions. It also helps us to make sense of the even greater variety of chemical compounds of the elements. In this topic we will focus mostly on two groups (columns) of the Periodic Table to illustrate the trends within the groups and several features of the elements and compounds in them.

In these groups you will meet familiar elements such as magnesium and chlorine, but also less familiar ones such as barium and bromine. For example, you will learn how the limewater test for carbon dioxide works, how milk of magnesia relieves indigestion and why farmers spread lime on their fields. You will learn that barium compounds are very poisonous but why you can safely eat a 'barium meal' before having an X-ray!

By learning how flame tests work, you will also understand how fireworks can have flames of different colours. Countless lives have been saved through the use of chlorine and its compounds in disinfectants and bleaches, and you will learn something of the chemistry involved in how they work.

### All the maths you need
- Use appropriate units in calculations including measurement of ionic radius and ionisation energy
- Use ratios to construct and balance equations
- Understand the connection between frequency and wavelength, and their use in representing the colour of light
- Understand proportional relationships such as the relationship between nuclear charge and electronegativity

## What have I studied before?
- The structure of an atom in terms of a nucleus and electrons in energy levels
- The Periodic Table as an arrangement of elements in order of increasing atomic number
- Elements in the same group of the Periodic Table have trends in properties
- How to represent chemical reactions with full and ionic equations
- The forces of attraction in covalent and ionic bonds, and between molecules
- Consideration of redox reactions in terms of electron transfer and changes in oxidation number

## What will I study later?
- The formation of ionic compounds from their elements considered as a sequence of steps (A level)
- These steps considered in terms of enthalpy changes in a Born–Haber cycle (A level)
- Understanding why some compounds are soluble but others are insoluble (A level)
- Extending the use of oxidation numbers when considering d-block elements (A level)
- Understanding the origin of colour changes in flame tests extended to the reactions of transition metals (A level)

## What will I study in this topic?
- Trends in physical properties of Group 2 and 7 elements
- Reactions of Group 2 and 7 elements and compounds
- Solubility and thermal stability of Group 2 compounds
- Redox reactions
- Tests for anions and cations

# 4.1 1 Trends in the Group 2 elements

By the end of this section, you should be able to...
- understand the reasons for the trend in ionisation energy down Group 2
- understand the reasons for the trend in reactivity of the Group 2 elements down the group

**fig A** From left to right: beryllium, magnesium, calcium, strontium and barium.

## Introduction to the Group 2 elements

There are six elements in Group 2, although you are not likely to see a sample of radium, as all of its isotopes are radioactive. The other five look almost the same in appearance – you can describe all of them, when pure, as bright silvery solids. However, when exposed to air, they combine with oxygen to form oxides as surface layers, which makes them appear dull. This photograph shows samples of five of them together.

In this section, we look at some trends in their properties.

## Trend in ionisation energy

We have looked at ionisation energy in a previous topic. You may remember that it is a fundamental property that affects physical and chemical properties, as we will see in later topics.

### First and second ionisation energies

It is helpful to recall the definition of **first ionisation energy** – it is the energy required to remove an electron from each atom in one mole of atoms in the gaseous state. A general equation for this process, using M to represent an atom of any element, is:

$$M(g) \rightarrow M^+(g) + e^-$$

In typical reactions, the Group 2 elements lose two electrons from each atom, so we should also consider the **second ionisation energy** – this is the energy required to remove an electron from each singly charged ion in one mole of positive ions in the gaseous state. A general equation for this process is:

$$M^+(g) \rightarrow M^{2+}(g) + e^-$$

The table shows the metallic radius and the values of the first and second ionisation energies for the Group 2 elements.

| Element | Metallic radius/nm | Ionisation energy/kJ mol$^{-1}$ | | |
|---|---|---|---|---|
| | | First | Second | First + second |
| Beryllium | 0.112 | 900 | 1757 | 2657 |
| Magnesium | 0.160 | 738 | 1451 | 2189 |
| Calcium | 0.197 | 590 | 1145 | 1735 |
| Strontium | 0.215 | 550 | 1064 | 1614 |
| Barium | 0.224 | 503 | 965 | 1468 |

**table A** The metallic radius and the values of the first and second ionisation energies for the Group 2 elements.

The energy needed for ionisation is to overcome the electrostatic attraction between the electron being removed and the protons in the nucleus.

## Factors to consider

The factors to consider when explaining trends in ionisation energy are:

- the nuclear charge (or the number of protons in the nucleus)
- the orbital in which the electron exists
- the shielding effect (sometimes called the 'screening effect') – the repulsion between filled inner shells and the electron being removed.

In a previous topic, we considered the subshell (or sublevel) from which the electron is being removed. For the Group 2 elements this is not necessary because in their reactions the electrons are always removed from an s subshell.

You should be able to understand why the trend is a decrease down the group for both values.

- As the nuclear charge increases, so the force of attraction for the electron being removed also increases – this means an increase in ionisation energy down the group.
- As each quantum shell is added, energy of the outermost electrons increases.
- As the number of filled inner shells increases, their force of repulsion on the electron being removed increases – this means a decrease in ionisation energy down the group.

So, one factor causes an increase and two factors cause a decrease. The combined effect of the last two factors outweighs the effect of the first factor, and this explains the decrease down the group.

## Trend in reactivity

We will look at the reactions of the Group 2 elements in the next section. In all of these reactions, the element changes into an $M^{2+}$ ion, and there is a general increase in reactivity down the group. This can be explained by the decrease in energy needed to remove the two electrons from each atom of the element.

### Learning tip

When writing an equation for an ionisation, you should always include the state symbol (g) after each atom and ion.

## Questions

1. Write equations to represent the first ionisation of beryllium and the second ionisation of barium.
2. Explain fully why beryllium is less reactive than barium.

### Key definitions

The **first ionisation energy** of an element is the energy required to remove an electron from each atom in one mole of atoms in the gaseous state.

The **second ionisation energy** of an element is the energy required to remove an electron from each singly charged positive ion in one mole of positive ions in the gaseous state.

# 4.1 2 Reactions of the Group 2 elements

By the end of this section, you should be able to...

- know the reactions of the Group 2 elements (magnesium to barium) with oxygen, chlorine and water
- know that these reactions become more vigorous down Group 2

## Reactions with oxygen

You should be familiar with the reaction that occurs when magnesium burns in air – there is a very bright flame and the formation of a white solid.

Similar observations can be made when other Group 2 elements burn in air, but they are qualitatively different – that is, you should be able to recognise that when calcium, strontium and barium are heated in air, the reactions are more vigorous. However, this may be hard to see if the metals are not fresh samples. If the burning metal is placed in a gas jar of oxygen, then the same reaction occurs, although more vigorously.

For all of these elements, the element needs to be heated for the reaction to start. However, even without heating, there is a slow reaction between the element and oxygen when the element is exposed to air. This forms a surface coating of oxide which helps to prevent the element from further reaction.

Barium is the most reactive. It is often stored under oil to keep it from reacting with oxygen and water vapour in the air.

The general equation for all of these reactions is:

$$2M(s) + O_2(g) \rightarrow 2MO(s)$$

The products are oxides containing $M^{2+}$ and $O^{2-}$ ions.

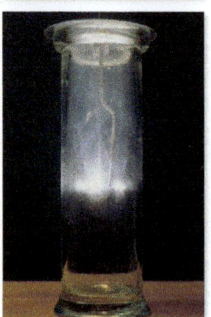

**fig A** (a) Burning magnesium in air. (b) Burning calcium in oxygen.

## Reactions with chlorine

The Group 2 elements combine with chlorine when heated in the gas. Just like the reactions with oxygen, the reactions with chlorine become more vigorous down the group, although this trend is harder to see than with the oxygen reactions. This is what magnesium burning in a flask of chlorine gas looks like:

**fig B** Magnesium burning in chlorine.

The general equation for all of these reactions is:

$$M(s) + Cl_2(g) \rightarrow MCl_2(s)$$

The products are chlorides containing $M^{2+}$ and $Cl^-$ ions.

## Reactions with water

The reaction between magnesium and water is very slow and does not proceed completely. Calcium, strontium and barium react with increasing vigour (i.e. down the group), which can be seen by the increase in effervescence.

In the figure below, you can see that a piece of magnesium in water is covered with bubbles of hydrogen gas, but that the reaction is not very vigorous.

**fig C** Magnesium reacting very slowly with water.

A piece of calcium in water is also covered with bubbles of hydrogen gas, but the reaction is more vigorous.

**fig D** Calcium reacting vigorously with water.

The general equation for all of these reactions is:

$$M(s) + 2H_2O(l) \rightarrow M(OH)_2(aq) + H_2(g)$$

The products are hydrogen gas and hydroxides containing $M^{2+}$ and $OH^-$ ions.

The equation for the reaction with calcium is:

$$Ca(s) + 2H_2O(l) \rightarrow Ca(OH)_2(s) + H_2(g)$$

Calcium hydroxide is only slightly soluble in water, so the liquid in this experiment goes cloudy as a precipitate of calcium hydroxide forms.

The equation for the reaction with barium is:

$$Ba(s) + 2H_2O(l) \rightarrow Ba(OH)_2(aq) + H_2(g)$$

Note the difference in the state symbol for the hydroxide in these equations, as barium hydroxide is soluble in water. You will learn more about the solubility of Group 2 hydroxides in the next section.

## Magnesium and steam

Magnesium reacts differently when heated in steam – it rapidly forms magnesium oxide (a white solid) and hydrogen gas in a vigorous reaction. The equation for this reaction is:

$$Mg(s) + H_2O(g) \rightarrow MgO(s) + H_2(g)$$

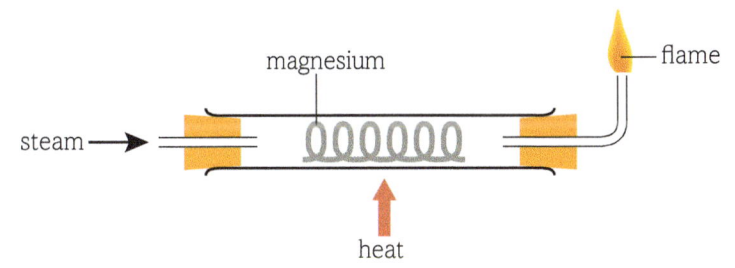

The hydrogen formed is burned as it leaves the tube. This is for safety reasons, to prevent the escape of a highly flammable gas into the laboratory.

## Reactions of beryllium and radium

You are not required to know any of the reactions of beryllium and radium.

However, you know that the trend is increasing reactivity down the group, so you should be able to predict that beryllium is less reactive than magnesium, and that radium is more reactive than barium.

### Learning tip

Practise writing equations for reactions of Group 2 elements with oxygen, chlorine and water.

## Questions

1. Write an equation for each of these reactions:
   (a) calcium with oxygen
   (b) strontium with chlorine
   (c) barium with water

2. Suggest why it is not a good idea to use water to put out a fire involving burning magnesium.

# 4.1 | 3 | Reactions of the Group 2 oxides and hydroxides, and trends in solubility

**By the end of this section, you should be able to...**

- know the reactions of the oxides of Group 2 elements with water and dilute acid, and their hydroxides with dilute acid
- know the trends in solubility of the hydroxides and sulfates of Group 2 elements

## Reactions of the oxides with water

The Group 2 oxides are classed as **basic oxides**, which means that they can react with water to form alkalis. These reactions occur when the oxides are added to water. The only observation to be made is that the solids react to form colourless solutions. The general equation for these reactions is:

$$MO(s) + H_2O(l) \rightarrow M(OH)_2(aq)$$

This equation can be simplified because there is no change to the $M^{2+}$ ion during the reactions.

$$O^{2-} + H_2O \rightarrow 2OH^-$$

This equation shows the formation of hydroxide ions, which is why the resulting solutions are alkaline.

## Trends in solubility of the hydroxides

The pH value of the alkaline solution formed depends partly on the relative amounts of oxide and water, but is also affected by differences in the solubility of the hydroxides.

For example, when magnesium oxide reacts with water, the magnesium hydroxide formed has a very low solubility in water. The solubility of the Group 2 hydroxides increases down the group. Therefore, the maximum alkalinity (pH value) of the solutions formed also increases down the group.

## Testing for carbon dioxide

You may remember from your previous study of chemistry that limewater is used to test for carbon dioxide – it goes cloudy (or milky) as a white precipitate forms. Limewater is the name used for a saturated aqueous solution of calcium hydroxide. Carbon dioxide reacts to form calcium carbonate, which is insoluble in water, and is the white precipitate. The equation for the reaction is:

$$CO_2 + Ca(OH)_2 \rightarrow CaCO_3 + H_2O$$

As carbon dioxide is bubbled through limewater, the amount of precipitate increases.

**fig A** When carbon dioxide is bubbled through limewater, the amount of precipitate increases.

## Milk of magnesia

For over a century, a suspension of magnesium hydroxide in water has been sold as an indigestion remedy called milk of magnesia. A bottle of this contains a saturated solution of magnesium hydroxide mixed with extra solid magnesium hydroxide, which acts as an antacid.

The human stomach contains hydrochloric acid that is needed to digest food, but sometimes there is too much of the acid and the person develops symptoms of indigestion. Taking milk of magnesia neutralises some of the hydrochloric acid and relieves the symptoms. The equation for the reaction is:

$$Mg(OH)_2 + 2HCl \rightarrow MgCl_2 + 2H_2O$$

Although hydroxide ions attack human tissue, the very low solubility of magnesium hydroxide means that the concentration of $OH^-$ ions in the medicine is also very low and does not pose a risk to health.

**fig B** A bottle of milk of magnesia.

## Reactions of the oxides and hydroxides with acids

All of the Group 2 oxides and hydroxides react with acids to form salts and water. These reactions can be described as neutralisation reactions.

During the reactions, the only observations to be made are that a white solid reacts to form a colourless solution. The reactions are exothermic, so you may use some of them in experiments to measure energy changes.

Here are some sample equations:

$$MgO + H_2SO_4 \rightarrow MgSO_4 + H_2O$$
$$CaO + 2HNO_3 \rightarrow Ca(NO_3)_2 + H_2O$$
$$Sr(OH)_2 + 2HCl \rightarrow SrCl_2 + 2H_2O$$
$$Ba(OH)_2 + 2HCl \rightarrow BaCl_2 + 2H_2O$$

## Use in agriculture

For centuries, farmers have used lime to control soil acidity so that a greater yield of crops can be obtained.

**fig C** Lime being spread on a field.

Lime is mostly calcium hydroxide (obtained from limestone, which is calcium carbonate), and neutralises excess acidity in the soil. Using nitric acid to represent the acid in soil, the equation for this reaction is:

$$Ca(OH)_2 + 2HNO_3 \rightarrow Ca(NO_3)_2 + 2H_2O$$

## Trends in solubility of the Group 2 sulfates

All Group 2 nitrates and chlorides are soluble, but the solubility of Group 2 sulfates decreases down the group.

- Magnesium sulfate is classed as soluble.
- Calcium sulfate is slightly soluble.
- Strontium sulfate and barium sulfate are insoluble.

You do not have to understand the reasons for this trend, but you do need to know how the very low solubility of barium sulfate is used in a test for sulfate ions in solution.

## Testing for sulfate ions

The presence of sulfate ions in an aqueous solution can be shown by adding a solution containing barium ions (usually barium chloride or barium nitrate). Any sulfate ions in the solution will react with the added barium ions to form a white precipitate of barium sulfate. The ionic equation for this reaction is:

$$Ba^{2+}(aq) + SO_4^{2-}(aq) \rightarrow BaSO_4(s)$$

There are other anions that could also form a white precipitate with barium ions, especially carbonate ions, so in the test there must be $H^+$ ions present to prevent barium carbonate from forming as a white precipitate. Dilute nitric acid or dilute hydrochloric acid is therefore added as part of the test.

As an example, here is how to test for the presence of sulfate ions in a solution of sodium sulfate.

- Add dilute nitric acid and barium nitrate solution.
- A white precipitate forms.

The equation for the reaction is

$$Ba(NO_3)_2(aq) + Na_2SO_4(aq) \rightarrow BaSO_4(s) + 2NaNO_3(aq)$$

## Barium meals

Solutions containing barium ions are poisonous to humans, yet barium sulfate is used in hospitals, where patients are sometimes given a barium 'meal'. This 'meal' contains barium sulfate, which is not poisonous because it is insoluble – although it contains barium ions, these ions are not free to move. Although bones show up well on X-rays, soft tissues do not. So, if the patient has a barium meal before an X-ray, these soft tissues will show up more clearly because of the dense white solid.

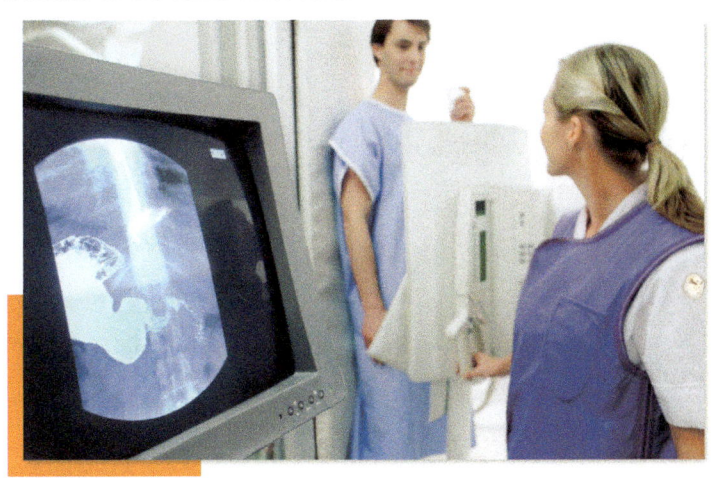

**fig D** How a barium meal can help show up soft tissues on an X-ray.

### Learning tip

There is no trend in reactivity with water for the Group 2 oxides, because they already contain metal ions, not metal atoms.

## Questions

1. Limewater is used to test for carbon dioxide. Why should limewater not be left exposed to air before using it in this test?

2. Why is an acid added when using barium chloride or barium nitrate solution to test for sulfate ions?

### Key definition

**Basic oxides** are oxides of metals that react with water to form metal hydroxides, and with acids to form salts and water.

## 4.1  4 ▶ Thermal stability of Group 2 compounds, and the comparison with Group 1

**By the end of this section, you should be able to...**

- understand the reasons for the trends in thermal stability of the nitrates and carbonates of the elements in Groups 1 and 2 in terms of the size and charge of the cations involved
- understand experimental procedures used to show patterns in the thermal decomposition of Group 1 and 2 nitrates and carbonates

### Factors affecting thermal stability

**Thermal stability** is a term that indicates how stable a compound is when it is heated. Does it not decompose at all (very thermally stable), or does it decompose as much as possible (not at all thermally stable), or somewhere in between?

You are familiar with ionic bonding in compounds and know that the bond strength is responsible for physical properties such as the melting temperature. For example, the high melting temperature of sodium chloride (NaCl) can be explained in terms of the strong electrostatic forces of attraction between the large numbers of oppositely charged $Na^+$ and $Cl^-$ ions. When sodium chloride melts, the ions change from being regularly arranged in a giant lattice to moving freely in a liquid. There is no decomposition occurring in this change of state.

The situation with Group 2 nitrates and carbonates is very different compared with a Group 1 chloride such as sodium chloride. There are three reasons for this.

1. The charge on a Group 2 cation is double that of a Group 1 cation (e.g. $Ca^{2+}$ compared with $Na^+$).
2. The size (ionic radius) of a Group 2 cation is smaller than that of the Group 1 cation in the same period.
3. The nitrate ($NO_3^-$) and carbonate ($CO_3^{2-}$) anions are more complex than the $Cl^-$ ion.

These differences mean that when Group 2 nitrates and carbonates are heated, they do not melt. Instead, they decompose. We need to look carefully at these factors to understand why this is.

- The larger, more complex nitrate ion can change into the smaller, more stable nitrite ion ($NO_2^-$) or oxide ion ($O^{2-}$) by decomposing and releasing oxygen gas and/or nitrogen dioxide gas.
- The larger, more complex carbonate ion can change into the smaller, more stable oxide ion ($O^{2-}$) by decomposing and releasing carbon dioxide gas, $CO_2$.
- The stabilities of the nitrate and carbonate anions are influenced by the charge and size of the cations present – smaller and more highly charged cations affect these anions more.

The table shows the charge and radius for each of the ions in Groups 1 and 2.

| Group 1 | | | Group 2 | | |
| --- | --- | --- | --- | --- | --- |
| Element | Charge on ion | Ionic radius/ nm | Element | Charge on ion | Ionic radius/ nm |
| Lithium | +1 | 0.074 | Beryllium | +2 | 0.027 |
| Sodium | +1 | 0.102 | Magnesium | +2 | 0.072 |
| Potassium | +1 | 0.138 | Calcium | +2 | 0.100 |
| Rubidium | +1 | 0.149 | Strontium | +2 | 0.113 |
| Caesium | +1 | 0.170 | Barium | +2 | 0.136 |

**table A**

You can see that the cation with the greatest influence (biggest charge and smallest size) on an anion is $Be^{2+}$, and the one with the least influence (smallest charge and largest size) is $Cs^+$.

## Thermal stability of nitrates

All of the nitrates of the Group 1 and Group 2 elements are white solids. When they are heated, they all decompose to nitrites or oxides, and give off nitrogen dioxide (brown fumes) and/or oxygen. If the nitrate contains water of crystallisation, then steam will also be observed.

If no brown fumes are observed, this indicates a lesser decomposition that can be represented by this word equation:

metal nitrate → metal nitrite + oxygen

Nitrates and nitrites are sometimes differentiated by using oxidation numbers – nitrate(V) for nitrate, and nitrate(III) for nitrite.

If brown fumes are observed, this indicates a greater decomposition that can be represented by this word equation:

metal nitrate → metal oxide + nitrogen dioxide + oxygen

The table shows typical observations obtained by heating samples of nitrates in test tubes over a Bunsen flame.

| Group 1 nitrate | | Group 2 nitrate | |
|---|---|---|---|
| Name | Result | Name | Result |
| Lithium nitrate | brown fumes | Beryllium nitrate | brown fumes |
| Sodium nitrate | no brown fumes | Magnesium nitrate | brown fumes |
| Potassium nitrate | no brown fumes | Calcium nitrate | brown fumes |
| Rubidium nitrate | no brown fumes | Strontium nitrate | brown fumes |
| Caesium nitrate | no brown fumes | Barium nitrate | brown fumes |

**table B** Results obtained from heating samples of nitrates in test tubes over a Bunsen flame.

This is what happens, in terms of decomposition, when samples of nitrates are heated in test tubes over a Bunsen flame.

| Group 1 nitrate | | Group 2 nitrate | |
|---|---|---|---|
| Name | Result | Name | Result |
| Lithium nitrate | greater decomposition | Beryllium nitrate | greater decomposition |
| Sodium nitrate | lesser decomposition | Magnesium nitrate | greater decomposition |
| Potassium nitrate | lesser decomposition | Calcium nitrate | greater decomposition |
| Rubidium nitrate | lesser decomposition | Strontium nitrate | greater decomposition |
| Caesium nitrate | lesser decomposition | Barium nitrate | greater decomposition |

**table C** Results, in terms of decomposition, when samples of nitrates are heated in test tubes over a Bunsen flame.

You can see that the greater decomposition occurs when:
- the cation has a 2+ charge (all of the Group 2 nitrates)
- the cation has a 1+ charge and is also the smallest Group 1 cation.

Here are some sample equations for the reactions that occur:

$4LiNO_3 \rightarrow 2Li_2O + 4NO_2 + O_2$ (lithium nitrate – the only Group 1 nitrate that decomposes in this way)

$2NaNO_3 \rightarrow 2NaNO_2 + O_2$ (all other Group 1 nitrates decompose in this way)

$2Be(NO_3)_2 \rightarrow 2BeO + 4NO_2 + O_2$ (all Group 2 nitrates decompose in this way)

## Thermal stability of carbonates

All of the carbonates of the Group 1 and Group 2 elements are white solids. When they are heated, they either do not decompose, or decompose to oxides and give off carbon dioxide.

As the gas given off is colourless and the carbonate and oxide are both white solids, there are no observations that can be made.

The table shows what happens when samples of carbonates are heated in test tubes over a Bunsen flame.

| Group 1 carbonate | | Group 2 carbonate | |
| --- | --- | --- | --- |
| Name | Result | Name | Result |
| Lithium carbonate | decomposition | Beryllium carbonate | decomposition |
| Sodium carbonate | no decomposition | Magnesium carbonate | decomposition |
| Potassium carbonate | no decomposition | Calcium carbonate | decomposition |
| Rubidium carbonate | no decomposition | Strontium carbonate | decomposition |
| Caesium carbonate | no decomposition | Barium carbonate | decomposition |

**table D** Results, in terms of decomposition, when samples of carbonates are heated in test tubes over a Bunsen flame.

You can see that the pattern is similar to that of the nitrates. The lithium compound and all of the Group 2 compounds behave differently from the other Group 1 compounds.

Lithium carbonate decomposes at lower temperatures than the other Group 1 carbonates:

$$Li_2CO_3 \rightarrow Li_2O + CO_2$$

Other Group 1 carbonates do not decompose on heating, except at very high temperatures.

All Group 2 carbonates decompose in the same way, but with increasing difficulty down the group. A typical equation for one of these decompositions is:

$$CaCO_3 \rightarrow CaO + CO_2$$

You can see a similar pattern as with the nitrates. Decomposition occurs when:

- the cation has a 2+ charge (all of the Group 2 carbonates)
- the cation has a 1+ charge and is also the smallest Group 1 cation (only lithium carbonate).

**fig A** Calcium oxide forms when calcium carbonate (limestone) is heated strongly. If the strong heating is continued, there is no further chemical change, but a very bright glow is seen. This is the origin of 'limelight' – it was formerly used in theatre lighting.

### Key definition

**Thermal stability** is a measure of the extent to which a compound decomposes when heated.

## Questions

### Learning tip

Look for the similar patterns in the thermal stability of Group 1 and Group 2 nitrates and carbonates. Although the reactions are different, the lithium compound behaves like the Group 2 compounds and not like the other Group 1 compounds.

1. What observations would be made when these compounds are heated in a test tube over a Bunsen flame?
   (a) calcium nitrate
   (b) sodium carbonate

2. Write a chemical equation for each of these reactions.
   (a) the decomposition of potassium nitrate
   (b) the decomposition of strontium carbonate

# 4.1 5 Flame tests and the test for ammonium ions

### By the end of this section, you should be able to...

- know the flame colours for Group 1 and 2 compounds
- understand experimental procedures used to show flame colours in compounds of Group 1 and 2 elements
- understand the formation of characteristic flame colours by Group 1 and 2 compounds in terms of electron transitions
- know how to identify ammonium ions

## Introduction to flame tests

A flame test seems a very simple chemical test to identify the presence of a cation in a compound. However, it is important to know how to do one carefully, to know how to interpret the results and to understand how the test works.

A flame test result can indicate the presence of some metals (in the form of cations) in Groups 1 and 2 of the Periodic Table. It does not work for all of them.

## How to do a flame test

We will assume that a small quantity of a solid is available in a glass dish. This is how to do a flame test.

- Wear safety glasses and a lab coat. Within a fume cupboard, light a Bunsen burner.
- Using a dropper, add a few drops of concentrated hydrochloric acid to the solid and mix together so that the metal compound begins to dissolve. (One reason for using hydrochloric acid is to convert any metal compound to a chloride – chlorides are more volatile than other salts so are more likely to give better results.)
- Dip a clean metal wire (platinum or nichrome) or silica rod into the mixture to obtain a sample of the compound.
- Hold the end of the wire or rod in the flame and observe the colour.

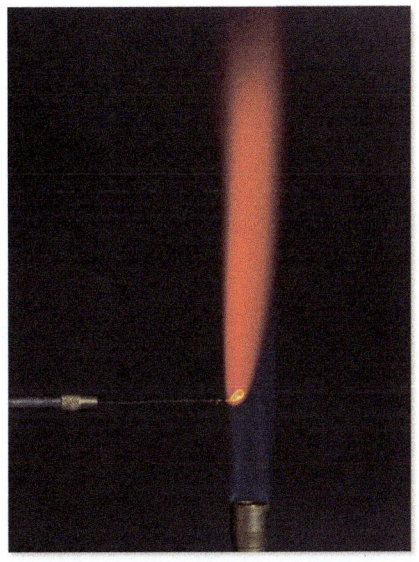

**fig A** A platinum wire loop with a sample of a strontium compound being held in a Bunsen flame.

## Problems with a flame test

There are two main problems with a flame test.

- Many compounds contain small amounts of sodium compounds as impurities, so the intense colour of sodium can mask other colours.
- Describing colours with words is subjective – people have different levels of colour vision, and a word description of a colour may mean different colours to different people.

## Colour descriptions

Some traditional descriptions of colours are problematic. For example, what is the colour 'brick red'? It may depend on the bricks used in a particular location, and these are not the same throughout the world!

Another example is the traditional colour description used for potassium. It is lilac, but can you tell the difference between lilac and lavender, magenta, mauve, pink, plum, puce, purple and violet?

The table shows some traditional colour descriptions obtained when Group 1 and 2 compounds are tested in this way.

| Metal cation | Formula | Colour |
|---|---|---|
| Lithium | $Li^+$ | red |
| Sodium | $Na^+$ | yellow/orange |
| Potassium | $K^+$ | lilac |
| Rubidium | $Rb^+$ | red/purple |
| Caesium | $Cs^+$ | blue/violet |

| Metal cation | Formula | Colour |
|---|---|---|
| Beryllium | $Be^{2+}$ | (no colour) |
| Magnesium | $Mg^{2+}$ | (no colour) |
| Calcium | $Ca^{2+}$ | (brick) red |
| Strontium | $Sr^{2+}$ | (crimson) red |
| Barium | $Ba^{2+}$ | (apple) green |

**table A** Colour descriptions obtained when Group 1 and 2 compounds undergo a flame test.

There are also some metal cations with characteristic flame colours that are not in Groups 1 or 2. For example, copper compounds produce a blue-green colour in a flame test.

## What causes the colours in flame tests?

The simple answer is electron transitions. However, we need a more complete explanation than this.

You know from previous sections that electrons occupy orbitals in specific energy levels in an atom. These are often represented using electron configurations such as 2.8.1 (for sodium).

Electrons can absorb energy and move to higher energy levels. Sometimes the term 'ground state' is used to describe an atom with all its electrons in their lowest possible energy levels. If an electron moves to a higher energy level, then the new situation can be described as an 'excited state'. This movement of an electron to an excited state occurs during a flame test.

However, this movement is immediately followed by the return of the electron to its ground state, which releases energy. If this energy corresponds to radiation in the visible light spectrum, then a characteristic colour appears – this is the flame test colour.

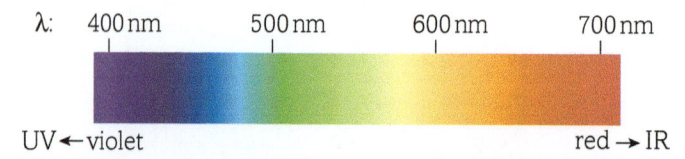

$\lambda$: 400 nm   500 nm   600 nm   700 nm

UV ← violet                         red → IR

**fig B** The wavelengths of the spectrum of visible light.

For example, the visible spectrum covers the part of the electromagnetic spectrum in the wavelength range 400–700 nm. The electron transition in sodium corresponds to a wavelength of about 590 nm, which is in the yellow-orange part of the spectrum, so this is the colour of a sodium compound in a flame test. The electron transition in magnesium corresponds to a wavelength outside the visible spectrum, so there is no flame colour for magnesium.

**fig C** The yellow-orange colour of street lights is the same as that in a flame test.

#### Learning tip
Remember that doing a flame test on a magnesium compound has nothing to do with burning magnesium metal in a Bunsen flame.

## The test for ammonium ions
A cation that does not give a colour in a flame test is the ammonium ion. The usual test for ammonium ions in a solid or solution is to add sodium hydroxide solution and warm the mixture. The addition of sodium hydroxide causes this reaction:

$$NH_4^+ + OH^- \rightarrow NH_3 + H_2O$$

and the warming releases ammonia gas. Ammonia can be recognised by its smell, but a simple chemical test is to use damp red litmus paper, which turns blue (ammonia is the only common alkaline gas). Alternatively, hydrogen chloride gas (from concentrated hydrochloric acid) reacts with ammonia to form white fumes of ammonium chloride:

$$NH_3 + HCl \rightarrow NH_4Cl$$

# Questions

1. Why is concentrated hydrochloric acid used in flame tests?
2. Explain why barium compounds give a characteristic flame colour but magnesium compounds do not.

# 4.2  1  General trends in Group 7

**By the end of this section, you should be able to...**

- understand the reasons for the trends in melting and boiling temperatures, physical state at room temperature, and electronegativity for the halogens
- understand the reasons for the trend in reactivity of Group 7 elements down the group

## Introduction to the Group 7 elements

Group 7 of the Periodic Table contains five elements. These elements are often known as halogens because they all form salts called halides. The term 'halogen' comes from Greek and means 'salt producer'.

When considering the properties of the group, the elements at the top and bottom of the group (fluorine and astatine) are often ignored. Fluorine is ignored because it sometimes behaves differently from chlorine, bromine and iodine, and astatine is ignored because (like radium in Group 2) it only exists as radioactive isotopes.

The table shows some information about the Group 7 elements.

| Element | State at room temperature | Melting temperature/°C | Boiling temperature/°C | Electronegativity |
|---|---|---|---|---|
| Fluorine | gas | −220 | −188 | 4.0 |
| Chlorine | gas | −101 | −35 | 3.0 |
| Bromine | liquid | −7 | 59 | 2.8 |
| Iodine | solid | 114 | 184 | 2.5 |
| Astatine | solid | 302 | 337 | 2.2 |

**table A** The physical state at room temperature, melting and boiling temperatures and electronegativity for the Group 7 elements.

## Trends in melting and boiling temperature

All of the halogens exist as diatomic molecules, so their melting and boiling temperatures depend on the strengths of the intermolecular forces of attraction between these molecules. We looked at these forces, known as London forces, in **Section 2.2.5**. Here is a reminder of how they arise.

As the two atoms in the diatomic molecule are identical, the pair of electrons forming the covalent bond between them is shared equally between the two atoms. This means that the halogen molecules are non-polar, at least on average. However, as the positive charges of the protons in the two nuclei are in fixed positions, but the electron density in a halogen molecule continuously fluctuates, sometimes the centres of positive and negative charge do not coincide. This situation results in a temporary dipole, often referred to as an instantaneous dipole.

We will use a rectangular shape to represent a halogen molecule. The two dots represent the nuclei of the two halogen atoms. When two molecules are close together, you would expect no interaction between the two because they are both non-polar.

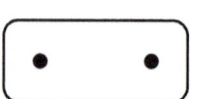

If the molecule on the left becomes an instantaneous dipole, then it will cause an induced dipole in the molecule on the right. This results in a force of attraction between the two molecules:

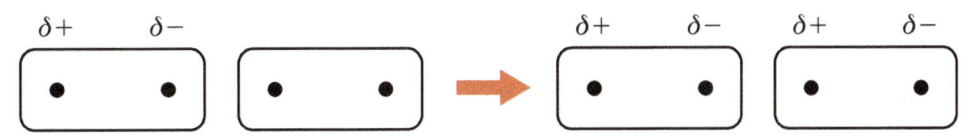

This force of attraction is described as an instantaneous dipole–induced dipole attraction, and these are the intermolecular forces of attraction that exist between halogen molecules. These weak forces increase as the number of electrons and the size of the electron cloud increases. So, the forces increase in strength down Group 7 as the number of electrons in the molecules increases. This explains the increase in both melting temperature and boiling temperature down Group 7.

## Equations for changes of state

You are used to writing equations for chemical changes – we use equations for physical changes less often. As well as using the correct state symbols, it is important to remember to write the formulae (not the symbols) of the halogens.

When bromine is left at room temperature, it gives off brown vapour, as its boiling point (59 °C) is not much higher than room temperature. The equation for this change is:

$Br_2(l) \rightarrow Br_2(g)$

When iodine is warmed, most of it changes directly into a vapour without melting – this change is called 'sublimation'. The equation for this change is:

$I_2(s) \rightarrow I_2(g)$

## Trend in electronegativity

We have already seen that electronegativity is the ability of an atom to attract the pair of electrons in a covalent bond. The 0–4 scale devised by Linus Pauling is still used. The electronegativity of an atom depends on:

- its nuclear charge – the bigger the nuclear charge, the higher the electronegativity
- the distance between the nucleus and the bonding pair of electrons – the shorter the distance, the higher the electronegativity
- the shielding effect of electrons in inner energy levels – the fewer energy levels, the higher the electronegativity.

The electronegativity of the Group 7 elements is the highest of any group in the Periodic Table. The electronegativity of fluorine is the highest of all elements.

## Trend in reactivity

Fluorine is an extremely reactive element, and reactivity decreases down Group 7. Because of their high electronegativity, most reactions of the halogens involve them acting as oxidising agents and gaining electrons to form negative ions or becoming the slightly negative ($\delta-$) part of a polar molecule. The decreasing reactivity down the group can therefore be explained by reference to the same factors used to explain the decreasing electronegativity down the group.

We will see examples of reactions in the next three sections.

### Learning tip
Write a brief summary to help you understand the difference between intermolecular forces involving permanent dipoles and temporary dipoles.

# Questions

1. Why does bromine have a higher boiling temperature than chlorine?
2. Why is fluorine the most electronegative element of all?

# 4.2 | 2 | Redox reactions in Group 7

By the end of this section, you should be able to...

- understand the trend in reactivity of Group 7 elements in terms of the redox reactions of chlorine, bromine and iodine with halide ions in aqueous solution, followed by the addition of an organic solvent
- understand, in terms of oxidation number, oxidation reactions with Group 1 and 2 metals and disproportionation reactions of chlorine
- make predictions about fluorine and astatine and their compounds, in terms of knowledge of trends in halogen chemistry

## Reactions with metals in Groups 1 and 2

There are 12 elements in Groups 1 and 2, and five elements in Group 7, so there are 60 possible reactions to consider! You do not need to know details of all of these reactions, but here are some useful generalisations that you should know.

- Reactions are most vigorous between elements at the bottom of Groups 1 and 2, and elements at the top of Group 7. So the most vigorous reaction should be between caesium (or francium) and fluorine, and the least vigorous between beryllium and iodine (or astatine).
- The products of these reactions are salts – ionic solids that are usually white.
- All of these reactions involve electron transfer to the halogen, so they are redox reactions in which the halogen acts as an oxidising agent.
- The oxidation number of the halogen decreases from 0 to $-1$, and the oxidation number of the metal increases from 0 to $+1$ or $+2$, depending on the group.

Here are a couple of sample equations:

- lithium reacting with chlorine     $2Li + Cl_2 \rightarrow 2LiCl$
- barium reacting with bromine     $Ba + Br_2 \rightarrow BaBr_2$

## Halogen/halide displacement reactions

A more reactive halogen can displace a less reactive halogen from one of its compounds. So:

- chlorine displaces bromine and iodine
- bromine displaces iodine but not chlorine
- iodine does not displace either chlorine or bromine.

These reactions occur in aqueous solution, so any reaction that occurs is indicated by a colour change. One problem in interpreting colour changes in these reactions is the similarity of some colours and the variation in colour with concentration. For example, bromine in its liquid state is red-brown, but bromine dissolved in water might be orange or yellow, depending on the concentration. Iodine dissolved in water may also appear brown at some concentrations.

When doing these reactions, it is a good idea to add an organic solvent (such as cyclohexane) after the reaction, and then shake the tube. Halogens are more soluble in cyclohexane than in water, so the halogen dissolves in the organic upper layer, where its colour can more easily be seen. The image shows the colours of the halogens dissolved in cyclohexane.

**fig A** The pale green colour of chlorine does not change much, the orange colour of bromine looks a bit darker, but the colour of iodine changes to purple or violet in cyclohexane.

Here are some sample equations:

- chlorine displacing bromine     $Cl_2 + 2NaBr \rightarrow 2NaCl + Br_2$
- bromine displacing iodine     $Br_2 + 2I^- \rightarrow 2Br^- + I_2$

These equations are examples of redox reactions. The reacting halogen decreases its oxidation number from 0 to $-1$, and the reacting halide increases its oxidation number from $-1$ to 0.

The decreasing reactivity of chlorine, bromine and iodine in the reactions above can be explained using the same factors as in the previous section. Chlorine is the most reactive of the three because:

- it is the smallest atom, so the incoming electron gets closer to and is more attracted by the protons in the nucleus
- it has the smallest number of complete inner energy levels of electrons, so the incoming electron experiences the least repulsion.

## Disproportionation reactions of chlorine

**Disproportionation** is a more unusual type of reaction, in which one element undergoes both oxidation and reduction at the same time. We will look at three examples of this type of reaction, all involving chlorine.

### Chlorine with water

When chlorine is added to water, it dissolves to form a solution that is sometimes called 'chlorine water' (just as 'bromine water' is used to refer to bromine dissolved in water). Some of the dissolved chlorine also reacts to form a mixture of two acids.

You are familiar with hydrochloric acid, but the other acid is chloric(I) acid (its old name is hypochlorous acid). Its formula is shown in the equation below as HClO, but HOCl is also commonly used. Both acids are colourless solutions, so there is no visible change during the reaction.

The disproportionation that occurs can be shown using oxidation numbers:

$$Cl_2 + H_2O \rightarrow HCl + HClO$$

| 0 | −1 | chlorine is reduced |
| 0 | +1 | chlorine is oxidised |

The addition of chlorine to disinfect water for drinking purposes has saved countless lives, and continues to do so today. It kills the pathogens responsible for water-borne diseases such as cholera.

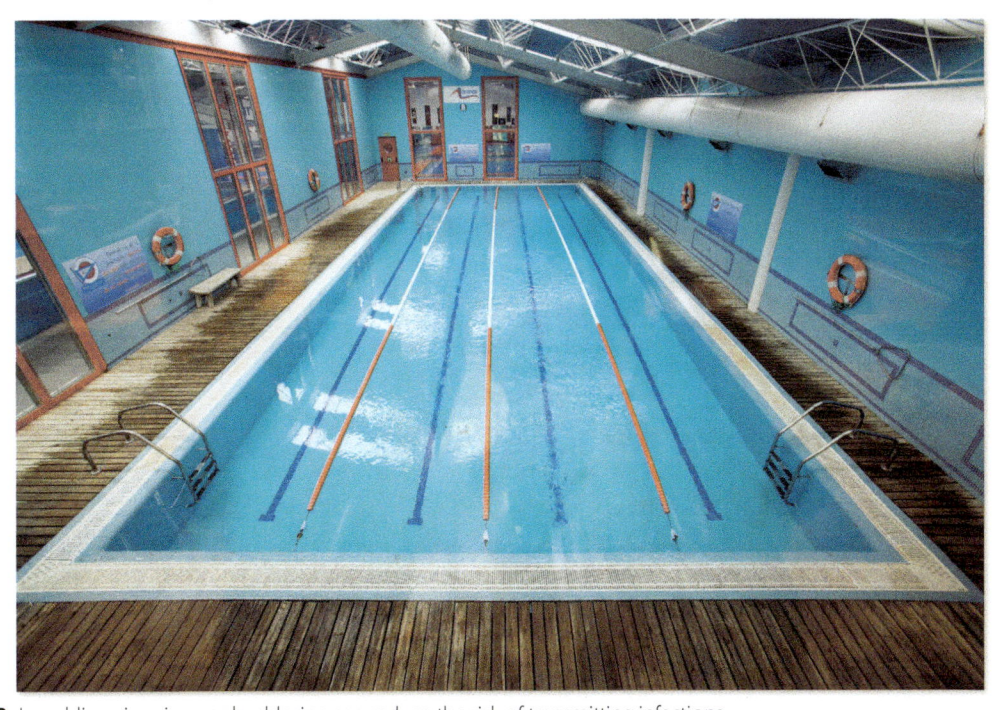

**fig B** In public swimming pools, chlorine can reduce the risk of transmitting infections.

### Chlorine with cold alkali

When chlorine is added to cold dilute aqueous sodium hydroxide, it reacts to form the salts of the acids in the equation above. These salts are sodium chloride and sodium chlorate(I), which is also known as sodium hypochlorite.

Again, the disproportionation that occurs can be shown using oxidation numbers:

$$Cl_2 + 2NaOH \rightarrow NaCl + NaClO + H_2O$$

| 0 | −1 | chlorine is reduced |
| 0 | +1 | chlorine is oxidised |

The sodium chlorate(I) formed is also a disinfectant, but it is mainly known for its bleaching action. It is used extensively in industry and is the active ingredient in household bleach.

### Chlorine with hot alkali

When chlorine is added to hot concentrated sodium hydroxide solution, it reacts to form sodium chloride and a different product, sodium chlorate(V).

Again, the disproportionation that occurs can be shown using oxidation numbers:

$$3Cl_2 + 6NaOH \rightarrow 5NaCl + NaClO_3 + 3H_2O$$

| | | |
|---|---|---|
| 0 | −1 | chlorine is reduced |
| 0 | +5 | chlorine is oxidised |

The sodium chlorate(V) formed is also used in bleaching, and as a weed killer.

Bromine and iodine react in similar ways.

## Reactions of fluorine and astatine

Fluorine and astatine are mentioned in some of the reactions in this topic, especially when it is not possible to make predictions about their reactions using trends in chlorine, bromine and iodine.

If you are asked to predict reactions of fluorine and astatine that you are not familiar with, then you can use information in this topic to help you. For example, you could write an equation to represent the reaction between sodium and astatine, based on your knowledge of the reaction between sodium and iodine.

#### Learning tip
Read the explanation for chlorine being more reactive than the other halogens (except for fluorine) at the end of the section on displacement reactions. Now try to explain why astatine is the least reactive halogen.

## Questions

1. (a) Write a chemical equation for the reaction between chlorine and potassium iodide.
   (b) Write an ionic equation for the reaction between bromine and sodium astatide.

2. Write equations and name the products of the reactions between bromine and:
   (a) cold dilute aqueous sodium hydroxide
   (b) hot concentrated sodium hydroxide solution.

#### Key definition
**Disproportionation** is the simultaneous oxidation and reduction of an element in a single reaction.

# 4.2 3 Reactions of halides with sulfuric acid

**By the end of this section, you should be able to...**

- understand the reactions of solid Group 1 halides with concentrated sulfuric acid, to illustrate the trend in reducing ability of the halide ions

## Redox reactions again

In previous sections you have come across several examples of the halogens acting as oxidising agents, and you know that this oxidising power decreases down the group.

In this sections, we will look at reactions of halide *ions*, not halogen *molecules*. It is important to realise that in these reactions, the halides act as reducing agents, and that the trend is different.

| | Halogen | | Halide | | |
|---|---|---|---|---|---|
| | Oxidising power | | Reducing power | | |
| High | fluorine | $F_2$ | fluoride | $F^-$ | Low |
| ↑ | chlorine | $Cl_2$ | chloride | $Cl^-$ | |
| | bromine | $Br_2$ | bromide | $Br^-$ | ↓ |
| | iodine | $I_2$ | iodide | $I^-$ | |
| Low | astatine | $At_2$ | astatide | $At^-$ | High |

**table A** Notice that the decreasing trend down the group in oxidising power of halogens goes with the increasing trend in reducing power of halides.

The reducing action of halide ions can be represented by this general half-equation:

$$2X^- \rightarrow X_2 + 2e^-$$

Sulfuric acid is of course an acid, but when concentrated it contains very few ions, and we can write an equation for its partial ionisation:

$$H_2SO_4 \rightleftharpoons H^+ + HSO_4^-$$

Note the reversible arrow. The position of this equilibrium lies well to the right. In the concentrated acid, this first ionisation is far from complete.

$$HSO_4^- \rightleftharpoons H^+ + SO_4^{2-}$$

Sulfuric acid, especially when concentrated, can act as an oxidising agent as well as an acid. When it acts as an oxidising agent, it is reduced, but the extent of its reduction and the products formed depend on the species being oxidised.

The three possible reduction products are:

- sulfur dioxide
- sulfur
- hydrogen sulfide.

Three different half-equations can be written to represent its oxidising action. Note the change in oxidation number of the sulfur in each case:

1. $H_2SO_4 + 2H^+ + 2e^- \rightarrow 2H_2O + SO_2$
   $\quad +6 \qquad\qquad\qquad\qquad\qquad +4$

2. $H_2SO_4 + 6H^+ + 6e^- \rightarrow 4H_2O + S$
   $\quad +6 \qquad\qquad\qquad\qquad\qquad 0$

3. $H_2SO_4 + 8H^+ + 8e^- \rightarrow 4H_2O + H_2S$
   $\quad +6 \qquad\qquad\qquad\qquad\qquad -2$

These half-equations may look complicated at first, but you should be able to see the pattern.
- In reaction 1, the decrease in oxidation number (+6 to +4) is 2, which is the same as the numbers of H⁺ ions and electrons in the equation.
- The pattern is similar in the other two equations. The decrease in oxidation number of the sulfur is the same as the numbers of H⁺ ions and electrons in the half-equation (6 in reaction 2, 8 in reaction 3).

## Observations and products

The table shows typical observations made, and the products formed, when concentrated sulfuric acid is added to three sodium halides.

| Halide | Observations | Products | |
|---|---|---|---|
| NaCl | misty fumes | hydrogen chloride | HCl |
| NaBr | misty fumes<br>brown fumes<br>colourless gas with choking smell | hydrogen bromide<br>bromine<br>sulfur dioxide | HBr<br>$Br_2$<br>$SO_2$ |
| NaI | misty fumes<br>purple fumes or black solid<br>colourless gas with choking smell<br>yellow solid<br>colourless gas with rotten egg smell | hydrogen iodide<br>iodine<br>sulfur dioxide<br>sulfur<br>hydrogen sulfide | HI<br>$I_2$<br>$SO_2$<br>S<br>$H_2S$ |

**table B** Observations and products formed when concentrated sulfuric acid is added to three sodium halides.

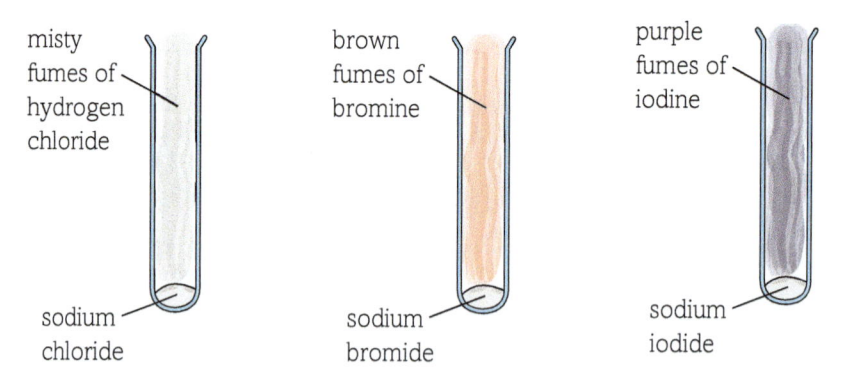

**fig A** From left to right: tubes containing hydrogen chloride (formed from sodium chloride), bromine (formed from sodium bromide) and iodine (formed from sodium iodide).

The tube on the left shows no colour. These are the misty fumes of hydrogen chloride, formed from sodium chloride. The tube in the middle contains brown fumes. This is bromine, formed from sodium bromide. The tube on the right contains a black solid. This is iodine, formed from sodium iodide.

With sodium chloride, the sulfuric acid behaves only as an acid, and not as an oxidising agent. This is because chloride ions have low reducing power.

With sodium bromide, the greater reducing power of bromide ions causes the sulfuric acid to be reduced, as in half-equation 1 above.

With sodium iodide, the much greater reducing power of iodide ions causes the sulfuric acid to be reduced, as in half-equations 1, 2 and 3.

## Constructing equations

Some of the equations you see, or are asked to write, may look very complicated and be hard to remember. It is better not to remember them, but to work them out by the addition of half-equations. This method is recommended for the redox equations of sodium bromide and sodium iodide.

### Sodium chloride

The reaction between sodium chloride and concentrated sulfuric can be represented by one equation, because no redox reactions are occurring:

$NaCl + H_2SO_4 \rightarrow NaHSO_4 + HCl$

## Sodium bromide

The formation of the misty fumes of hydrogen bromide can be represented by an equation analogous to the one for sodium chloride:

$NaBr + H_2SO_4 \rightarrow NaHSO_4 + HBr$

The table of observations shows that only one redox reaction occurs – the formation of sulfur dioxide in half-equation 1. The two relevant half-equations are:

$2Br^- \rightarrow Br_2 + 2e^-$

and

$H_2SO_4 + 2H^+ + 2e^- \rightarrow 2H_2O + SO_2$

Adding these together, then cancelling the $2e^-$ on each side, gives:

$2Br^- + H_2SO_4 + 2H^+ \rightarrow 2H_2O + SO_2 + Br_2$

You could combine the ions on the left to give:

$2HBr + H_2SO_4 \rightarrow 2H_2O + SO_2 + Br_2$

So, this equation represents the oxidation of the misty fumes of hydrogen bromide.

## Sodium iodide

The formation of the misty fumes of hydrogen iodide can be represented by an equation analogous to the one for sodium chloride:

$NaI + H_2SO_4 \rightarrow NaHSO_4 + HI$

The table of observations shows that three redox reactions occur, so the situation is more complicated. You could construct an equation showing the formation of sulfur dioxide in the same way as for sodium bromide.

Here is the result of applying the same method to the formation of sulfur in half-equation 2. The two relevant half-equations are:

$2I^- \rightarrow I_2 + 2e^-$

and

$H_2SO_4 + 6H^+ + 6e^- \rightarrow 4H_2O + S$

Before you add these together, you need to multiply the first one by three, so that the $6e^-$ on each side will cancel, giving:

$6I^- + H_2SO_4 + 6H^+ \rightarrow 4H_2O + S + 3I_2$

You could combine the ions on the left to give:

$6HI + H_2SO_4 \rightarrow 4H_2O + S + 3I_2$

So, this equation represents the oxidation of the misty fumes of hydrogen iodide.

You should now be able to use the same method to construct an equation to represent the oxidation of the misty fumes of hydrogen iodide to form hydrogen sulfide.

### Learning tip

Look carefully at all of the equations in this section. Make sure you can identify whether they are redox reactions and, if so, what the extent of the reaction is.

# Questions

1. Explain what kind of reaction occurs when concentrated sulfuric acid is added to sodium fluoride.
2. Use half-equations to construct an overall equation for the reaction between iodide ions and concentrated sulfuric acid that results in the formation of hydrogen sulfide.

# 4.2 4 Other reactions of halides

By the end of this section, you should be able to...

- understand precipitation reactions of the aqueous anions Cl⁻, Br⁻ and I⁻ with aqueous silver nitrate solution, followed by aqueous ammonia solution
- understand reactions of hydrogen halides with ammonia and with water (to produce acids)

## Testing for halide ions in solution

These tests depend on the very low solubility of silver halides in water and their different solubility in aqueous ammonia.

The reagent is silver nitrate solution, but dilute nitric acid is added beforehand to make sure that any other anions (especially carbonate ions) are removed, as they would also form precipitates.

If a precipitate is obtained, it is then usual to add some ammonia solution – this can be dilute or concentrated.

The table shows the results obtained.

|  | Chloride ions | Bromide ions | Iodide ions |
| --- | --- | --- | --- |
| Add silver nitrate solution | white precipitate | cream precipitate | yellow precipitate |
| Add dilute aqueous ammonia | soluble | insoluble | insoluble |
| Add concentrated aqueous ammonia | soluble | soluble | insoluble |

**table A** Results obtained from precipitation reactions of the aqueous anions Cl⁻, Br⁻ and I⁻ with aqueous silver nitrate solution, followed by aqueous ammonia solution.

This test cannot be used to detect fluoride ions in aqueous solution, because silver fluoride is soluble.

The photograph shows the results of these tests.

From left to right:

- tube 1 shows the white precipitate formed from a chloride
- tube 2 shows the result of adding dilute aqueous ammonia to the white precipitate
- tube 3 shows the cream precipitate formed from a bromide
- tube 4 shows the result of adding concentrated aqueous ammonia to the cream precipitate
- tube 5 shows the yellow precipitate formed from an iodide
- tube 6 shows the result of adding concentrated aqueous ammonia to the yellow precipitate – it has not dissolved.

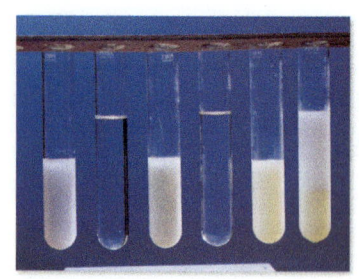

**fig A** Testing for halide ions in solution.

The general ionic equation for the formation of the precipitates is:

$$Ag^+(aq) + X^-(aq) \rightarrow AgX(s)$$

You can write a specific equation, such as:

$$AgNO_3(aq) + NaCl(aq) \rightarrow AgCl(s) + NaNO_3(aq)$$

The table of results suggests that a halide ion could be identified without using aqueous ammonia. However, the colours of the three precipitates are similar. Even when all three are seen together, it is not easy to be sure. In a single test, where only one precipitate is seen, this would be even harder.

Aqueous ammonia is useful because the precipitates have different solubilities in it.
- Silver chloride dissolves readily in both dilute and concentrated aqueous ammonia.
- Silver bromide dissolves readily in concentrated aqueous ammonia, but not in dilute aqueous ammonia.
- Silver iodide dissolves in neither.

Dissolving of the precipitates occurs because of the formation of a complex ion. In the case of silver chloride, the equation for the reaction is:

$$AgCl(s) + 2NH_3(aq) \rightarrow [Ag(NH_3)_2]^+(aq) + Cl^-(aq)$$

## Hydrogen halides acting as acids

All of the hydrogen halides are colourless gases and exist as polar diatomic molecules.

## Reactions with water

They readily react with water to form acidic solutions, all of which are colourless.

The table shows these reactions.

| Hydrogen halide | Acid formed | Equation |
|---|---|---|
| Hydrogen fluoride | hydrofluoric acid | $HF + H_2O \rightleftharpoons H_3O^+ + F^-$ |
| Hydrogen chloride | hydrochloric acid | $HCl + H_2O \rightarrow H_3O^+ + Cl^-$ |
| Hydrogen bromide | hydrobromic acid | $HBr + H_2O \rightarrow H_3O^+ + Br^-$ |
| Hydrogen iodide | hydriodic acid | $HI + H_2O \rightarrow H_3O^+ + I^-$ |

**table B** Reactions of hydrogen halides with water to form acidic solutions.

### Learning tip

Note the inconsistent spelling of hydriodic acid (not hydroiodic acid).

Note also the reversible arrow in the first equation. Although hydrogen fluoride is very soluble in water, it is only a weak acid, whereas the other acids formed are strong.

## Reactions with ammonia

Hydrogen halides all react with ammonia gas to form salts, all of which are white ionic solids.

For example, when ammonia and hydrogen chloride gases are mixed together, they react to form ammonium chloride:

$$NH_3(g) + HCl(g) \rightarrow NH_4Cl(s)$$

You may be familiar with the use of this reaction to illustrate diffusion and the different rates of diffusion of ammonia and hydrogen chloride.

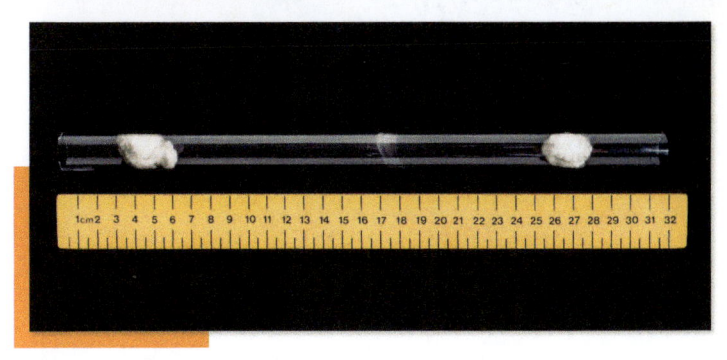

**fig B** This experiment illustrates the different rates of diffusion of ammonia (from the left) and hydrogen chloride (from the right).

Ammonia and hydrogen chloride gases are given off from the cotton wool pieces soaked in concentrated aqueous ammonia and concentrated hydrochloric acid. These colourless gases move through the tube until they meet and react to form ammonium chloride. Ammonia gas molecules ($M_r$ = 17.0) move more quickly and therefore travel further in the same time as hydrogen chloride molecules ($M_r$ = 36.5).

## Fluorides and astatides

Fluorides and astatides have been mentioned in some of the reactions in this chapter, especially when it is not possible to make predictions about their reactions using trends in the chlorides, bromides and iodides. For example, you might predict that silver nitrate would form a white precipitate with sodium fluoride because sodium chloride does, but this is not the case.

If you are asked to predict reactions of fluorides and astatides that you are not familiar with, you can use information in this chapter to help you. For example, you could write an equation to show the formation of an acid when hydrogen astatide is added to water, based on the equation for hydrogen iodide.

### Did you know?

None of the reactions in this section involve redox. This is because the oxidation number of every halogen in a halogen compound in this section remains unchanged at −1.

## Questions

1. Write equations to show:
   (a) the formation of the cream precipitate in the test for bromide ions
   (b) the precipitate dissolving in concentrated aqueous ammonia.

2. (a) Describe what happens when ammonia and hydrogen bromide gases are mixed together.
   (b) Name the product of the reaction and write an equation for its formation.

# THINKING BIGGER

## NOT-SO-LIFELESS NITROGEN

Nitrogen is the most common gas in the Earth's atmosphere present at 78% in the form $N_2$. However, its ability to exist in at least eight different oxidation states means that its chemistry is as diverse as it is vital.

## NITRATE DISAPPEARS IN 'IMPOSSIBLE' REACTION

A simple reaction which chemists previously believed impossible has been brought about by a team of chemists in the US. The unprecedented reaction is the transformation of nitrate ions ($NO_3^-$) into ammonium ions ($NH_4^+$) by simple organic molecules such as methane or ethanol.

The reaction was discovered by Alan Hutson and Ayusman Sen of Pennsylvania State University. They say that if a catalyst could be found to promote it, the discovery could lead one day to a practical way of removing not only nitrogen oxides but also unburnt hydrocarbons from car exhausts (*Journal of the American Chemical Society*, vol. 116, p. 4527).

In the reaction, the nitrogen atom gains electrons: in other words, it undergoes reduction. In nature, the reduction of nitrate to nitrite ($NO_2^-$), to nitric oxide (NO), and even to nitrogen gas ($N_2$), happens quite normally. In these conversions, up to four electrons ($e^-$) are required. However, the direct reduction from $NO_3^-$ to $NH_4^+$ requires eight electrons:

Reaction 1: $NO_3^- + 10H^+ + 8e^- \rightarrow NH_4^+ + 3H_2O$

Even metal hydrides, which are extremely electron rich and are therefore among the strongest reducing agents known, cannot accomplish this transformation in one step. However, Hutson and Sen have brought it about simply by dissolving the nitrate in sulfuric acid, and treating it with simple organic compounds. This is all the more surprising because sulfuric acid is usually thought of as an oxidising agent – a compound which removes electrons, rather than adding them.

The chemists report that a number of commonplace organic chemicals will do the trick. They include hydrocarbon gases such as methane ($CH_4$) and ethane ($C_2H_6$), simple alcohols such as ethanol ($C_2H_5OH$) and propanol ($C_3H_7OH$), and acetic (ethanoic) acid ($CH_3COOH$). None of these chemicals is regarded as a reducing agent, let alone a strong reducing one.

In a typical reaction, Hutson and Sen dissolved sodium nitrate in concentrated sulfuric acid (96% grade), added the organic compound and heated the combination to around 170 °C. For the gases methane and ethane they used a steel pressure vessel to contain the reactants. They used nuclear magnetic resonance to prove that ammonia had been formed.

In some reactions the chemists found that 100% of the nitrate was converted to ammonia – in the case of ethanol this had happened after only 15 minutes. With methane as the reducing agent, they observed only a trace of ammonia after 20 hours' heating, but when they added a little mercury(II) sulfate to the solution the yield of ammonia rose to 100%. Hutson and Sen believe that the mercury acts as a catalyst in a reaction that converts methane to methanol, and that it is the methanol that acts as the reducing agent.

Hutson and Sen are now investigating the route by which the reduction occurs. They believe the first step is the conversion of $NO_3^-$ to $NO_2^-$ by the sulfuric acid. What happens next is not clear.

Where else will I encounter these themes?

Let us start by considering the nature of the writing in the article.

1. In the title of this article from the *New Scientist* magazine, the author uses the word 'impossible' in inverted commas. What do you think the author is trying to convey using this term? Do you think the same term is likely to be used in the original research paper in the *Journal of the American Chemical Society*?
2. How might the results of this research be validated by the scientific community?

Now we will look at the chemistry in, or connected to, this article. Don't worry if you are not ready to give answers to these questions yet. You may like to return to the questions once you have covered other topics in the book.

3. Explain why reaction 1 is considered to be a reduction process.
4. Assign oxidation numbers to nitrogen in the following species:
   a. $NO_2$   b. $N_2O$   c. $HNO_3$   d. $NH_3$   e. $NaNO_2$   f. $N_2H_4$
5. Write a balanced half-equation for the reduction of $NO_3^-$ ions to NO in acidic solution.
6. The author mentions that 'concentrated sulfuric acid is usually thought of as an oxidising agent'. Write a balanced equation for the reaction between concentrated sulfuric acid and sodium bromide identifying the changes in oxidation state that occur. Explain why sulfuric acid is behaving as an oxidising agent in this case.'
7. Comment on the difficulty of assigning a single oxidation number to carbon in a molecule such as ethanoic acid. It might help if you draw out the displayed formula of ethanoic acid.
8. In the experiment cited in the original paper, 0.12 mmol $NaNO_3$ and 0.17 mmol $C_2H_5OH$ were heated in a reaction vessel to 180 °C for 15 minutes in 96% sulfuric acid. Calculate the mass of $NaNO_3$ and $C_2H_5OH$ used giving your answers in mg. (1 mmol = 1 × 10$^{-3}$ mol = one thousandth of a mole)

> **Command word**
> When you are asked to comment on something, you need to think about several variables from the data or information you are given and form your own judgement.

> Make sure that you are comfortable using the two equations below. You will use these equations again and again in chemistry whenever quantities are used to make sure that you know them!
>
> Number of moles of substance
> $$= \frac{\text{mass of substance in g}}{\text{mass of 1 mole of that substance}}$$
>
> Concentration in mol dm$^{-3}$
> $$= \frac{\text{number of moles}}{\text{volume of solution in dm}^3}$$

# Activity

Design a poster (A1 or A2 size) showing the nitrogen cycle. The poster should include all of the following:
- The chemical formulae of all of the nitrogen species involved, showing the oxidation number of nitrogen in each case.
- The main processes responsible for the transitions between oxidation states (this will include biotic and abiotic processes) e.g. denitrifying bacteria.
- The most important nitrogen sinks (reservoirs).

> Your poster will be a helpful resource when it comes to revising topics including redox and balancing ionic equations.

### Did you know?
The gas we now call nitrogen was known to Antoine Lavoisier (considered by many to be the father of chemistry) as 'azote' from the Greek meaning 'without life'. Ironically, nitrogen turns out to be one of the most important elements *for* life and in some habitats is the life-limiting element.

● From an article in *New Scientist* magazine

# 4 Exam-style questions

1. This question is about the chemistry of the elements in Groups 1 and 2 of the Periodic Table.
   (a) When magnesium burns in air it forms a mixture of magnesium oxide and magnesium nitride.
   Give the formula of both magnesium oxide and magnesium nitride. [2]
   (b) (i) Write an equation for the reaction between calcium and water. Include state symbols. [2]
   (ii) Explain how the reactivity of the Group 2 metals with water changes down the group. [3]
   (iii) State how the solubility of the hydroxides of the Group 2 metals changes down the group. [1]
   (c) Explain the trend in thermal stability of the carbonates of the Group 2 metals. [3]
   (d) The nitrates of most of the Group 1 metals decompose on heating as shown in the equation
   $$MNO_3 \rightarrow MNO_2 + \tfrac{1}{2}O_2$$
   where M represents a Group 1 metal.
   However, lithium nitrate decomposes further.
   Write an equation for the thermal decomposition of lithium nitrate. State symbols are not required. [2]
   **[Total: 13]**

2. Radium (Ra) and astatine (At) are radioactive elements in Groups 2 and 7 respectively in the Periodic Table.
   (a) Predict the appearance at room temperature of
   (i) radium [1]
   (ii) astatine. [2]
   (b) Predict the reactivity of astatine compared to that of iodine. Justify your answer. [3]
   (c) Write an equation for the reaction of astatine with hydrogen. State symbols are not required. [1]
   (d) Write an equation for the reaction between radium and chlorine. State symbols are not required. [1]
   (e) The product of the reaction between radium and water, and the pH of the resulting solution is most likely to be
   **A** radium oxide and 1
   **B** radium oxide and 14
   **C** radium hydroxide and 1
   **D** radium hydroxide and 12 [1]
   **[Total: 9]**

3. This question is about halides.
   (a) Group 1 halides react with concentrated sulfuric acid in different ways.
   Sodium chloride reacts to give hydrogen chloride as the only gaseous product.
   Sodium bromide reacts to give both hydrogen bromide and bromine as gaseous products.
   (i) Write an equation for the reaction of both sodium chloride and sodium bromide with concentrated sulfuric acid. [2]
   (ii) Explain why the two reactions produce different gaseous products. [3]
   (iii) Describe how ammonia gas can be used to show that hydrogen chloride has been given off in the reaction between sodium chloride and concentrated sulfuric acid. [2]
   (b) Hydrogen halides react with water.
   (i) Write an equation for the reaction between hydrogen bromide and water. [1]
   (ii) Explain the effect that the solution formed has on methyl orange indicator. [3]
   (c) Describe how you show that an aqueous solution contains both chloride and iodide ions [4]
   **[Total: 15]**

4. This question is about some compounds of the elements in Group 2 of the Periodic Table.
   (a) Water was slowly added to a solid lump of calcium oxide. The lump got very hot and broke apart to form another white solid.
   Excess water was added to this white solid and the mixture was then filtered to produce a colourless solution. The solution turned milky when carbon dioxide was bubbled through it.
   (i) State why the lump of calcium oxide got very hot when water was added. [1]
   (ii) Name the colourless solution formed when an excess of water was added to the white solid and predict its pH. [2]
   (iii) Write an equation for the reaction between this colourless solution and carbon dioxide. Include state symbols. [2]
   (b) Calcium oxide can be formed in several ways.
   One way is to heat calcium in oxygen. Another is to heat calcium nitrate.
   (i) Write an equation for each method of forming calcium oxide. State symbols are not required. [2]
   (ii) Explain the trend in decomposition temperatures of the Group 2 nitrates. [4]
   **[Total: 11]**

5 Sodium chlorate(V), NaClO₃ can be used as a weed killer. It can be made by reacting chlorine with a hot, concentrated aqueous solution of sodium hydroxide. The equation for the reaction is

$3Cl_2(g) + 6NaOH(aq) \rightarrow 5NaCl(aq) + NaClO_3(aq) + 3H_2O(l)$

(a) Write an ionic equation for this reaction. State symbols are not required. [2]
(b) Comment on the changes of oxidation number of chlorine. [3]
(c) Write an equation for the reaction between chlorine and cold, dilute aqueous sodium hydroxide. Include state symbols. [2]

[Total: 7]

6 The trend in reactivity of the halogens can be shown by displacement reactions.

(a) Complete the table to show the products of the reaction of each halogen with halide ions. If the reagents do not react, write **no reaction**. [3]

| Reagents | An aqueous solution containing chloride ions | An aqueous solution containing bromide ions | An aqueous solution containing iodide ions |
|---|---|---|---|
| chlorine, Cl₂ | | | |
| bromine, Br₂ | | | |
| iodine, I₂ | | | |

(b) Write an equation for the reaction between chlorine and an aqueous solution containing astatide ions, At⁻. Include state symbols. [2]
(c) The colours of silver halides and their reactions with aqueous ammonia may be used to identify the presence of halides ions in aqueous solution.
Using this information, describe how you could perform a test to show the presence of bromide ions in an aqueous solution. [4]

[Total: 9]

7 Magnesium oxide and barium oxide both react with water.

$MgO(s) + H_2O(l) \rightarrow \mathbf{X}(aq)$

$BaO(s) + H_2O(l) \rightarrow \mathbf{Y}(aq)$

(a) (i) Identify **X** and **Y**. [2]
(ii) Predict the approximate pH of a saturated aqueous solution of **X** and of **Y**. [2]
(iii) Justify your answers to (a)(ii). [2]
(b) When carbon dioxide is bubbled into an aqueous solution of **Y**, a white precipitate is formed.
Write an equation for the reaction that takes place. Include state symbols. [2]
(c) Magnesium oxide and barium oxide can be distinguished from each other by using a flame test. State what you would expect to see when a flame test is separately carried out on each compound. [2]

[Total: 10]

# TOPIC 5
# Formulae, equations and amounts of substance

## Introduction

Studying chemistry at this level involves many skills – these include describing observations, giving explanations, planning and interpreting experiments. Although all of these skills are discussed in this topic, the most important feature of this part of the course involves using numbers – lots of numbers! Some students have a natural feel for numbers, while others find them off-putting. Wherever you are on the spectrum of mathematical competence, it is important to spend time mastering as much as possible of this topic.

Using numbers accurately is an important skill in so many occupations – here are just a few examples:

- In medicine, calculating the dosage of a drug to maximise benefit to the patient and minimise unwanted side-effects
- In business, predicting whether capital investment will produce an income sufficient to justify the investment
- In industry, considering whether the costs and polluting effects of a new method of making a substance justify making the change
- In ship salvage (recently of the Costa Concordia), working out the volume of air needed to float a capsized ship

As far as your chemistry course is concerned, this topic may not appear to be the most exciting, but it is the foundation of success in other topics, such as energetics and kinetics.

## All the maths you need

- Use appropriate units and conversions, including for masses, gas and solution volumes
- Use standard and ordinary form, decimal places, and significant figures
- Use ratios, fractions and percentages, including in yield and atom economy calculations
- Select values and calculate means, including in titrations
- Consider uncertainties in measurements, including in titrations
- Rearrange expressions and substitute values

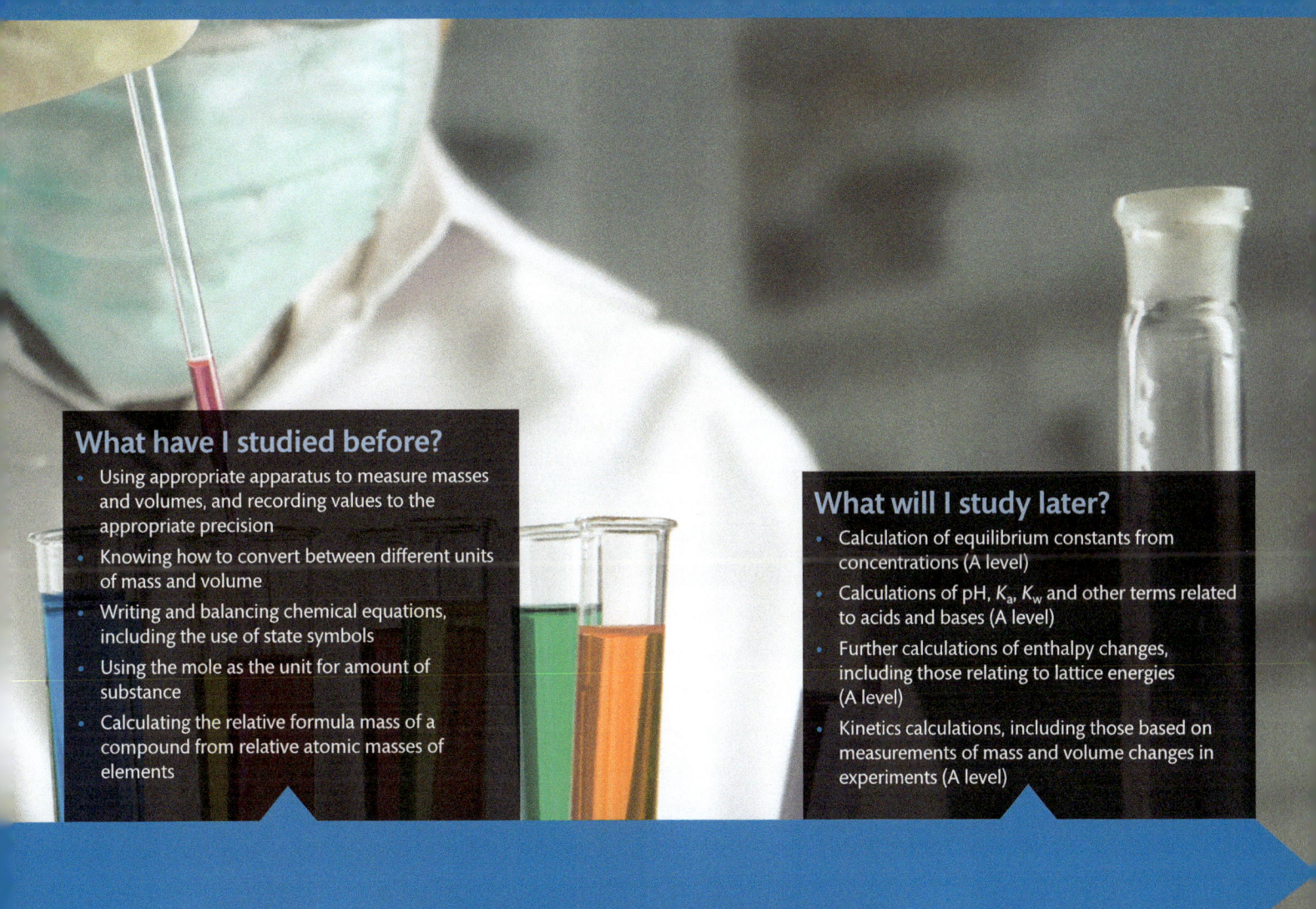

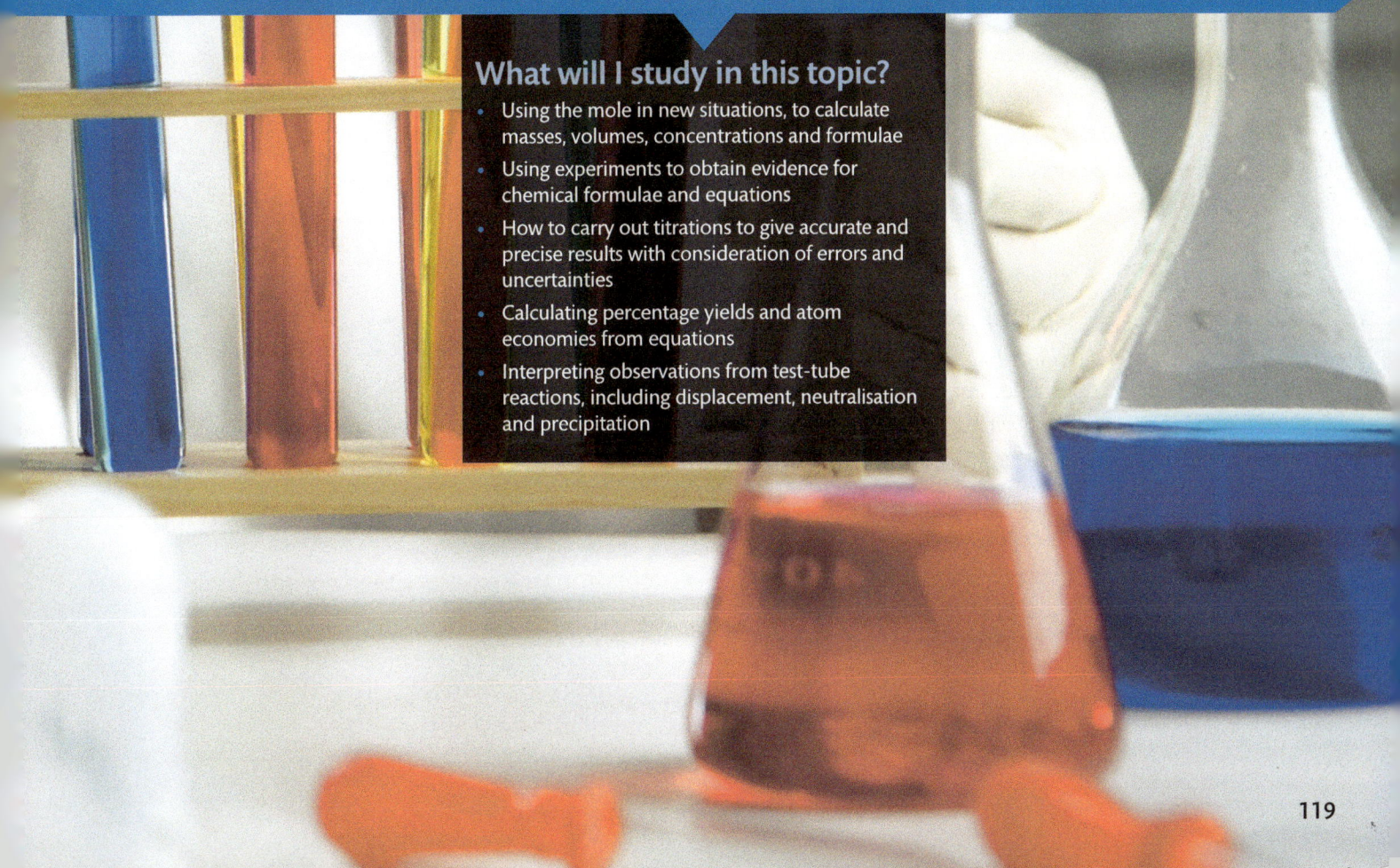

## What have I studied before?

- Using appropriate apparatus to measure masses and volumes, and recording values to the appropriate precision
- Knowing how to convert between different units of mass and volume
- Writing and balancing chemical equations, including the use of state symbols
- Using the mole as the unit for amount of substance
- Calculating the relative formula mass of a compound from relative atomic masses of elements

## What will I study later?

- Calculation of equilibrium constants from concentrations (A level)
- Calculations of pH, $K_a$, $K_w$ and other terms related to acids and bases (A level)
- Further calculations of enthalpy changes, including those relating to lattice energies (A level)
- Kinetics calculations, including those based on measurements of mass and volume changes in experiments (A level)

## What will I study in this topic?

- Using the mole in new situations, to calculate masses, volumes, concentrations and formulae
- Using experiments to obtain evidence for chemical formulae and equations
- How to carry out titrations to give accurate and precise results with consideration of errors and uncertainties
- Calculating percentage yields and atom economies from equations
- Interpreting observations from test-tube reactions, including displacement, neutralisation and precipitation

# 5.1    1    Empirical formulae

By the end of this section, you should be able to...
- know what is meant by the term empirical formula
- calculate empirical formulae from experimental data

## Introduction to empirical formulae

The three letters 'mol' appear in several words used in chemistry. You will know some of these from your previous studies, especially 'molecule', which is a group of two or more atoms joined together by covalent bonds. You know that a molecule of water can be represented by the formula $H_2O$, so this is the molecular formula of water. You have already come across the term 'relative molecular mass', which you will probably remember is 18 (or more accurately, 18.0) for water.

It can be tricky to understand the term 'mole' (abbreviated to 'mol' when considering a specific number of moles). We will introduce and explain this term gradually over the next few sections.

The term **empirical** indicates that some information has been found by experiment. An empirical formula shows the smallest whole number ratio of the atoms of each element in a compound.

## One example of an experimental method

A simple example involves determining the formula of an oxide of copper, which is a black solid. This oxide can be converted to copper by removing the oxygen.

Here are the stages of the experiment.

- Place a known mass of copper oxide in the tube.
- Heat it in a stream of hydrogen gas (or natural gas, which is mostly methane).
- The hydrogen atoms in the gas react with the oxygen atoms in the copper oxide and form steam.
- The solid gradually changes colour from black to orange-brown, which is the colour of copper.
- The excess gas is burned off at the end of the tube for safety reasons.

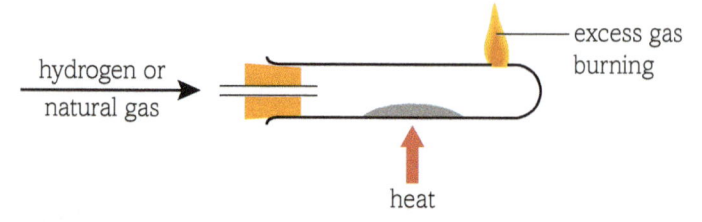

**fig A** Apparatus used to convert copper oxide to copper by removing the oxygen.

- After cooling, remove and weigh the solid copper.
- It is good practice to heat the solid again in the stream of gas to check whether its mass changes. Heating to a constant mass suggests that the conversion to copper is complete.

## Calculating empirical formulae

The calculation method involves these steps.

- Divide the mass, or percentage composition by mass, of each element by its relative atomic mass.
- If necessary, divide the answers from this step by the smallest of the numbers.
- This answer gives numbers that should be in an obvious whole-number ratio, such as 1 : 2 or 3 : 2.
- These whole numbers are used to write the empirical formula.

The numbers may not be in an exact ratio because of experimental error, but you should be able to decide what the nearest ratio is.

Use at least two significant figures in the calculation (preferably three), and beware of inappropriate rounding. For example, if the ratio is 1.2 : 1.2 : 1.9, do not convert this to a 1 : 1 : 2 ratio.

It may help you to organise the calculation using a table, but this is not essential.

### Calculation using masses

Assume that these are the results of the experiment above.

    mass of copper oxide = 4.28 g
    mass of copper = 3.43 g
    the mass of oxygen removed is 4.28 − 3.43 = 0.85 g

This is the calculation table for these results.

|  | Cu | O |
|---|---|---|
| mass of element/g | 3.43 | 0.85 |
| relative atomic mass | 63.5 | 16.0 |
| division by $A_r$ | 0.0540 | 0.0531 |
| ratio | 1 | 1 |

Here, the ratio is obviously 1 : 1, so the empirical formula is CuO.

### Calculation using percentage composition by mass

The calculation is done in the same way.

The calculation table uses the results for a compound containing three elements. A compound has the percentage composition by mass C = 38.4%, H = 4.8%, Cl = 56.8%.

|  | C | H | Cl |
|---|---|---|---|
| % of element | 38.4 | 4.8 | 56.8 |
| relative atomic mass | 12.0 | 1.0 | 35.5 |
| division by $A_r$ | 3.2 | 4.8 | 1.6 |
| ratio | 2 | 3 | 1 |

The empirical formula is $C_2H_3Cl$.

## Calculation when the oxygen value is not provided

Analysis of the results for some compounds does not include values for oxygen because it is often difficult to obtain an experimental value. So, sometimes you need to remember to calculate the percentage of oxygen by subtraction.

A compound has the percentage composition by mass of Na = 29.1%, S = 40.5%, with the remainder being oxygen.

The percentage of oxygen = 100 − (29.1 + 40.5) = 30.4%.

This is the calculation table.

|  | Na | S | O |
|---|---|---|---|
| % of element | 29.1 | 40.5 | 30.4 |
| relative atomic mass | 23.0 | 32.1 | 16.0 |
| division by $A_r$ | 1.27 | 1.26 | 1.90 |
| division by the smallest | 1 | 1 | 1.5 |
| ratio | 2 | 2 | 3 |

The empirical formula is $Na_2S_2O_3$.

## Calculation using combustion analysis

Many organic compounds contain carbon and hydrogen, or carbon, hydrogen and oxygen. When a known mass of such a compound is completely burned, it is possible to collect and measure the masses of carbon dioxide and water formed. The calculation is more complex because there are extra steps. Here is an example.

A 1.87 g sample of an organic compound was completely burned, forming 2.65 g of carbon dioxide and 1.63 g of water.

In this type of calculation, the first steps are to calculate the masses of carbon and hydrogen in the carbon dioxide and water.

- The relative molecular mass of carbon dioxide is 44.0 but, because the relative atomic mass of carbon is 12.0, the proportion of carbon in carbon dioxide is always 12.0 ÷ 44.0.
- Similarly, the proportion of hydrogen in water is always (2 × 1.0) ÷ 18.0.

All of the carbon in the carbon dioxide comes from the carbon in the organic compound. Similarly, all of the hydrogen in the water comes from the hydrogen in the organic compound. So, in this example:

$$\text{mass of carbon} = \frac{2.65 \times 12.0}{44.0} = 0.723\,g$$

$$\text{mass of hydrogen} = \frac{1.63 \times 2.0}{18.0} = 0.181\,g$$

These two masses add up to 0.904 g.

The original mass of the organic compound was 1.87 g, so the difference must be the mass of oxygen present in the organic compound (1.87 − 0.904 = 0.966 g).

Here is the calculation table.

|  | C | H | O |
|---|---|---|---|
| mass of element/g | 0.723 | 0.181 | 0.966 |
| relative atomic mass | 12.0 | 1.0 | 16.0 |
| division by $A_r$ | 0.0603 | 0.181 | 0.0604 |
| ratio | 1 | 3 | 1 |

The empirical formula is $CH_3O$.

### Learning tip

Divide by the relative atomic mass, *not* the atomic number or the relative molecular mass.

So, for oxygen, do not divide by 8 or 32.

$C_6H_{12}O_6$

*If subscripts in a formula will reduce, it is NOT an empirical formula!*

**fig B** This is glucose. Can you see that the empirical formula is $CH_2O$?

## Questions

1. A compound has the percentage composition by mass Ca = 24.4%, N = 17.1% and O = 58.5%. What is its empirical formula?

2. Combustion analysis of 2.16 g of an organic compound produced 4.33 g of carbon dioxide and 1.77 g of water. What is its empirical formula?

### Key definition

An **empirical formula** shows the smallest whole-number ratio of atoms of each element in a compound.

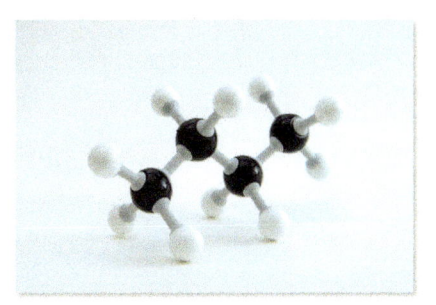

**fig C** Can you work out the empirical formula of this molecule? Black represents carbon and white represents hydrogen.

# 5.1 | 2 | Molecular formulae

**By the end of this section, you should be able to...**

- know what is meant by the term molecular formula and be able to calculate molecular formulae from experimental data
- know that the molar mass of a substance is the mass per mole of the substance in $g\,mol^{-1}$
- calculate the molar mass of a gas or volatile liquid in SI units, using the ideal gas equation $pV = nRT$

## Introduction to molecular formulae

In the previous section, you learned how to calculate empirical formulae. Sometimes the empirical formula of a compound is the same as its molecular formula – carbon dioxide ($CO_2$) and water ($H_2O$) are familiar examples. The **molecular formula** of a compound shows the actual numbers of the atoms of each element in the compound.

Examples of compounds with different empirical and molecular formulae are:

| hydrogen peroxide | empirical formula is HO | molecular formula is $H_2O_2$ |
| butane | empirical formula is $C_2H_5$ | molecular formula is $C_4H_{10}$ |

To determine the molecular formula of a compound, you need:

- the empirical formula
- the relative molecular mass.

For example, if you had already found that the empirical formula of a compound was HO and you then found that its relative molecular mass was 34, you could compare the relative molecular mass of the empirical formula (17) with 34 and deduce that the molecular formula was double the empirical formula.

## Using terminology correctly

In the last section, we said we would introduce moles gradually. In fact, without realising it, you have already used the idea of moles. When you divided a mass of an element by the relative atomic mass of the element, you were actually calculating the amount in moles of that element.

We have not got as far as understanding the proper meaning of a mole yet, but here is a useful expression that allows you to work with moles:

$$\text{amount in moles} = \frac{\text{mass of substance in grams}}{\text{relative atomic/molecular mass}}$$

The abbreviation of the unit is mol. You can write the amount in words (e.g. two moles), but it is better to use a value and the unit (e.g. 2.0 mol).

## Relative molecular mass

The term 'relative molecular mass' was introduced in an earlier section, and we can still use it in this section. However, it does have one disadvantage – it can only be used for molecules (compounds containing covalent bonds). It is not correct to state that the relative molecular mass of sodium hydroxide is 40.0 because sodium hydroxide is an ionic compound and does not contain any molecules.

## Relative formula mass

An alternative term, 'relative formula mass', can be used for both molecular and ionic compounds. There is no problem in referring to the relative formula mass of the ionic compound sodium hydroxide.

## Molar mass

Another way around this problem is to use the term **molar mass**, which is the mass per mole of any substance (molecular or ionic). Its symbol is $M$ (not $M_r$) and it has the units $g\,mol^{-1}$ (grams per mole).

So, this is the expression you can use:

$$\text{amount in mol} = \frac{\text{mass of substance in g}}{\text{molar mass in } g\,mol^{-1}} \quad \text{or} \quad n = \frac{m}{M}$$

The table shows examples of working out the amounts in moles of some substances.

| Substance | $O_2$ | $CH_4$ | $H_2O$ | $NH_4NO_3$ |
|---|---|---|---|---|
| mass in g | 5.26 | 4.0 | 100 | 14.7 |
| molar mass, $M$ in $g\,mol^{-1}$ | 32.0 | 16.0 | 18.0 | 80.0 |
| amount in mol | 0.164 | 0.25 | 5.56 | 0.184 |

## Calculating molecular formulae

You have already practised calculating an empirical formula from experimental data. Now look at how to calculate a molecular formula from an empirical formula.

### Example 1

In this example, the empirical formula is given. A compound has the empirical formula CH and a relative molecular mass of 104.

The 'formula mass' of the empirical formula is 13.0, so as 104 is 8 times 13.0, the molecular formula of the compound is $C_8H_8$.

### Example 2

In this example, you first have to work out the empirical formula. A compound contains the percentage composition by mass Na = 34.3%, C = 17.9%, O = 47.8%, and has a molar mass of 134 $g\,mol^{-1}$.

|  | Na | C | O |
|---|---|---|---|
| % of element | 34.3 | 17.9 | 47.8 |
| relative atomic mass | 23.0 | 12.0 | 16.0 |
| division by $A_r$ | 1.49 | 1.49 | 2.99 |
| ratio | 1 | 1 | 2 |

The empirical formula is $NaCO_2$.

The 'formula mass' of the empirical formula is 23.0 + 12.0 + (2 × 16.0) = 67.0, so as 134 is 2 times 67.0, then the molecular formula of the compound is $Na_2C_2O_4$.

## Using $pV = nRT$

$pV = nRT$ is the ideal gas equation, and can be used for gases (or volatile liquids above their boiling temperatures) to find the amount of a substance in moles. If the mass of the substance is also known, then the molar mass of the substance can be calculated. This gives the extra information needed to work out a molecular formula from an empirical formula.

## SI units

You need to be careful when using this equation that the units are the correct ones. It is always safest to work in SI units.

The SI units you should use are:

- $p$ = pressure in pascals (Pa)
- $V$ = volume in cubic metres ($m^3$)
- $T$ = temperature in kelvin (K)
- $n$ = amount of substance in moles (mol)
- $R$ = the gas constant – this appears in the data book provided for use in the examinations but has the value $8.31\,J\,mol^{-1}K^{-1}$.

You may find in a question that the units quoted are not in these SI units. If this is the case, then you will need to convert them to SI units. The table shows the conversions you are likely to need.

| Conversion | How to do it |
|---|---|
| kPa → Pa | multiply by $10^3$ |
| $cm^3$ → $m^3$ | divide by $10^6$ or multiply by $10^{-6}$ |
| $dm^3$ → $m^3$ | divide by $10^3$ or multiply by $10^{-3}$ |
| °C → K | add 273 |

### Example 1

In this example, you just have to calculate the molar mass of the gas. It may help you (until you have had more practice) to write a list of the values with any necessary conversions.

A 0.280 g sample of a gas has a volume of 58.5 $cm^3$, measured at a pressure of 120 kPa and a temperature of 70 °C. Calculate the molar mass of the gas.

$p$ = 120 kPa = 120 × $10^3$ Pa

$V$ = 58.5 $cm^3$ = 58.5 × $10^{-6}\,m^3$

$T$ = 70 °C = 343 K

$R$ = 8.31 $J\,mol^{-1}K^{-1}$

So, $n = \dfrac{pV}{RT} = \dfrac{120 \times 10^3 \times 58.5 \times 10^{-6}}{8.31 \times 343} = 0.00246\,mol$

$M = \dfrac{m}{n} = \dfrac{0.280}{0.00246} = 114\,g\,mol^{-1}$

### Example 2

In this example, you first have to calculate the empirical formula, then the amount in moles, then the molar mass, then the molecular formula.

This is the question:

A compound has the percentage composition by mass C = 52.2%, H = 13.0%, O = 34.8%. A sample containing 0.173 g of the compound had a volume of 95.0 cm³ when measured at 105 kPa and 45 °C.

What is the molecular formula of this compound?

Step 1: the empirical formula

|  | C | H | O |
|---|---|---|---|
| % of element | 52.2 | 13.0 | 34.8 |
| relative atomic mass | 12.0 | 1.0 | 16.0 |
| division by $A_r$ | 4.35 | 13.0 | 2.175 |
| ratio | 2 | 6 | 1 |

The empirical formula is $C_2H_6O$.

Step 2: the amount in moles

$$n = \frac{pV}{RT} = \frac{105 \times 10^3 \times 95.0 \times 10^{-6}}{8.31 \times 318} = 0.00377 \text{ mol}$$

Step 3: the molar mass

$$M = \frac{m}{n} = \frac{0.173}{0.00377} = 45.9 \text{ g mol}^{-1}$$

Step 4: – the molecular formula

The 'formula mass' of the empirical formula is $(2 \times 12.0) + (6 \times 1.0) + 16.0 = 46.0$

As this is the same as the molar mass, the empirical and molecular formulae are the same.

So, the molecular formula is $C_2H_6O$.

### Learning tip

The word 'amount' is often used loosely to refer to a mass or volume of a substance, but you should *not* use it in this way. Use the term only for the amount of a substance in moles.

## Questions

1. A 2.82 g sample of a gas has a volume of 1.26 dm³, measured at a pressure of 103 kPa and a temperature of 55 °C. Calculate the molar mass of the gas.

2. A compound has the percentage composition by mass C = 40.0%, H = 6.7%, O = 53.3%. A sample containing 0.146 g of the compound had a volume of 69.5 cm³ when measured at 98 kPa and 63 °C. What is the molecular formula of this compound?

### Key definitions

A **molecular formula** shows the actual number of atoms of each element in a molecule.

The **molar mass** is the mass per mole of a substance. It has the symbol *M* and the units g mol⁻¹.

# 5.2  1  Calculations using moles and the Avogadro constant

**By the end of this section, you should be able to...**

- know that the mole (mol) is the unit for amount of a substance
- use the Avogadro constant in calculations

## What is a mole?

So far we have referred to the mole and have used it in a simple form of calculation, but we have not properly explained what it is.

### The definition of a mole

A **mole** is the amount of substance that contains the same number of particles as the number of carbon atoms in exactly 12 g of the $^{12}C$ isotope.

This definition is tricky to understand, but keep on reading and all should become clear.

**Fig A** shows what one mole of four different elements looks like.

### Counting atoms

As you know, atoms are very tiny particles that cannot be seen by the human eye. However, we can look at the sand on a beach and know that we are looking at billions and billions of atoms of, mostly, silicon and oxygen in the compound silicon dioxide ($SiO_2$). Quoting the actual numbers of atoms involved in a reaction – whether in a test tube or on an industrial scale – would involve extremely large numbers that would be very difficult to handle.

You are already familiar with the use of relative atomic masses to compare the relative masses of atoms. In a water molecule, the oxygen atom has a mass that is 16 times the mass of a hydrogen atom. You know this because in the Periodic Table the relative atomic masses are H = 1.0 and O = 16.0. Note the word 'relative'. These values do not tell us the actual mass in grams of an oxygen atom or a hydrogen atom – only that an oxygen atom has a mass 16 times that of a hydrogen atom.

Now consider using these numbers (16.0 and 1.0) with the familiar unit g (grams). You can more easily visualise 16.0 g of oxygen than a single atom of oxygen. Doing this is effectively scaling up on a very large scale. The number of oxygen atoms in 16.0 g of oxygen is the same as the number of hydrogen atoms in 1.0 g of hydrogen.

**fig A** From left to right: tin (Sn), magnesium (Mg), iodine (I) and copper (Cu). Each sample contains $6.02 \times 10^{23}$ atoms.

## What is Avogadro's constant?

Amedeo Avogadro (1776–1856) was an Italian chemist whose name lives on in the form of the **Avogadro constant** and Avogadro's law.

We are introducing him here because the scaling-up factor from atoms to grams is named after him.

Its value is approximately 602 000 000 000 000 000 000 000 mol$^{-1}$, which is far better shown using standard form as $6.02 \times 10^{23}$ mol$^{-1}$.

The symbol $L$ is used for the Avogadro constant (using $A$ might be confusing because of the use of $A_r$ for relative atomic mass). $L$ comes from the surname of Johann Josef Loschmidt (1821–1895), an Austrian chemist who was a contemporary of Avogadro. He made many contributions to our understanding of the same area of knowledge.

You can easily apply this value to other elements. For example, there are $6.02 \times 10^{23}$ atoms of He in 4.0 g of helium, $6.02 \times 10^{23}$ atoms of Na in 23.0 g of sodium, and so on.

**fig B** The Italian chemist, Amedeo Avogradro.

## Calculations using moles

### What to remember when doing calculations

You can use the mole to count atoms, molecules, ions, electrons and other species. So, it is important to include an exact description of the species being referred to. Consider the examples of hydrogen, oxygen and water.

One mole of water has a mass of 18.0 g, but what is the mass of one mole of hydrogen or oxygen? It depends on whether you are referring to atoms or molecules, so you need to make this clear.

Remember that the symbol n is used for the amount of substance in mol.

Consider the substances in the table. The masses are the same, but the amounts are different.

|  | 1 g of oxygen atoms (O) | 1 g of oxygen molecules ($O_2$) | 1 g of ozone molecules ($O_3$) |
|---|---|---|---|
| n (amount in mol) | 1 ÷ 16.0 = 0.0625 mol | 1 ÷ 32.0 = 0.0313 mol | 1 ÷ 48.0 = 0.0208 mol |

You should always clearly identify the species that you are referring to if there is any possibility that there could be more than one meaning. This is not usually necessary for compounds.

It is good practice to refer to both the formula and name. Examples include:

- the amount, in moles, of O in 9.4 g of oxygen atoms (0.59 mol)
- the amount, in moles, of $O_2$ in 9.4 g of oxygen molecules (0.29 mol)
- the amount, in moles, of $O_3$ in 9.4 g of ozone molecules (0.20 mol)
- the amount, in moles, of $CO_2$ in 9.4 g of carbon dioxide (0.21 mol)
- the amount, in moles, of $SO_4^{2-}$ in 9.4 g of sulfate ions (0.098 mol).

### The equation for calculating moles

The equation for calculating moles is:

$$\text{amount of substance in moles} = \frac{\text{mass in grams}}{\text{molar mass}} \quad \text{or} \quad n = \frac{m}{M}$$

You will use this expression in many calculations during your chemistry studies. It is sometimes rearranged as:

$$M = \frac{m}{n} \quad \text{and} \quad m = n \times M$$

### Example 1

What is the amount of substance in 6.51 g of sodium chloride?

$$n = \frac{m}{M} = \frac{6.51}{58.5} = 0.111 \text{ mol}$$

### Example 2

What is the mass of 0.263 mol of hydrogen iodide?

$$m = n \times M = 0.263 \times 127.9 = 33.6 \text{ g}$$

### Example 3

A sample of 0.284 mol of a substance has a mass of 17.8 g. What is the molar mass of the substance?

$$M = \frac{m}{n} = \frac{17.8}{0.284} = 62.7 \text{ g mol}^{-1}$$

### Calculations using the Avogadro constant

You will need to use the value of $L$ in these types of calculation.

- Calculate the number of particles in a given mass of a substance.
- Calculate the mass of a given number of particles of a substance.

### Example 1

How many $H_2O$ molecules are there in 1.25 g of water?

$$n = \frac{1.25}{18.0} = 0.0694 \text{ mol}$$

number of molecules = $6.02 \times 10^{23} \times 0.0694 = 4.18 \times 10^{22}$

### Example 2

What is the mass of 100 million atoms of gold?

$$n = \frac{100 \times 10^6}{6.02 \times 10^{23}} = 1.66 \times 10^{-16} \text{ mol}$$

$$m = 1.66 \times 10^{-16} \times 197.0 = 3.27 \times 10^{-14} \text{ g}$$

(Lots of atoms, but only a tiny mass!)

### Learning tip

When using moles, always make clear what particles you are referring to – atoms, molecules, ions or electrons. It is also a good idea to state the formula.

## Questions

1. What is the amount of substance in each of the following?
   (a) 8.0 g of sulfur, S
   (b) 8.0 g of sulfur dioxide, $SO_2$
   (c) 8.0 g of sulfate ions, $SO_4^{2-}$

2. How many particles are there of the specified substance?
   (a) atoms in 2.0 g of sulfur, S
   (b) molecules in 4.0 g of sulfur dioxide, $SO_2$
   (c) ions in 8.0 g of sulfate ions $SO_4^{2-}$

### Key definitions

A **mole** is the amount of substance that contains the same number of particles as the number of carbon atoms in exactly 12 g of the carbon-12 isotope.

The **Avogadro constant** is the number of atoms of $^{12}C$ in exactly 12 g of $^{12}C$.

# 5.2  2  Writing chemical equations

**By the end of this section, you should be able to...**

- write balanced full and ionic equations, including state symbols, for chemical reactions

## What to remember when writing equations

This section is a useful summary of what you are expected to do when you write equations.

### Writing formulae for names

You may be given the formulae of unfamiliar compounds in a question. You will be expected to work out the formulae of many of them and to remember others. You also need to show elements with the correct formulae. It is not possible to provide a definitive list, but here are some examples.

You need to remember that:

- oxygen is $O_2$ and not $O$
- hydrogen is $H_2$ and not $H$
- nitrogen is $N_2$ and not $N$
- water is $H_2O$
- sodium hydroxide is NaOH
- nitric acid is $HNO_3$.

You should be able to work out that:

- iron(II) sulfate is $FeSO_4$
- iron(III) oxide is $Fe_2O_3$
- calcium carbonate is $CaCO_3$.

### Writing an equation from a description

You will need to convert words into formulae and decide which ones are reactants and which are products.

Consider this description: when carbon dioxide reacts with calcium hydroxide, calcium carbonate and water are formed. This wording makes it clear that carbon dioxide and calcium hydroxide are the reactants, and calcium carbonate and water are the products.

Now you have to write the formulae in the correct places:

$$CO_2 + Ca(OH)_2 \rightarrow CaCO_3 + H_2O$$

The next step is balancing. You need to add up the numbers of all the atoms to make sure that, for each element, the totals are the same on the left and on the right of the equation. In this example, there is one carbon, one calcium, two hydrogen and four oxygen atoms on each side, so the equation is already balanced.

Here is an example where the first equation you write is not balanced.

The description is: hydrogen peroxide decomposes to water and oxygen.

The formulae are:

$$H_2O_2 \rightarrow H_2O + O_2$$

This is already balanced for hydrogen, but not for oxygen. With careful practice, which sometimes involves trial and error, you should be able to write the **coefficients** needed to balance the equation (coefficient is the technical term for the numbers you write in front of a species when balancing an equation). In this case, the balanced equation is:

$$2H_2O_2 \rightarrow 2H_2O + O_2$$

Most equations are balanced using whole-number coefficients, but using fractions is usually acceptable. This is especially the case in organic chemistry.

Consider this unbalanced equation for the complete combustion of butane:

$$C_4H_{10} + O_2 \rightarrow CO_2 + H_2O$$

The balanced equation can be either:

$$2C_4H_{10} + 13O_2 \rightarrow 8CO_2 + 10H_2O$$

or

$$C_4H_{10} + 6\tfrac{1}{2}O_2 \rightarrow 4CO_2 + 5H_2O$$

Using 6.5 instead of $6\tfrac{1}{2}$ is also acceptable.

### Using state symbols

Many chemical equations include state symbols. The symbols are:

- (s) = solid
- (l) = liquid
- (g) = gas
- (aq) = aqueous (dissolved in water).

It is important to distinguish between (l) and (aq). A common error is to write $H_2O(aq)$ instead of $H_2O(l)$.

Although it is good practice to include state symbols in all equations, in some cases they are essential, while in others they are often omitted. For example, when using Avogadro's constant in the previous section, it was important to identify which species were gases. However, later in this book you will write equations for organic reactions, where state symbols are often ignored.

Here is another description: when aqueous solutions of silver nitrate and calcium chloride are mixed, a white precipitate of silver chloride forms. As this precipitate settles, a solution of calcium nitrate becomes visible.

After writing the correct formulae, balancing the equation and including state symbols, the equation is:

$$2AgNO_3(aq) + CaCl_2(aq) \rightarrow 2AgCl(s) + Ca(NO_3)_2(aq)$$

### Arrows in equations

Most equations are shown with a conventional (left to right) arrow. However, some important reactions are reversible, which means that the reaction can go in the forward and reverse directions. The symbol $\rightleftharpoons$ is used for these. You can find guidance later in this book about when this reversible arrow should be used.

Sometimes a conventional arrow is lengthened to allow information about the reaction to be shown above the arrow (and sometimes below it). This information may be about reaction conditions, such as temperature, pressure and the use of a catalyst. In organic chemistry, where reaction schemes are important, it may simply be a label indicating Step 1 or Stage 2 in a sequence of reactions or in a reaction scheme.

## Ionic equations

### Simplifying full equations

Ionic equations show any atoms and molecules involved, but only the ions that react together, and not the **spectator ions** (these are the ones that are there before and after the reaction but are not involved in the reaction).

This is the easiest process to follow:

1. Start with the full equation for the reaction.
2. Replace the formulae of ionic compounds by their separate ions.
3. Delete any ions that appear identically on both sides.

### Example 1

What is the simplest ionic equation for the neutralisation of sodium hydroxide solution by dilute nitric acid?

The full equation is:

$$NaOH(aq) + HNO_3(aq) \rightarrow NaNO_3(aq) + H_2O(l)$$

You now have to consider which of these species is ionic and replace them by ions:

$$Na^+(aq) + OH^-(aq) + H^+(aq) + NO_3^-(aq) \rightarrow Na^+(aq) + NO_3^-(aq) + H_2O(l)$$

After deleting the identical ions, the equation becomes:

$$H^+(aq) + OH^-(aq) \rightarrow H_2O(l)$$

### Example 2

What is the simplest ionic equation for the reaction that occurs when solutions of lead(II) nitrate and sodium sulfate react together to form a precipitate of lead(II) sulfate and a solution of sodium nitrate?

The full equation is:

$$Pb(NO_3)_2(aq) + Na_2SO_4(aq) \rightarrow PbSO_4(s) + 2NaNO_3(aq)$$

Replacing the appropriate species by ions gives:

$$Pb^{2+}(aq) + 2NO_3^-(aq) + 2Na^+(aq) + SO_4^{2-}(aq) \rightarrow PbSO_4(s) + 2Na^+(aq) + 2NO_3^-(aq)$$

After deleting the identical ions, the equation becomes:

$$Pb^{2+}(aq) + SO_4^{2-}(aq) \rightarrow PbSO_4(s)$$

These remaining ions are not deleted because they are not shown identically. Before the reaction they were free-moving ions in two separate solutions. After the reaction they are joined together in a precipitate.

## Example 3

Carbon dioxide reacts with calcium hydroxide solution to form water and a precipitate of calcium carbonate.

The full equation is:

$$CO_2(g) + Ca(OH)_2(aq) \rightarrow CaCO_3(s) + H_2O(l)$$

Replacing the appropriate species by ions gives:

$$CO_2(g) + Ca^{2+}(aq) + 2OH^-(aq) \rightarrow CaCO_3(s) + H_2O(l)$$

Note that carbon dioxide and water are molecules, so their formulae are not changed. In this example, no ions are shown identically on both sides, so this is the simplest ionic equation.

## Example 4

Solutions of sodium chloride and potassium bromide are mixed together. What is the simplest ionic equation for any reaction that occurs?

A full equation would be:

$$NaCl(aq) + KBr(aq) \rightarrow NaBr(aq) + KCl(aq)$$

Replacing the appropriate species by ions gives:

$$Na^+(aq) + Cl^-(aq) + K^+(aq) + Br^-(aq) \rightarrow Na^+(aq) + Br^-(aq) + K^+(aq) + Cl^-(aq)$$

After deleting the identical ions, there are none left! This is because the reactants are soluble in water and do not react with each other.

This example has been included to make a point. It is possible to write an equation that looks convincing, but it is meaningless if no reaction occurs.

## Ionic half-equations

Ionic half-equations are written for reactions involving oxidation and reduction, and they usually show what happens to only one reactant. A simple example is the reaction that occurs at the negative electrode during the electrolysis of aqueous sulfuric acid:

$$2H^+(aq) + 2e^- \rightarrow H_2(g)$$

You will learn much more about ionic half-equations later in this book.

> **Learning tip**
>
> Consider carefully what the correct symbol for a species should be – it may be different in different reactions. For example, water is never $H_2O(aq)$, but it may be $H_2O(s)$, $H_2O(l)$ or $H_2O(g)$, depending on the temperature.

## Questions

1. Sodium thiosulfate ($Na_2S_2O_3$) solution reacts with dilute hydrochloric acid to form a precipitate of sulfur, gaseous sulfur dioxide and a solution of sodium chloride. Write an equation, including state symbols, for this reaction.

2. Solutions of ammonium sulfate and sodium hydroxide are warmed together to form sodium sulfate solution, water and ammonia gas. Write the simplest ionic equation for this reaction.

> **Key definitions**
>
> **Coefficients** are the numbers written in front of species when balancing an equation.
>
> **Spectator ions** are the ions in an ionic compound that do not take part in a reaction.

# 5.2　3　Calculations using reacting masses

**By the end of this section, you should be able to...**

- calculate reacting masses from chemical equations, and vice versa, using the concepts of amount of substance and molar mass

## Introduction to reacting masses

You can use the ideas from previous sections about amounts of substance and equations to do calculations involving the masses of reactants and products in equations.

A balanced equation for a reaction shows the same number of each species (atoms, molecules, ions or electrons) on both sides of the equation. It is also balanced for the masses of each species. This means that we can make predictions about the masses of reactants needed to form a specified mass or amount of a product, or the other way round.

Consider this equation used in the manufacture of ammonia:

$N_2 + 3H_2 \rightarrow 2NH_3$

This shows that one molecule of nitrogen reacts with three molecules of hydrogen to form two molecules of ammonia. This statement can be made about the amounts involved:

1 mol of $N_2$ reacts with 3 mol of $H_2$ to form 2 mol of $NH_3$.

This statement can be made about the masses involved:

28.0 g of $N_2$ reacts with 6.0 g of $H_2$ to form 34.0 g of $NH_3$.

These amounts and masses can be different, provided that the ratio does not change.

## Calculating reacting masses from equations

Using a balanced equation, predictions can be made about masses.

### Example 1

The equation for a reaction is:

$SO_3 + H_2O \rightarrow H_2SO_4$

What mass of sulfur trioxide is needed to form 75.0 g of sulfuric acid?

- Step 1: calculate the molar masses of all substances you are told about and asked about – in this case, sulfur trioxide and sulfuric acid.

    $M(SO_3) = 80.1 \text{ g mol}^{-1}$ and $M(H_2SO_4) = 98.1 \text{ g mol}^{-1}$

- Step 2: calculate the amount of sulfuric acid.

    $n = \dfrac{m}{M} = \dfrac{75.0}{98.1} = 0.765 \text{ mol}$

- Step 3: use the reaction ratio in the equation to work out the amount of sulfur trioxide needed.

    as the ratio is 1 : 1, the amount is the same, so $n(SO_3) = 0.765 \text{ mol}$

- Step 4: calculate the mass of sulfur trioxide.

    $m = n \times M = 0.765 \times 80.1 = 61.2 \text{ g}$

## Example 2

The equation for a reaction is:

$$2NH_3 + H_2SO_4 \rightarrow (NH_4)_2SO_4$$

What mass of ammonia is needed to form 100 g of ammonium sulfate?

- Step 1: $M(NH_3) = 17.0\,g\,mol^{-1}$ and $M((NH_4)_2SO_4) = 132.1\,g\,mol^{-1}$
- Step 2: $n((NH_4)_2SO_4) = \dfrac{100}{132.1} = 0.757\,mol$
- Step 3: $n(NH_3) = 2 \times 0.757 = 1.51\,mol$ (note the 2 : 1 ratio in the equation)
- Step 4: $m(NH_3) = n \times M = 1.51 \times 17.0 = 25.7\,g$

## Working out equations from reacting masses

You might assume that the equations for all reactions are already known. However, there are sometimes two or more possible reactions for the same reactants, and the reacting masses can be used to identify which of the reactions is occurring.

## Example 1

Sodium carbonate exists as the pure (anhydrous) compound but also as three **hydrates**. Careful heating can decompose these hydrates to one of the other hydrates or to the anhydrous compound. The measurement of reacting masses can allow you to determine the correct equation for the decomposition.

Question
A 16.7 g sample of a hydrate of sodium carbonate ($Na_2CO_3.10H_2O$) is heated at a constant temperature for a specified time until a reaction is complete. A mass of 3.15 g of water is obtained. What is the equation for the reaction occurring?

Method
- Step 1: calculate the molar masses of the relevant substances:

  $M(Na_2CO_3.10H_2O) = 286.1\,g\,mol^{-1}$ and $M(H_2O) = 18.0\,g\,mol^{-1}$

- Step 2: calculate the amounts of these substances:

  $Na_2CO_3.10H_2O$

  $n = \dfrac{m}{M} = \dfrac{16.7}{286.1} = 0.0584\,mol$

  water

  $n = \dfrac{m}{M} = \dfrac{3.15}{18.0} = 0.175\,mol$

- Step 3: use these amounts to calculate the simplest whole-number ratio for these substances:

  $Na_2CO_3.10H_2O$ and $H_2O$ are in the ratio $0.0584 : 0.175$ or $1 : 3$

- Step 4: use this ratio to work out the equation for the reaction:

  $Na_2CO_3.10H_2O \rightarrow Na_2CO_3.7H_2O + 3H_2O$

We have worked out the $Na_2CO_3.7H_2O$ formula by considering the ratio of the other two formulae.

## Example 2

Copper forms two oxides. Both of them can be converted to copper by heating with hydrogen.

Question
An oxide of copper is heated in a stream of hydrogen to constant mass. The masses of copper and water formed are Cu = 17.6 g and $H_2O$ = 2.56 g. What is the equation for the reaction occurring?

Method
- Step 1: $M(Cu) = 63.5\,g\,mol^{-1}$ and $M(H_2O) = 18.0\,g\,mol^{-1}$
- Step 2: $n(Cu) = \dfrac{17.6}{63.5} = 0.277\,mol$ and $n(H_2O) = \dfrac{2.56}{18.0} = 0.142\,mol$
- Step 3: ratio is $0.277 : 0.142 = 2 : 1$
- Step 4: the equation has 2 mol of Cu and 1 mol of $H_2O$, so the products must be $2Cu + H_2O$

So the equation is:

$$Cu_2O + H_2 \rightarrow 2Cu + H_2O \text{ and not } CuO + H_2 \rightarrow Cu + H_2O$$

#### Learning tip

One important part of both of these calculation methods is the use of the relevant ratio from the equation. Practise deciding which substances should be used for the ratio and which way round to use the ratio.

# Questions

1. A fertiliser manufacturer makes a batch of 20 kg of ammonium nitrate. What mass of ammonia does he need?

2. A sample of an oxide of iron was reduced to iron by heating with hydrogen. The mass of iron obtained was 4.35 g and the mass of water was 1.86 g. Deduce the equation for the reaction that occurs.

**fig A** Chemistry on this scale needs careful calculations so that no reactants are wasted.

#### Key definition

**Hydrates** are compounds containing water of crystallisation, represented by formulae such as $CuSO_4.5H_2O$.

# 5.2  4  Avogadro's law and gas volume calculations

**By the end of this section, you should be able to...**

- know what Avogadro's law is
- calculate reacting volumes of gases from chemical equations, using Avogadro's law

## Avogadro's law

### What is Avogadro's law?

By the early years of the nineteenth century, scientists had discovered something remarkable about the way gases react with each other: the volumes of gases that react are in a simple whole-number ratio, provided that the volumes are measured under the same conditions of temperature and pressure.

For example, when hydrogen and oxygen react to form water, the ratio is 2 : 1. This is also true for the volume of any gaseous product, so when nitrogen and hydrogen react to form ammonia, the ratio is 1 : 3 : 2.

In 1811, Amedeo Avogadro put forward a hypothesis. A hypothesis is a suggestion that is supported by some evidence. As more supporting evidence is found, and there is little or no contradictory evidence, the hypothesis may later be referred to as a law. Avogadro's hypothesis became **Avogadro's law**.

In English, the text of the original hypothesis reads: 'Equal volumes of gases under the same conditions of temperature and pressure contain the same numbers of molecules'.

This is very useful, because it allows us to use volumes, rather than masses, to do calculations for gas reactions. For gases, volumes are much easier to measure than masses. One mole of any gas occupies exactly the same volume.

**fig A** The wording of Avogadro's original hypothesis, commemorated on an Italian stamp.

### Avogadro's law and equations

Before Avogadro's law, it was known that the ratios of the volumes of reacting gases are in a simple whole-number ratio. After Avogadro's law, it was clear that this ratio also applies to the numbers of molecules that react. You can see the link.

1 volume of nitrogen reacts with 3 volumes of hydrogen to form 2 volumes of ammonia:

$$N_2(g) + 3H_2(g) \rightarrow 2NH_3(g)$$

So, when we compare the volumes of reacting gases, we are indirectly comparing the numbers of molecules that react.

## Calculations using reacting volumes of gases

If you are given an equation that shows reacting gases (even if one or more solids or liquids is involved), you can work out the volumes involved. In these examples, assume that all volumes are measured at the same temperature and pressure.

### Example 1

Hydrogen and chlorine react to form hydrogen chloride according to this equation:

$$H_2(g) + Cl_2(g) \rightarrow 2HCl(g)$$

100 cm$^3$ of hydrogen is mixed with 100 cm$^3$ of chlorine, and reacted. What volume of hydrogen chloride is formed?

The answer is 200 cm$^3$. This answer is not obtained just by adding the two values of 100 cm$^3$ together. It is worked out using the 1 : 1 : 2 ratio in the equation.

## Example 2

Carbon dioxide and carbon react to form carbon monoxide according to this equation:

$$CO_2(g) + C(s) \rightarrow 2CO(g)$$

$1.00\,dm^3$ of carbon dioxide is reacted with an excess of carbon to form carbon monoxide. What volume of carbon monoxide is formed?

The answer is $2.00\,dm^3$. We must ignore the carbon because it is not a gas, so the only ratio to consider is the 1:2 ratio for carbon dioxide and carbon monoxide.

## Example 3

Nitrogen monoxide and oxygen react to form nitrogen dioxide according to this equation:

$$2NO(g) + O_2(g) \rightarrow 2NO_2(g)$$

$50\,cm^3$ of nitrogen monoxide are mixed with $25\,cm^3$ of oxygen, and reacted. What volume of nitrogen dioxide is formed?

The answer is $50\,cm^3$, not $75\,cm^3$. We must use the ratio in the equation, which is 2:1:2, so the volume of nitrogen dioxide formed is the same as the volume of nitrogen monoxide we started with.

## Example 4

Sulfur dioxide and oxygen react to form sulfur trioxide according to this equation:

$$2SO_2(g) + O_2(g) \rightarrow 2SO_3(g)$$

$200\,cm^3$ of sulfur dioxide are mixed with $200\,cm^3$ of oxygen, and reacted. What volume of sulfur trioxide is formed?

This calculation is a bit trickier. The volumes given in the question are equal and so do not match the ratio in the equation (2:1:2). You should be able to see that all of the sulfur dioxide will react, but it can only react with half of the volume of oxygen available.

The answer is $200\,cm^3$, but $100\,cm^3$ of unreacted oxygen gas is left over.

## Example 5

Methane and oxygen react to form carbon dioxide and water according to this equation:

$$CH_4(g) + 2O_2(g) \rightarrow CO_2(g) + 2H_2O(g)$$

$100\,cm^3$ of methane is mixed with $100\,cm^3$ of oxygen, and reacted. What volume of gas is formed?

This calculation is a bit trickier. Firstly, the volumes given in the question are equal and so do not match the ratio in the equation (1:2:1:2). There is also the problem of being asked about the volume of gas formed. There are two gases, so you need to calculate the total.

You should be able to see that all of the oxygen will react, but it can only react with half of the available volume of methane. So, $50\,cm^3$ of methane reacts with $100\,cm^3$ of oxygen to form $50\,cm^3$ of carbon dioxide and $100\,cm^3$ of water (in the form of steam).

The answer to the question is $50 + 100 = 150\,cm^3$. There is also $50\,cm^3$ of unreacted methane gas left over.

## Example 6

Hydrogen sulfide and oxygen react to form sulfur dioxide and water according to this equation:

$$2H_2S(g) + 3O_2(g) \rightarrow 2SO_2(g) + 2H_2O(l)$$

$250\,cm^3$ of hydrogen sulfide is mixed with $600\,cm^3$ of oxygen, and reacted. What is the volume of the resulting gaseous mixture?

A careful look at the equation shows that the water formed is a liquid, so it cannot figure in this calculation. The ratio is 2:3:2 (ignoring the water). You should be able to see that all of the hydrogen sulfide will react, but with only $375\,cm^3$ of the oxygen, because of the 2:3 ratio. This will form $250\,cm^3$ of sulfur dioxide.

Note that the question asks about the resulting gas mixture – this includes any gaseous reactants that are not completely used up.

It might help to list the original and final amounts of gases. These are:

- $H_2S(g)$: originally $250\,cm^3$, finally zero
- $O_2(g)$: originally $600\,cm^3$, finally $600 - 375 = 225\,cm^3$
- $SO_2(g)$: originally zero, finally $250\,cm^3$.

So, the volume of the resulting gas mixture = $0 + 225 + 250 = 475\,cm^3$.

### Learning tip

Practise finding the relevant reacting ratio for a given equation and identifying which reactant may only partly react because the volumes used are not in the reacting ratio.

## Questions

In these questions, assume that all volumes are measured at the same temperature and pressure.

1. Hydrogen sulfide and sulfur dioxide react at room temperature to form sulfur and water, as shown in this equation.

$$2H_2S(g) + SO_2(g) \rightarrow 2H_2O(l) + 3S(s)$$

$1\,dm^3$ of hydrogen sulfide and $1\,dm^3$ of sulfur dioxide are mixed together and the reaction occurs until complete. What is the final volume of the gas?

2. A mixture of $50\,cm^3$ of propane and $150\,cm^3$ of oxygen is ignited to form carbon dioxide and water. The equation for the reaction is:

$$C_3H_8(g) + 5O_2(g) \rightarrow 3CO_2(g) + 4H_2O(l)$$

What is the volume of the final reaction mixture?

### Key definition

**Avogadro's law** states that equal volumes of gases under the same conditions of temperature and pressure contain the same numbers of molecules.

# 5.2 / 5 Molar volume calculations

By the end of this section, you should be able to...

- calculate reacting volumes of gases from chemical equations using the concept of molar volume of gases

## Molar volume

The work of Avogadro and others led to the idea of **molar volume** – the volume of gas that contains one mole of that gas. The molar volume is approximately the same for all gases, but its value varies with temperature and pressure. The value most often used is for gases at room temperature and pressure (sometimes abbreviated as RTP). Room temperature is 298 K (or 25 °C) and standard pressure is 101 325 kPa (1 atm). The value is usually quoted as 24.0 dm³ or 24 000 cm³. Its symbol is $V_m$.

$$V_m = 24.0 \text{ dm}^3 \text{ mol}^{-1} \text{ at RTP}$$

and

$$V_m = 24\,000 \text{ cm}^3 \text{ mol}^{-1} \text{ at RTP}$$

## Calculations using molar volume

### Calculations involving a single gas

If you are asked about a single gas, the calculation is straightforward. Assume that in these examples, all volumes are measured at RTP.

You use the expression $V_m = 24.0 \text{ dm}^3 \text{ mol}^{-1}$ which you might want to consider using in these alternative forms:

$$V_m = 24.0 = \frac{\text{volume in dm}^3}{\text{amount in mol}} \quad \text{or} \quad V_m = 24\,000 = \frac{\text{volume in cm}^3}{\text{amount in mol}}$$

You will need to rearrange the expression depending on the actual question. Make sure that you use only dm³ or only cm³ in the calculation.

### Example 1

What is the amount, in moles, of CO in 3.80 dm³ of carbon monoxide?

$$\text{Answer} = \frac{3.80}{24.0} = 0.158 \text{ mol}$$

### Example 2

What is the amount, in moles, of $CO_2$ in 500 cm³ of carbon dioxide?

$$\text{Answer} = \frac{500}{24\,000} = 0.0208 \text{ mol}$$

### Example 3

What is the volume of 0.356 mol of hydrogen?

$$\text{Answer} = 24.0 \times 0.356 = 8.54 \text{ dm}^3 \quad \text{or} \quad 24\,000 \times 0.356 = 8540 \text{ cm}^3$$

### Calculations involving gases and solid or liquids

If you are given an equation that involves one or more gases and a solid or a liquid, and you use information about the amount (in mol) of a solid or liquid, you can combine two calculation methods that we have already used.

The basis of this type of calculation is that for gases you can interconvert between amount and volume, while for solids and liquids you can interconvert between amount and mass.

- Step 1: is to calculate the amount in moles from either the mass or the volume, depending on which one is given.
- Step 2: is to use the relevant reaction ratio in the equation to calculate the amount of another substance.
- Step 3: is to convert this amount to a mass or a volume, depending on what the question asks.

## Example 1

A piece of magnesium with a mass of 1.0 g is added to an excess of dilute hydrochloric acid. What volume of hydrogen gas is formed? The equation for the reaction is:

$$Mg(s) + 2HCl(aq) \rightarrow MgCl_2(aq) + H_2(g)$$

You are not given any information about the hydrochloric acid, and you are not asked anything about magnesium chloride. You can use the mole expression to calculate the amount of magnesium:

$$n(Mg) = \frac{1.0}{24.3} = 0.0412 \, mol$$

Next, you can see that the $Mg : H_2$ ratio in the equation is 1 : 1, which means that 0.0412 mol of hydrogen is formed.

Finally, convert this amount to a volume and you have the answer.

$$Volume = 24\,000 \times 0.0412 = 988 \, cm^3$$

## Example 2

Calcium carbonate reacts with nitric acid to form calcium nitrate, water and carbon dioxide, as shown in the equation:

$$CaCO_3(s) + 2HNO_3(aq) \rightarrow Ca(NO_3)_2(aq) + H_2O(l) + CO_2(g)$$

In a reaction, 100 cm³ of carbon dioxide is formed. What mass of calcium carbonate is needed for this?

As in Example 1, you are not told anything about nitric acid, or asked anything about calcium nitrate or water. You can use the molar volume expression to calculate the amount of carbon dioxide.

$$amount = \frac{100}{24\,000} = 0.00417 \, mol$$

Next, you can see that the $CaCO_3 : CO_2$ ratio in the equation is 1 : 1, which means that 0.00417 mol of calcium carbonate is needed.

Finally, convert this amount to a mass and you have the answer.

$$m = n \times M = 0.00417 \times 100.1 = 0.417 \, g$$

## Example 3

Ammonium sulfate reacts with sodium hydroxide solution to form sodium sulfate, water and ammonia, as shown in the equation:

$$(NH_4)_2SO_4(s) + 2NaOH(aq) \rightarrow Na_2SO_4(aq) + 2H_2O(l) + 2NH_3(g)$$

What volume of ammonia is formed by reacting 2.16 g of ammonium sulfate with excess sodium hydroxide solution?

You are not given any information about the sodium hydroxide, and you are not asked anything about sodium sulfate or water.

You can use the mole expression to calculate the amount of ammonium sulfate:

$$n((NH_4)_2SO_4) = \frac{2.16}{132.1} = 0.01635 \, mol$$

Next, you can see that the $(NH_4)_2SO_4 : NH_3$ ratio in the equation is 1 : 2, which means that 0.0327 mol of ammonia is formed.

Finally, convert this amount to a volume and you have the answer.

$$volume = 24\,000 \times 0.0327 = 785 \, cm^3$$

### Learning tip

Practise using the three-step method for calculating masses from volumes, and volumes from masses, in a reaction.

## Questions

In these questions, assume that all volumes are measured at STP.

1. A flask contains 2 dm³ of butane. What is the amount, in moles, of gas in the flask?

2. 10.0 g of copper(II) oxide is heated with hydrogen according to this equation:

   $$CuO(s) + H_2(g) \rightarrow Cu(s) + H_2O(l)$$

   What volume of hydrogen gas is needed to react with the copper(II) oxide, and what mass of copper is formed?

### Key definition

**Molar volume** is the volume occupied by 1 mol of any gas.

# 5.3 1 Concentrations of solutions

By the end of this section, you should be able to...
- understand how to do calculations using mass concentration, molar concentration and using concentration and mass
- calculate solution concentrations, in $mol\,dm^{-3}$ and $g\,dm^{-3}$

## Calculations using mass concentration

If you know the mass of a **solute** that you dissolve in a **solvent** (usually water), and the volume of the **solution** formed, then it is straightforward to calculate the **mass concentration**.

You use the expression:

$$\text{mass concentration in } g\,dm^{-3} = \frac{\text{mass of solute in g}}{\text{volume of solution in } dm^3}$$

In this section, we only use values based on g and $dm^3$. You may also see other units, such as $g\,cm^{-3}$ and $kg\,m^{-3}$. You may have come across another unit, milligrams per cubic decimetre ($mg\,dm^{-3}$), although you are more likely to recognise its abbreviation, ppm (parts per million).

As with other similar expressions, you will need to rearrange it, depending on the wording of the question. You also need to remember to convert $cm^3$ to $dm^3$ (by dividing by 1000).

### Example 1
$200\,cm^3$ of a solution contains $5.68\,g$ of sodium bromide. What is its mass concentration?

$$\text{mass concentration} = \frac{m}{V} = \frac{5.68}{0.200} = 28.4\,g\,dm^{-3}$$

### Example 2
The concentration of a solution is $15.7\,g\,dm^{-3}$. What mass of solute is there in $750\,cm^3$ of solution?

$$m = \text{mass concentration} \times V = 15.7 \times 0.750 = 11.8\,g$$

### Example 3
A chemist uses $280\,g$ of a solute to make a solution of concentration $28.4\,g\,dm^{-3}$. What volume of solution does he make?

$$V = \frac{m}{\text{mass concentration}} = \frac{280}{28.4} = 9.86\,dm^3$$

## Calculations using molar concentration

**Molar concentration** (or molarity) is a much more widely used term than mass concentration. If only the term 'concentration' is mentioned, then you should assume that it refers to molar concentration.

Its units are $mol\,dm^{-3}$, and this is often denoted by the use of square brackets. So, if a solution of hydrochloric acid has a concentration of $0.150\,mol\,dm^{-3}$, this can be shown as $[HCl] = 0.150\,mol\,dm^{-3}$. The symbol c is sometimes used.

You need to be able to use these two expressions together:

amount = mass/molar mass    or    $n = \frac{m}{M}$

and

(molar) concentration = $\frac{\text{amount}}{\text{volume}}$    or    $c = \frac{n}{V}$

As before, you will need to rearrange them. Look at the question wording to decide which one to use first.

### Example 1
A chemist makes $500\,cm^3$ of a solution of nitric acid of concentration $0.800\,mol\,dm^{-3}$. What mass of $HNO_3$ does he need?

Step 1
You are given values of $V$ and $c$, so you can use the second expression to calculate a value for $n$.

$$n = c \times V = 0.800 \times 0.500 = 0.400\,mol$$

Step 2
You can now use the first expression to calculate the mass of nitric acid.

$$m = n \times M = 0.400 \times 63.0 = 25.2\,g$$

### Example 2
A student has $50.0\,g$ of sodium chloride. What volume of a $0.450\,mol\,dm^{-3}$ solution can she make?

Step 1
You are given the value of $m$ and can work out $M$ from the Periodic Table, so you can calculate $n$.

$$n = \frac{m}{M} = \frac{50.0}{58.5} = 0.855\,mol$$

Step 2
You can now use the second expression to calculate the volume of solution.

$$V = \frac{n}{c} = \frac{0.855}{0.450} = 1.90\,dm^3$$

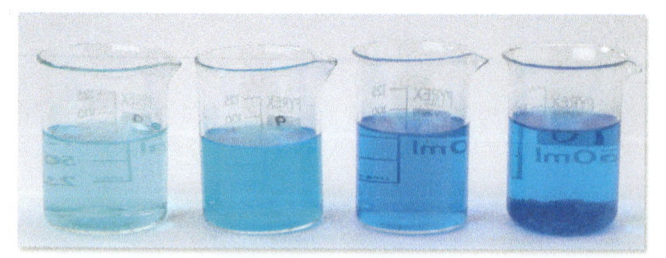

**fig A** These containers indicate the relative solute concentrations by the different colour intensities. Unfortunately, most solutions we use are not coloured, so we cannot rely on different colour intensities to indicate different concentrations.

## Calculations from equations using concentration and mass

In this type of calculation you can use an equation to calculate the mass of a reactant or product if you are given the volume and molar concentration of another substance, and the other way round.

The expressions you need are the same as those you have just used, but you also need the equation for the reaction so that you can see the reacting ratio.

## Example 1

An excess of magnesium is added to $100\,cm^3$ of $1.50\,mol\,dm^{-3}$ dilute hydrochloric acid. The equation for the reaction is:

$$Mg + 2HCl \rightarrow MgCl_2 + H_2$$

What mass of hydrogen is formed?

**Step 1**
You are given the values of $V$ and $c$, so you can use the second expression to calculate the value of n for hydrochloric acid.

$$n = 0.100 \times 1.50 = 0.150\,mol$$

**Step 2**
The ratio for $HCl : H_2$ is 2 : 1, so $n(H_2) = 0.0750\,mol$

**Step 3**
For hydrogen, $m = n \times M = 0.0750 \times 2.0 = 0.150\,g$

## Example 2

A mass of $47.8\,g$ of magnesium carbonate reacts with $2.50\,mol\,dm^{-3}$ hydrochloric acid. The equation for the reaction is:

$$MgCO_3 + 2HCl \rightarrow MgCl_2 + H_2O + CO_2$$

What volume of acid is needed?

**Step 1**
You are given the value of m and can work out $M$ from the Periodic Table, so you can calculate $n$.

$$n = \frac{47.8}{84.3} = 0.567\,mol$$

**Step 2**
The ratio for $MgCO_3 : HCl$ is 1:2, so $n(HCl) = 2 \times 0.567 = 1.134\,mol$

**Step 3**

$$\text{for HCl, } V = \frac{1.134}{2.50} = 0.454\,dm^3$$

### Learning tip

Remember that the volumes referred to in this section are of solutions, not solvents. If you dissolve a solute in $100\,cm^3$ of a solvent, the volume of the solution is not exactly $100\,cm^3$.

## Questions

1 $50.0\,g$ of sodium hydroxide is dissolved in water to make $1.50\,dm^3$ of solution. What is the molar concentration of the solution?

2 $150\,cm^3$ of $0.125\,mol\,dm^{-3}$ lead(II) nitrate solution is mixed with an excess of potassium iodide solution. The equation for the reaction that occurs is:

$$Pb(NO_3)_2(aq) + 2KI(aq) \rightarrow PbI_2(s) + 2KNO_3(aq)$$

What mass of lead(II) iodide is formed?

### Key definitions

A **solute** is a substance that is dissolved.

A **solvent** is a substance that dissolves a solute.

A **solution** is a solute dissolved in a solution.

The **mass concentration** of a solution is the mass (in g) of the solute divided by the volume of the solution.

The **molar concentration** of a solution is the amount (in mol) of the solute divided by the volume of the solution.

# 5.3.2 Making standard solutions

**By the end of this section, you should be able to...**

- understand how a primary standard can be used to prepare a standard solution with an accurately known concentration

## What are standard solutions and primary standards?

**Section 5.3.3** is about titrations. One substance needed in a titration is a standard solution, which we will look at in this section.

A **standard solution** is a solution whose concentration is accurately known. One obvious way to prepare a standard solution is to take a known mass of a substance and dissolve it in water to make a known volume of solution.

These substances are known as **primary standards**. Ideally, they should:

- be solids with high molar masses
- be available in a high degree of purity
- be chemically stable (neither decompose nor react with substances in the air)
- not absorb water from the atmosphere
- be soluble in water
- react rapidly and completely with other substances when used in titrations.

Unfortunately, several substances that are often used in titrations are not suitable as primary standards. For example, hydrochloric acid does not exist as a solid, but only as HCl(g) and HCl(aq). Sodium hydroxide is a solid, but it absorbs water vapour and carbon dioxide from the atmosphere. A sample of sodium hydroxide can be accurately weighed, but it is not certain what is being weighed, as there will be unknown masses of water and sodium carbonate mixed with it. You may remember seeing a white powder around the neck of a bottle of sodium hydroxide solution – this is sodium carbonate.

## Making a standard solution of sulfamic acid

Sulfamic acid is probably unfamiliar to you. It is a readily available primary standard for use in acid–base titrations that has the necessary characteristics. Its formula can be shown in more than one way, including $NH_2SO_3H$, and it has a molar mass of $97.1 \text{ g mol}^{-1}$.

We will look in detail at a method used to obtain an accurately known value for the concentration of this solution. The weighing method used is known as 'weighing by difference'. In your practical work, you may be told to use a modified version.

### Calculating roughly how much to weigh

You need some idea of the approximate concentration and volume of the solution to be made. Typical values are $0.1 \text{ mol dm}^{-3}$ and $250 \text{ cm}^3$.

Using the calculation method in the previous section, the amount of sulfamic acid,
$n = c \times V = 0.1 \times 0.25 = 0.025 \text{ mol}$, so the mass needed, $m = n \times M = 0.025 \times 97.1 = 2.4 \text{ g}$.

Note that even though we are going to do a very accurate weighing, we only need to know an approximate mass at this stage.

## Apparatus

The apparatus you need comprises:

- safety glasses and a lab coat
- an accurate balance (we will assume one reading to 3 decimal places)
- a weighing bottle (or weighing boat)
- a spatula
- a 250 cm³ beaker
- a 250 cm³ volumetric flask
- a wash bottle containing deionised water (or distilled water)
- a small funnel
- a glass stirring rod.

## Method

1. Add between 2.3 and 2.5 g of sulfamic acid to the weighing bottle and weigh accurately.
2. Transfer as much as possible of the acid to a clean beaker and reweigh the weighing bottle.
3. Add about 100 cm³ of deionised water to the beaker and stir until all of the sulfamic acid has dissolved.
4. Remove the stirring rod, washing traces of solution from the rod into the beaker using the wash bottle.
5. Place a funnel in the neck of the volumetric flask and pour the solution from the beaker into the flask.
6. Rinse the inside of the beaker several times with the wash bottle and transfer the rinsings to the flask.
7. Add deionised water to the flask and make up exactly to the graduation mark.
8. Stopper the flask and invert it several times to make a uniform solution.

You can now calculate an accurate value for the concentration of the solution, using these example values.

mass of weighing bottle + sulfamic acid = 19.542 g

mass of weighing bottle + any traces of sulfamic acid = 17.151 g

mass of sulfamic acid added = 2.391 g

$n(NH_2SO_3H) = \dfrac{2.391}{97.1} = 0.02462 \text{ mol}$

$c = \dfrac{0.02462}{0.250} = 0.0985 \text{ mol dm}^{-3}$

### Learning tip

Look carefully at each step of the method. How does each step contribute to the accuracy of the final value of the concentration?

## Questions

1. A student used the method described but made two mistakes.
   (a) In Step 6, he poured the rinsings down the sink instead of transferring them to the flask.
   (b) In Step 7, he added water above the graduation mark.
   Explain how each mistake affected the calculated concentration.

2. A student used the method to make 500 cm³ of a solution of sodium carbonate ($M = 106.0 \text{ g mol}^{-1}$). These are her weighings:

   mass of weighing bottle + sodium carbonate = 23.382 g

   mass of weighing bottle = 18.218 g

   Calculate the concentration of the solution made.

### Key definitions

A **standard solution** is a solution whose concentration is accurately known.

**Primary standards** are substances used to make a standard solution by weighing.

**fig A** When making a standard solution in a volumetric flask, it is important to ensure that the lowest part of the liquid (the meniscus) is aligned with the graduation mark.

# 5.3 3 Doing titrations

By the end of this section, you should be able to...
- understand how to carry out a titration to obtain accurate results

## What is a titration?

A titration is a practical method with the aim of measuring the volumes of two solutions that react together, and using the results to calculate the concentration of one of the solutions. Because the method involves measuring volumes, it is sometimes known as volumetric analysis.

You may come across different titration types (redox titrations and complexometric titrations), but in this book we will consider only acid–base titrations. Many, but not all, bases are soluble in water, which makes them alkalis. They are often referred to as acid–alkali titrations.

All of the acids and bases you will use in titrations are colourless, as are the products, so there is nothing to see when the reaction is complete. This problem is solved by using an indicator, a substance which has different colours in acids and bases/alkalis.

In the method described in this section, we will assume that a solution of sodium hydroxide of approximate concentration $0.1\,mol\,dm^{-3}$ is available and that the standard solution of sulfamic acid referred to in the previous section will be used. The aim is to calculate an accurate value for the concentration of the sodium hydroxide solution. The method will be described in some detail, and we will focus on avoiding common errors in technique.

In some titrations, it is important to choose which way round to do the titration (which solution goes in the pipette and which solution in the burette). In other titrations, it does not matter.

## Outline of the titration method

Here is an outline of the titration method that introduces some key terms.

- Add the acid to the alkali until the **equivalence point** of the titration and the **end point** of the indicator is reached.
- Record the lowest part of the **meniscus**.
- Record the **titre**.
- Repeat the titration until **concordant titres** are obtained.

## Apparatus

The apparatus likely to be used comprises:
- a conical flask (usually $250\,cm^3$)
- a burette (usually $50\,cm^3$) and stand
- a pipette (usually $25\,cm^3$) and pipette filler
- a wash bottle containing deionised water (or distilled water)
- a small funnel
- a white tile.

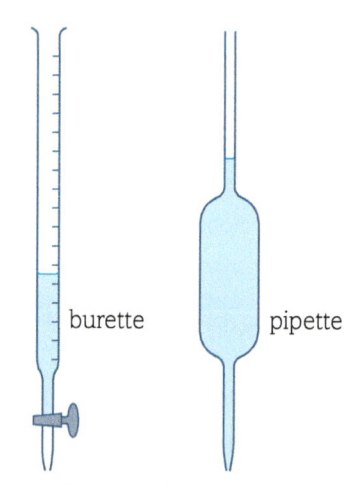

**fig A** Whoever has used the burette has not filled the space between the tap and the tip!

## Method

1. Rinse the conical flask with deionised water and place it on a white tile.
2. Using a pipette filler, rinse the pipette with deionised water and then with some of the sodium hydroxide solution.
3. Use the pipette to transfer $25.0\,cm^3$ of the sodium hydroxide solution to the conical flask.
4. Add about 3 drops of methyl orange indicator.
5. Rinse the burette with deionised water and then with some of the sulfamic acid solution.
6. Fill the burette with the sulfamic acid solution and set it up in the stand above the conical flask.
7. Record the burette reading.
8. Add the sulfamic acid solution to the conical flask until the indicator just changes colour, and again record the burette reading.
9. Empty and rinse the conical flask with deionised water, and repeat the titration until concordant titres have been obtained.

## 5.3 Equations and calculations

## Titration techniques

To obtain accurate results in a titration, it is important to work carefully. The diagram shows some important techniques and the reasons for them.

| TECHNIQUE | REASON |
|---|---|
| Use a white tile. | Although not essential, this provides a constant white background that enables the indicator colour change to be seen more clearly. |
| Add about 3 drops of indicator. | Many acid-base indicators are weak acids and so have an effect on the end point of titration. This would be a problem if different volumes of indicator were used in the repeated titrations. |
| Fill the burette so that the space between the tap and the tip is full of solution. | If this is not done, then as the level in the burette goes down, some of the liquid will fill this space and not enter the conical flask. |
| Set up the burette with its tip inside the neck of the conical flask. | This is to minimise the risk of some of the solution from the burette ending up outside the conical flask. |
| Record the burette reading to the nearest half of a small division (0.05 cm$^3$) using a light background to see the bottom of the meniscus. | This is to increase the accuracy of the reading of the meniscus. |
| Add the sulfamic acid solution from the burette steadily at first, then much more slowly as the end point is approached, then drop by drop when very close to the end point, swirling all the time. | This is to decrease the chance of overshooting the end point – adding too much sulfamic acid solution. Swirling is done to ensure continuous mixing of the two solutions. |
| Stop adding the solution from the burette when the indicator *just* changes colour. | This is to increase the accuracy of the titre. Adding more solution does not change the colour further (although it may make it more intense) but it will decrease the accuracy. |
| Repeat to obtain concordant results. | With all the techniques used to increase the accuracy of the titres, the titres should be the same, but concordant means that they should be within 0.20 cm$^3$ of each other. |
| Rinse the pipette and burette with both deionised water and with the solution to be used. Rinse the conical flask with deionised water only. | If the conical flask were rinsed with the sodium hydroxide solution, there would be an unknown extra amount of substance being titrated, which would introduce an error. |

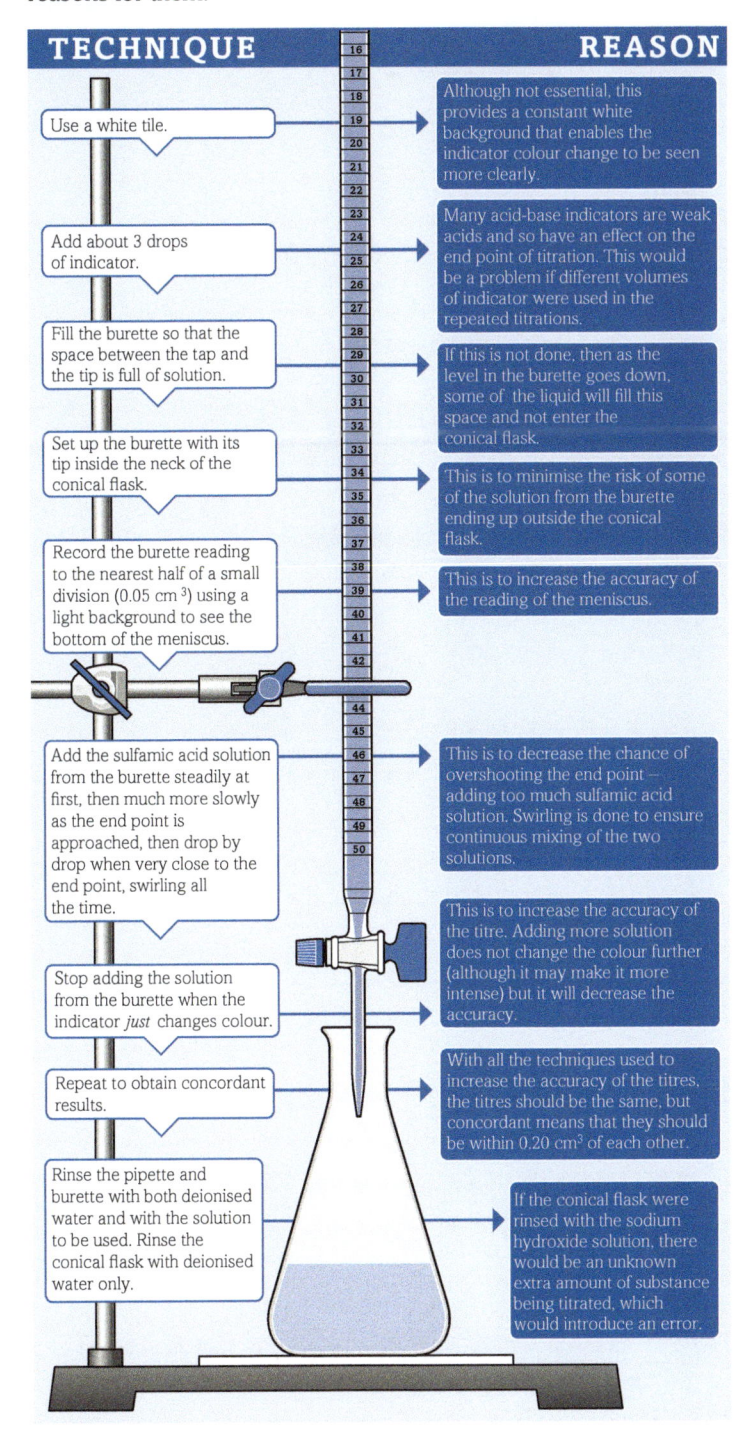

**fig B** Titration techniques and the reasons for them.

## Choosing an indicator

Two common indicators are:
- methyl orange
- phenolphthalein

Sometimes it is important to use one of these and not the other, but in other titrations it does not matter which one you use.

The table shows the colours of these indicators and which combination of acid and base they should be used with.

| Indicator | Colour in acid | Colour in alkali | Acid–base combination |
|---|---|---|---|
| methyl orange | red | yellow | strong acid – weak base and strong acid – strong base |
| phenolphthalein | colourless | pink | weak acid – strong base and strong acid – strong base |

**table A** Information about methyl orange and phenolphthalein.

Examples of strong acids are:
- hydrochloric acid
- nitric acid.

Examples of strong bases are:
- sodium hydroxide
- potassium hydroxide.

The commonest weak base is ammonia.

The commonest weak acid is ethanoic acid.

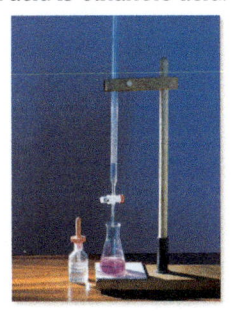

**fig C** This shows the burette after the titration has been done. Note the white tile and the colour of the phenolphthalein indicator – the pink colour shows that an acid has been neutralised and there is an excess of an alkali.

### Learning tip

The best way to understand all the features of successful titrations is to do several yourself!

## Questions

1. A student does not fill the burette space between the tap and the tip in a titration. Explain the effect of this mistake on the value of the titre.

2. A student rinses out the conical flask with deionised water, then with the solution used in the pipette. Explain the effect of this mistake on the value of the titre.

### Key definitions

The **equivalence point** is the point at which there are exactly the right amounts of substances to complete the reaction.

The **end point** is the point at which the indicator just changes colour. Ideally, the end point should coincide with the equivalence point.

The **meniscus** is the curving of the upper surface in a liquid in a container. The lowest (horizontal) part of the meniscus should be read.

The **titre** is the volume added from the burette during a titration.

**Concordant titres** are those that are close together (usually within 0.20 cm$^3$ of each other).

# 5.3 4 Calculations from titrations

By the end of this section, you should be able to...

- calculate solution concentrations, in mol dm⁻³, for simple acid–base titrations using a range of acids, alkalis and indicators

## Calculating the average (mean) titre

The terms 'average' and 'mean' can have more than one mathematical meaning. However, for practical purposes, consider them to be the same – a set of values added to give a total. The total is then divided by the number of values.

Before calculating an average titre, the concordant ones must be selected, i.e. those that are within $0.20\,cm^3$ of each other. This table shows some typical titration results and a student's choice of concordant titres.

| Titration number | 1 | 2 | 3 | 4 |
|---|---|---|---|---|
| final burette reading/cm³ | 24.15 | 25.30 | 24.60 | 23.25 |
| initial burette reading/cm³ | 1.20 | 2.70 | 1.90 | 0.60 |
| titre/cm³ | 22.95 | 22.60 | 22.70 | 22.65 |
| concordant titres | ✗ | ✓ | ✓ | ✓ |

**table A** A set of titration results and a student's choice of concordant titres.

In this example, titrations 2, 3 and 4 are all within $0.20\,cm^3$ of each other and so have been correctly ticked as concordant. It is not surprising that titration 1 is not concordant – it is normal practice to do the first titration more quickly to obtain a rough titre, so the end point is more likely to be overshot. This saves time in the long run because in the other titrations the liquid in the burette can be added quickly at first, until the end point is close, then much more slowly.

This is how to calculate the average.

$$\text{average} = \frac{22.60 + 22.70 + 22.65}{3} = 22.65\,cm^3$$

## Calculating a concentration

### Example 1

Use the example from **Section 5.3.3** to give a typical set of titration results.

volume of sodium hydroxide solution used = $25.0\,cm^3$

volume of sulfamic acid solution used = $22.65\,cm^3$

concentration of sulfamic acid solution = $0.0985\,mol\,dm^{-3}$

The equation for the reaction is:

$$NaOH(aq) + NH_2SO_3H(aq) \rightarrow NH_2SO_3Na(aq) + H_2O(l)$$

You may be shown a shorter method than the following one, but this method, using moles, is more versatile and can be used in more difficult examples.

Step 1: calculate the amount of one substance, in this case the sulfamic acid.

$$n(NH_2SO_3H) = c \times V = 0.0985 \times \frac{22.65}{1000} = 0.00223\,mol$$

Step 2: calculate the amount of the other substance using the reacting ratio in the equation.

as the reacting ratio is 1 : 1, $n(NaOH) = 0.00223\,mol$

Step 3: calculate the concentration of sodium hydroxide solution.
$$c = \frac{n}{V} = \frac{0.00223}{0.0250} = 0.0892 \text{ mol dm}^{-3}$$

## Example 2

A titration is done to calculate the concentration of a solution of nitric acid, using a standard solution of sodium carbonate. The equation for the reaction is:

$$Na_2CO_3 + 2HNO_3 \rightarrow 2NaNO_3 + H_2O + CO_2$$

The titration results are:

volume of sodium carbonate solution used = 25.0 cm³

volume of nitric acid solution used = 27.25 cm³

concentration of sodium carbonate solution = 0.108 mol dm⁻³

Step 1: calculate the amount of sodium carbonate.
$$n = 0.108 \times \frac{25.0}{1000} = 0.00270 \text{ mol}$$

Step 2: calculate the amount of the other substance using the reacting ratio in the equation (1 : 2 in this example).
$$n = 0.00270 \times 2 = 0.00540 \text{ mol}$$

Step 3: calculate the concentration of the nitric acid.
$$c = \frac{0.00540}{0.02725} = 0.198 \text{ mol dm}^{-3}$$

## Example 3

A titration is done to calculate the concentration of a solution of hydrochloric acid, using the sodium hydroxide solution from Example 1. The equation for the reaction is:

$$NaOH + HCl \rightarrow NaCl + H_2O$$

The titration results are:

volume of sodium hydroxide solution used = 25.0 cm³

volume of hydrochloric acid used = 22.68 cm³

concentration of sodium hydroxide solution = 0.0892 mol dm⁻³

This time, we will use the shorter method of calculation, using the expression

$$V_1 \times \frac{M_1}{n_1} = V_2 \times \frac{M_2}{n_2}$$

where $V$ is the volume, $M$ is the molar concentration and $n$ is the coefficient in the equation. The subscript 1 refers to the first substance in the equation, and the subscript 2 to the second substance.

In this example, the unknown is $M_2$, so rearranging the expression and substituting the values gives the answer.

$$M_2 = \frac{V_1 \times M_1 \times n_2}{V_2 \times n_1} = \frac{25.0 \times 0.0892 \times 1}{22.68 \times 1} = 0.0983 \text{ mol dm}^{-3}$$

### Learning tip

Practise the calculation method. Make sure that you use the reacting ratio the right way round.

## Questions

1  A student records these readings during a titration:

| | | | | |
|---|---|---|---|---|
| final burette reading/cm³ | 28.0 | 27.8 | 27.4 | 27.0 |
| initial burette reading/cm³ | 2.7 | 2.8 | 2.8 | 2.8 |
| titre/cm³ | 25.3 | 25.0 | 24.6 | 24.2 |

average titre = 24.4 cm³

What mistakes has the student made in recording these results?

2  A student does a titration using this reaction:

$$2KOH + H_2SO_4 \rightarrow K_2SO_4 + 2H_2O$$

She records these results:

volume of KOH solution = 25.0 cm³

volume of $H_2SO_4$ = 19.83 cm³

concentration of $H_2SO_4$ = 0.0618 mol dm⁻³

What is the concentration of the KOH solution, in mol dm⁻³?

## 5.4.1 Mistakes, errors, accuracy and precision

**By the end of this section, you should be able to…**
- understand the difference between a mistake and an error
- understand the difference between accuracy and precision
- comment on sources of error in experimental procedures

### Using the correct terminology

As with many aspects of chemistry (and other sciences), it can be difficult to find the correct words to use when considering the results of experiments and the calculations based on these results. This is because in the non-scientific world, words are often used with less care. You have already come across the idea that 'amount' has a specific meaning in chemistry (amount of substance, in moles) and should not be used to refer to mass or volume.

We look at terminology in this section.

- Mistakes and errors are not the same thing.
- Accuracy and precision have different meanings.
- Systematic and random errors have different causes.

In this and the next few sections, we look at the differences in meaning between some of the more important terms.

To do this, we will refer back to previous sections, which involve measuring masses and volumes using different methods and pieces of apparatus.

### Mistakes are not errors

Put very simply, an error is something that even a skilled operator would find difficult to avoid, and is a consequence of the way the apparatus has been constructed and how readings can be made using it.

A mistake is something that a skilled operator can avoid by being careful.

Here are some examples of mistakes.

1. A chemist weighs a beaker on a balance without making sure the balance is tared (set to zero) beforehand. The reading on the balance could be very different from the actual mass of the beaker, so the reading should not be used, although this careless chemist may not realise this.

2. A student sees a burette reading of 27.35 cm$^3$ but writes it down as 23.75 cm$^3$. This is the student's mistake, and has nothing to do with the apparatus.

3. A student fills a burette using a funnel and forgets to remove the funnel before adding the liquid. During the addition, some of the liquid in the funnel drips into the burette, and this causes an incorrect burette reading to be recorded. This is due to the student's faulty technique. Again, it has nothing to do with the actual apparatus, only his careless use of it.

### Accuracy and precision, and systematic and random errors

The most important terms to consider in this section are:

- error
- accuracy
- precision.

# Errors and uncertainties 5.4

The last two terms are often confused. We will try to understand the difference between them by considering some burette readings and an archery competition!

A teacher and some students do a titration using the same solutions. The teacher works carefully and obtains an average titre of 24.27 cm³, which we will assume is the 'correct' value.

Consider the following titres, which were recorded by students doing the same titration as the teacher.

| Student | Titres/cm³ | | | | Average of all titres |
|---|---|---|---|---|---|
| | 1 | 2 | 3 | 4 | |
| A | 24.80 | 24.85 | 24.90 | 24.80 | 24.84 |
| B | 24.95 | 24.80 | 23.25 | 23.80 | 24.20 |
| C | 24.20 | 24.30 | 24.25 | 24.20 | 24.24 |

**table A** Titres recorded by Students A, B and C doing the same titration as their teacher.

All the average titres have been calculated correctly.

Now for the archery competition! The aim of an archery competition is to win by managing to land all the arrows as close as possible to the centre of the target. These diagrams show how the three students have fared with their titre values.

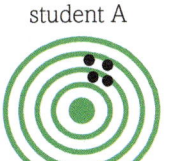

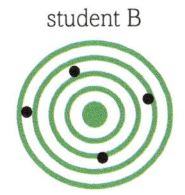

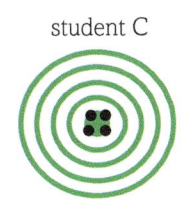

student A   student B   student C

The table provides commentary on these values.

| Student | Comments |
|---|---|
| A | Student A has titres that are all concordant (within 0.10 cm³ of each other), but the average is 0.57 cm³ higher than the correct value. This suggests that the titrations have been carefully done, but that there is probably something about the apparatus that is responsible for the large difference from the correct titre value. This is called a systematic error. The titre values are precise but not accurate. |
| B | Student B has no concordant titres (they are all over the place) but the average is within 0.07 cm³ of the correct value. This suggests that the titrations have been carelessly done, but the student has been lucky because the average happens to be close to the correct value. This is called a random error. Even though each individual titre is not accurate, and all four of them are not precise, the average titre is accurate. |
| C | Student C has titres that are all concordant, and the average is within 0.03 cm³ of the correct value. This suggests that the titrations have been carefully done, and that the apparatus used is of the same standard as that used by the teacher to obtain the correct value. The titre values are both accurate and precise. |

**table B** Comments explaining how Students A, B and C have fared with their titre values.

### Learning tip

Use the archery competition analogy to try to understand the difference between accuracy and precision.

## Questions

1. An experimental method requires a 25.0 cm³ pipette to be used to measure a volume of liquid in different experiments. A student uses a 25 cm³ measuring cylinder instead in each case. Explain whether the student has made a mistake or has introduced a random error or a systematic error.

2. Consider these titres recorded in cm³.

    | Student 1 | 27.60 | 27.70 | 27.70 | 27.65 |
    | Student 2 | 26.15 | 26.85 | 26.60 | 26.30 |
    | Student 3 | 26.40 | 26.50 | 26.50 | 26.40 |

    The teacher's average titre was 26.50 cm³, which can be assumed to be correct.

    Explain whether the students' titres indicate accuracy, precision, both or neither.

### Key definitions

An **error** is the difference between an experimental value and the accepted or correct value.

**Accuracy** is a measure of how close values are to the accepted or correct value.

**Precision** is a measure of how close values are to each other.

# 5.4 2 Measurement errors and measurement uncertainties

**By the end of this section, you should be able to...**

- understand the difference between random and systematic errors
- understand how measurement uncertainties depend on the apparatus used

## Random and systematic errors

Consider the accepted meanings of some of the terms briefly mentioned in the previous section.

| Term | Meaning |
|---|---|
| random error | This is an error caused by unpredictable changes in conditions such as temperature or pressure, or by a difference in recording that is difficult to get exactly right. |
| systematic error | This is an error caused by the apparatus, and leads to the recorded value being either too low or too high. |

### Random errors

Here are examples of random errors.

- The volume of a gas collected in a syringe is measured in different experiments done on the same day. The atmospheric pressure and the temperature of the laboratory may vary during the day, and this will cause an unpredictable change in the value recorded.
- The mass of an object is measured using the same balance but at different times during the day. The values may differ slightly, as a result of changes in temperature, draughts or condensation of water vapour on the balance pan.
- An experiment is done with a flask in an electric water bath. The thermostat is set to 50 °C, but as the heater automatically switches on, the actual temperature rises slightly above 50 °C, and then falls to slightly less than 50 °C after the thermostat switches off.

Repeating an experiment should lead to a more accurate final value being recorded because these random fluctuations are less important when values are averaged.

### Systematic errors

Here are examples of systematic errors.

- A 25.0 cm³ pipette has been wrongly calibrated during its manufacture, so that the graduation mark is lower than it should be. This means that no matter how carefully it is used, the volume added will always be less than 25.0 cm³.
- The amount of liquid in a thermometer is more than it should be, so that the height of the liquid at all temperatures is higher than it should be. This means that the temperature recorded will always be greater than the correct temperature.
- A measuring cylinder has markings on it from 0 to 10 cm³, but the diameter of the cylinder is smaller than the manufacturer intended. This means that when the liquid level shows 10 cm³, the volume is less than 10 cm³.

Repeating an experiment using the same apparatus will not lead to a more accurate value being recorded.

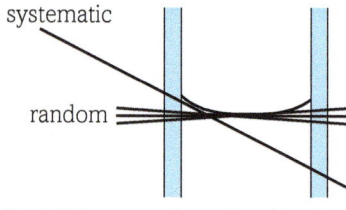

**fig A** When recording a liquid level, the operator should view the meniscus at eye level (horizontally). This diagram shows that minor variations from the horizontal can be considered as random errors. If the meniscus is consistently viewed from above, then the error can be described as systematic.

### Learning tip

Think about how using a burette could involve both random and systematic errors.

# Errors and uncertainties 5.4

## Measurement uncertainty

When using apparatus, there is always a potential error, known as **measurement uncertainty**. The size of the measurement uncertainty is determined by the precision of the apparatus.

## Balances

Digital balances can have various degrees of precision. A balance that measures to three decimal places is more precise that one that measures to only one decimal place. This means that the measurement uncertainty involved in using the three-decimal place balance is lower than in using the one-decimal place balance. This is illustrated in **table A**. The greater the degree of precision of the balance, the smaller the measurement uncertainty in the recorded mass.

| Number of decimal places | Measurement uncertainty | Example |
|---|---|---|
| 1 | ±0.05 g | A reading of 17.1 g could be between 17.05 and 17.15 g |
| 2 | ±0.005 g | A reading of 17.10 g could be between 17.095 g and 17.105 g |
| 3 | ±0.0005 g | A reading of 17.100 g could be between 17.0995 g and 17.1005 g |

**table A** Examples of measurement uncertainties when using a balance.

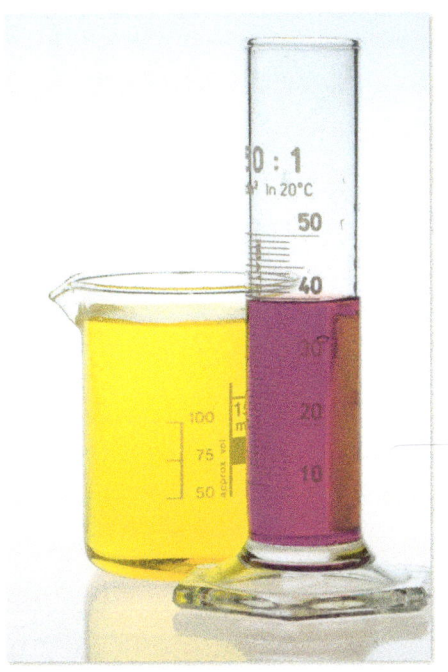

**fig B** The scale on a beaker is there only as a guide and should not be used to record values to be used in calculations. A measuring cylinder is better for this purpose.

## Glassware

Most laboratory glassware is manufactured to Class A and Class B (or Grade A and Grade B) standards. Class A apparatus is considerably more expensive than Class B apparatus, and the extra cost reflects the way in which the apparatus is calibrated, rather than the quality of the actual apparatus.

You will almost certainly use Class B apparatus, which is still capable of giving measurements to a high degree of accuracy. If you look carefully, you should be able to see the manufacturer's claim about level of precision marked on the apparatus. You should see a temperature quoted (usually 20 °C). At higher temperatures, the glass and the solution it contains will expand to different extents, and the value of the volume will no longer be accurate within the given tolerances.

The table shows information about typical pieces of Class B glassware that are used to measure volumes.

| Apparatus | Capacity | Measurement uncertainty |
|---|---|---|
| burette | 50 cm$^3$ | ±0.05 cm$^3$ (but the burette is read twice in a titration, so the total measurement uncertainty is ±0.10 cm$^3$) |
| pipette | 25 cm$^3$ | ±0.06 cm$^3$ |
| volumetric flask | 250 cm$^3$ | ±0.3 cm$^3$ |

**table B** Examples of measurement uncertainties in glassware.

## Questions

1. A student carries out a titration and obtains a titre of 23.40 cm$^3$. She repeats the titration the next day using different apparatus and obtains a titre of 24.50 cm$^3$.
   (a) What is the total measurement uncertainty of the first titre?
   (b) Identify a potential source of random error that may have affected her results.
   (c) Identify a potential source of systematic error that may have affected her results.

2. A student uses a pipette and a burette to add 50.0 cm$^3$ of a liquid to a flask in different experiments. Use the table of measurement uncertainties (**table B** above) to calculate which piece of apparatus is the better one to use in this experiment.

### Key definitions

**Random errors** are errors caused by unpredictable variations in conditions.

**Systematic errors** are errors that are constant or predictable, usually because of the apparatus used.

A **measurement uncertainty** is the potential error involved when using a piece of apparatus to make a measurement.

### Learning tip

Remember that, as long as you have used the piece of apparatus correctly, you cannot say for certain that any measurement made with the apparatus is accurate. It is only possible to know if there has been an error if you happen to know the true value of the measurement made. The best you can say is that there is a potential error involved, which is given by the *measurement uncertainty* by using the apparatus.

# 5.4 3 Percentage measurement uncertainty

By the end of this section, you should be able to...
- understand how to calculate percentage measurement uncertainties and total percentage measurement uncertainties
- understand how to minimise percentage measurement uncertainties

## Percentage measurement uncertainties

Each piece of apparatus you use to record a value (such as mass, volume, temperature or time) has a measurement uncertainty associated with it, which depends on the way it has been manufactured and calibrated.

The actual measurement uncertainty may be fixed, but in many cases the **percentage uncertainty** when you use the apparatus depends on the value you measure. This mostly depends on whether you use the apparatus to record only one value or two values, and on how big the value is compared to the capacity of the apparatus.

### Glassware

The table shows typical percentage uncertainties for common items of glassware.

| Apparatus | Capacity | Uncertainty | Percentage uncertainty |
|---|---|---|---|
| burette | 50 cm$^3$ | ±0.05 cm$^3$ | Note that two burette values are read in a titration, so the total measurement uncertainty is ±0.10 cm$^3$. If the titre is 22.50 cm$^3$, then the percentage uncertainty is: $$\pm\frac{0.10 \times 100}{22.50} = \pm 0.44\%$$ |
| pipette | 25 cm$^3$ | ±0.06 cm$^3$ | The reading is taken only once, and for the same volume each time, so the percentage uncertainty is always: $$\pm\frac{0.06 \times 100}{25} = \pm 0.24\%$$ |
| volumetric flask | 250 cm$^3$ | ±0.3 cm$^3$ | The reading is taken only once, and for the same volume each time, so the percentage uncertainty is always: $$\pm\frac{0.3 \times 100}{250} = \pm 0.12\%$$ |

**table A** Typical percentage uncertainties for a burette, pipette and volumetric flask.

### Balances

The percentage uncertainty in using a balance depends on:
- the precision of the balance, i.e. the number of decimal places to which the balance can be read
- the mass being weighed, as the percentage uncertainty will be greater for a smaller mass.

## Example 1
- mass of a marble chip = 3.57 g

The measurement uncertainty in a two-decimal place balance is ±0.005 g. We have to count this uncertainty twice, since there is a ±0.005 g uncertainty when calibrating the balance to zero, as well as a ±0.005 g uncertainty when measuring the mass.
So the percentage uncertainty is:

$$\pm \frac{2 \times 0.005 \times 100}{3.57} = \pm 0.28\%$$

## Example 2
- mass of weighing bottle + solid = 20.354 g
- mass of weighing bottle = 19.816 g
- mass of solid = 0.538 g

The measurement uncertainty in a three-decimal place balance is ±0.0005 g. We have to count this uncertainty twice for each measurement and we have two measurements, so we need to count it four times. So the percentage uncertainty is:

$$\pm \frac{4 \times 0.0005 \times 100}{0.538} = \pm 0.37\%$$

The percentage uncertainty is greater in Example 2, even though the balance reads to one more decimal place. This is because the balance is used twice and also because the mass being weighed is much smaller.

### Adding measurement uncertainties

If a final answer has been obtained using more than one piece of apparatus, then the approximate total measurement uncertainty is obtained by adding together the individual uncertainties.

For example, if a concentration of 0.118 mol dm$^{-3}$ has been calculated from a titration that has involved balance and glassware uncertainties, then the uncertainties might be:

| | |
|---|---|
| balance | ±0.09% |
| volumetric flask | ±0.12% |
| pipette | ±0.24% |
| burette | ±0.47% |
| total percentage uncertainty | ±0.92% |

This means that there is an uncertainty in the concentration of ±0.001, so the final value can be quoted like this:

concentration = 0.118 ± 0.001 mol dm$^{-3}$

which means that the exact value is in the range 0.117–0.119 mol dm$^{-3}$.

### Minimising error and uncertainty

How can errors and uncertainties be minimised in an experiment? This depends on a number of factors, some of which are easier to control than others.

For example, in a thermochemistry experiment carried out in the laboratory using standard equipment, there will always be transfer of heat energy to the surroundings, which creates a random error in the measurement of the temperature change. In this type of experiment, using a balance that reads to three decimal places (instead of one that reads to two decimal places) will have no significant effect on the overall uncertainty of the final value. Minimising heat energy losses will have a much greater effect.

Where a measuring instrument can be used for a range of values, the percentage uncertainty can be minimised by using a higher value rather than a lower one. For example, when using a balance, weighing a sample of 5 g will lead to a much lower percentage uncertainty than weighing a 0.5 g sample. In a titration, it is not a good idea to have a titre value of 10 cm$^3$ instead of 30 cm$^3$, as the larger titre has the lower percentage uncertainty.

**fig A** This balance is tared (set to zero) and ready to use. The pan is enclosed – this is necessary on a four-decimal place balance because of fluctuations in the reading caused by draughts of air.

### Learning tip
Practise calculating measurement uncertainties and percentage uncertainties for different pieces of apparatus.

## Questions

1. A student uses a one-decimal place balance to weigh a piece of zinc. He records a mass of 2.8 g. What is the percentage uncertainty?

2. A thermometer has a measurement uncertainty of 1 °C. A student uses it to measure a temperature rise and records these values:

   start temperature = 15 °C

   final temperature = 28 °C

   What is the percentage uncertainty in the temperature rise?

### Key definition
The **percentage uncertainty** in an experiment is the actual measurement uncertainty multiplied by 100 and divided by the value recorded.

# 5.5 1 The yield of a reaction

**By the end of this section, you should be able to...**
- understand the meanings of the terms theoretical yield, actual yield and percentage yield
- calculate percentage yields using chemical equations and experimental results

## Theoretical yield, actual yield and percentage yield

In the laboratory, when you are making a product, you naturally want to obtain as much of it as possible from the reactants you start with. In industry, where reactions occur on a much larger scale, and there is economic competition between manufacturers, it is even more important to maximise the product of a reaction.

There are some reasons why the mass of a reaction product may be less than the maximum possible.
- The reaction is reversible and so may not be complete.
- There are side-reactions that lead to other products that are not wanted.
- The product may need to be purified, which may result in loss of product.

## Terminology relating to 'yield'

We normally use the term 'yield' with other words, such as:
- **theoretical yield**
- **actual yield**
- **percentage yield.**

In the laboratory, theoretical yield and actual yield may be measured in grams, but in industry, kilograms and tonnes are more likely to be used.

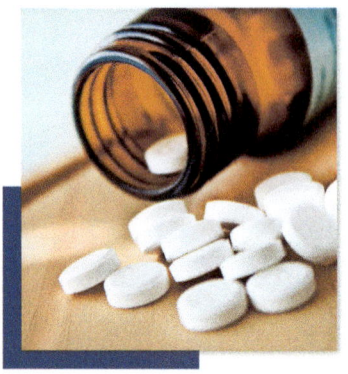

**fig A** Pharmaceutical companies are always looking for ways to increase the percentage yield when manufacturing a drug.

## Theoretical yield

We calculate theoretical yield using the equation for the reaction, and we use a method you are familiar with from previous sections. It is always assumed that the reaction goes to completion, with no losses.

## Example 1

Copper(II) carbonate is decomposed to obtain copper(II) oxide. The equation for the reaction is:
$$CuCO_3 \rightarrow CuO + CO_2$$

What is the theoretical yield of copper(II) oxide obtainable from 5.78 g of copper(II) carbonate?

Step 1: calculate the amount of starting material.
$$n(CuCO_3) = \frac{5.78}{123.5} = 0.0468 \text{ mol}$$

Step 2: use the reacting ratio to calculate the amount of desired product.
$$n(CuO) = 0.0468 \text{ mol}$$

Step 3: calculate the mass of desired product.
$$m = 0.0468 \times 79.5 = 2.97 \text{ g}$$

## Example 2

Magnesium phosphate can be prepared from magnesium by reacting it with phosphoric acid. The equation for the reaction is:
$$3Mg + 2H_3PO_4 \rightarrow Mg_3(PO_4)_2 + 3H_2$$

What is the theoretical yield of magnesium phosphate obtainable from 5.62 g of magnesium?

Step 1: $n(Mg) = \dfrac{5.62}{24.3} = 0.231 \, mol$

Step 2: $n(Mg_3(PO_4)_2) = \dfrac{0.231}{3} = 0.0770 \, mol$

Step 3: $m = 0.0770 \times 262.9 = 20.2 \, g$

## Actual yield

This is the actual mass obtained by weighing, not by calculation.

## Percentage yield

Percentage yield is calculated using

$$\dfrac{\text{actual yield} \times 100}{\text{theoretical yield}} = \text{percentage yield}$$

This calculation may be done independently or in conjunction with the calculation of theoretical yield.

## Example 1

The theoretical yield in a reaction is 26.7 tonnes. The actual yield is 18.5 tonnes. What is the percentage yield?

$$\text{Percentage yield} = \dfrac{18.5 \times 100}{26.7} = 69.3\%$$

## Example 2

A manufacturer uses this reaction to obtain methanol from carbon monoxide and hydrogen:

$$CO + 2H_2 \rightarrow CH_3OH$$

He obtains 4.07 tonnes of methanol starting from 4.32 tonnes of carbon monoxide. What is the percentage yield?

First, calculate the theoretical yield.

Step 1: $n(CO) = \dfrac{4.32 \times 10^6}{28.0} = 1.54 \times 10^5 \, mol$

Step 2: $n(CH_3OH) = 1.54 \times 10^5 \, mol$ (because of 1:1 ratio)

Step 3: $m = 1.54 \times 10^5 \times 32.0 = 4.94 \times 10^6 \, mol$

$$\text{Percentage yield} = \dfrac{4.07 \times 10^6 \times 100}{4.94 \times 10^6} = 82.4\%$$

### Learning tip

When doing calculations using kilograms and tonnes, remember that:

$1 \, kg = 1 \times 10^3 \, g$ and
$1 \, tonne = 1 \times 10^6 \, g$

# Questions

1. A student prepares a sample of copper(II) sulfate crystals, $CuSO_4.5H_2O$, weighing 7.85 g. She starts with 4.68 g of copper(II) oxide. What is the percentage yield?

2. A manufacturer makes some ethanoic acid using this reaction:

    $$CH_3OH + CO \rightarrow CH_3COOH$$

    Starting with 50.0 kg of methanol, he obtains 89.2 kg of ethanoic acid. What is the percentage yield?

### Key definitions

The **theoretical yield** in a reaction is the maximum possible mass of a product, assuming complete reaction and no losses.

The **actual yield** in a reaction is the actual mass obtained.

The **percentage yield** is 100 × the actual yield divided by the theoretical yield.

# 5.5 | 2 | Atom economy

By the end of this section, you should be able to...

- calculate atom economies using chemical equations and experimental results

### Background to atom economy

In the previous section, we looked at the percentage yield of a reaction – the closer the value is to 100%, the better. A higher percentage means that less of the starting materials is lost or ends up as unwanted products.

Percentage yield is not the only factor to take into consideration when assessing the suitability of an industrial process. In addition to those mentioned in the previous section, other factors include:

- the scarcity of non-renewable raw materials
- the cost of raw materials
- the quantity of energy needed.

### How atom economy works

There are two main processes used to manufacture phosphoric acid. To make the comparison easier to follow, a single summary equation is shown for each process.

Process 1    $Ca_3(PO_4)_2 + 3H_2SO_4 \rightarrow 2H_3PO_4 + 3CaSO_4$

Process 2    $P_4 + 5O_2 + 6H_2O \rightarrow 4H_3PO_4$

There are advantages and disadvantages of both processes. However, what you can see from these equations is that all of the atoms in the starting materials for Process 2 end up in the desired product. In Process 1, many of the atoms end up in a second, unwanted product.

### Barry Trost's contribution

An American chemist, Barry Trost, developed the idea of **atom economy** as an alternative way of assessing chemical reactions, especially in industrial processes. He believed that it was important to consider how many atoms from the reactants end up in the desired product.

The expression we use to calculate atom economy (sometimes described as percentage atom economy) is:

$$\text{atom economy} = \frac{\text{molar mass of the desired product}}{\text{sum of the molar masses of all products}} \times 100$$

You can see that you do not need a calculator to work out the atom economy of Process 2. There is only one product, so it must be 100%.

$$\text{For Process 1, atom economy} = \frac{(98.0 \times 2) \times 100}{(98.0 \times 2) + (136.1 \times 3)} = 32.4\%$$

So, you can see that less than one-third of the mass of the starting materials end up in the desired product, which does not look good if the $CaSO_4$ is a waste product that has to be disposed of. If it has a use, then the manufacturer can sell it, which would partly offset the low atom economy of the process. Even if the percentage yield were as high as 100%, the atom economy is still only 32.4%.

# Yield and atom economy 5.5

fig A Barry Trost – a pioneer of the concept of atom economy.

## Reaction types and atom economy

It is possible to make some generalisations about certain types of reaction.

- Addition reactions have 100% atom economy.
- Elimination and substitution reactions have lower atom economies.
- Multistep reactions have lower atom economies.

## Examples of calculations

### Example 1

Sodium carbonate is an important industrial chemical manufactured by the Solvay process. The overall equation for the process is:

$$CaCO_3 + 2NaCl \rightarrow Na_2CO_3 + 2NaCl$$ *[annotated: CaCl₂]*

A manufacturer starts with 75.0 kg of calcium carbonate and obtains 76.5 kg of sodium carbonate. Calculate the percentage yield and atom economy for this reaction.

Theoretical yield = 79.5 kg

Percentage yield = $\frac{76.5 \times 100}{79.5}$ = 96.2%

Atom economy = $\frac{106 \times 100}{106.0 + (58.5 \times 2)}$ = 47.5%

*[annotated: 81.1    39.5%]*

### Example 2

Hydrazine ($N_2H_4$) can be used as a rocket fuel and is manufactured using this reaction:

$$2NH_3 + NaOCl \rightarrow N_2H_4 + NaCl + H_2O$$

What is the atom economy for this reaction?

Atom economy = $\frac{32.0 \times 100}{32.0 + 58.5 + 18.0}$ = 29.5%

### Example 3

A manufacturer of ethene wants to convert some of the ethene into 1,2-dichlorethane. He considers two possible reactions:

Reaction 1  $H_2C=CH_2 + Cl_2 \rightarrow ClCH_2CH_2Cl$

Reaction 2  $2H_2C=CH_2 + 4HCl + O_2 \rightarrow 2ClCH_2CH_2Cl + 2H_2O$

Explain, without doing a calculation, which reaction would be a good choice on the basis of atom economy.

The answer is Reaction 1, because there is only one product, so all the atoms in the reactants end up in the desired product and the atom economy is 100%.

Reaction 2 has a lower atom economy because some of the atoms in the reactants form water, which has no value as a product.

#### Learning tip

Remember that percentage yield indicates how efficient a reaction is at converting the reactants to the products. Atom economy indicates the percentage of atoms from the starting materials that end up in the desired product.

## Questions

1. Ethanol can be manufactured by the hydration of ethene:

    $$C_2H_4 + H_2O \rightarrow C_2H_5OH$$

    What is the atom economy of this process?

2. Ethene can be manufactured by the dehydration of ethanol:

    $$C_2H_5OH \rightarrow C_2H_4 + H_2O$$

    What is the atom economy of this process?

#### Key definition

**Atom economy** is the molar mass of the desired product divided by the sum of the molar masses of all the products, expressed as a percentage.

# 5.6  1  Displacement reactions

**By the end of this section, you should be able to...**

- relate ionic and full equations, with state symbols, to observations from simple test tube reactions, for displacement reactions

## What is a displacement reaction?

As you learn about more chemical reactions, you will know that they are often classified into different types of reaction. You will recognise reaction types such as neutralisation, combustion, oxidation and several others.

In this section, we look at a reaction type called displacement. In simple terms, it is a reaction in which one element replaces another element in a compound. In **Topic 6**, Organic chemistry of this book, you will find examples of substitution reactions, and these could be described using similar words as those used for displacement reactions.

In this book, we use the term substitution reaction in organic chemistry and **displacement reaction** in inorganic chemistry.

## Displacement reactions involving metals

Here are the equations for two displacement reactions of metals:

1  $Mg(s) + CuSO_4(aq) \rightarrow Cu(s) + MgSO_4(aq)$
2  $2Al(s) + Fe_2O_3(s) \rightarrow 2Fe(s) + Al_2O_3(s)$

What do these reactions have in common?

- Both involve one metal reacting with the compound of a different metal.
- Both produce a metal and a different metal compound.
- Both are redox reactions.
- You can see that the metal element on the reactants side has taken the place of the metal in the metal compound on the reactants side.

What are the differences between the reactions?

- Reaction 1 takes place in aqueous solution, while Reaction 2 involves only solids.
- Reaction 1 occurs without the need for energy to be supplied, while Reaction 2 requires a very high temperature to start it.
- Reaction 1 is likely to be done in the laboratory, while Reaction 2 is done for a specific purpose in industry.

## Metal displacement reactions in aqueous solution

Take a closer look at Reaction 1, shown above. When magnesium metal is added to copper(II) sulfate solution, the blue colour of the solution becomes paler. If an excess of magnesium is added, the solution becomes colourless, as magnesium sulfate forms. The magnesium changes in appearance from silvery to brown as copper forms.

The equation can be rewritten as an ionic equation:

$Mg(s) + Cu^{2+}(aq) + SO_4^{2-}(aq) \rightarrow Cu(s) + Mg^{2+}(aq) + SO_4^{2-}(aq)$

Cancelling the ions that appear identically on both sides gives:

$Mg(s) + Cu^{2+}(aq) \rightarrow Cu(s) + Mg^{2+}(aq)$

# Types of reaction 5.6

Now the reaction can be seen as a redox reaction. Electrons are transferred from magnesium atoms to copper(II) ions, so magnesium atoms are oxidised (loss of electrons) and copper(II) ions are reduced (gain of electrons).

This reaction is just one example of many similar reactions in which a more-reactive metal displaces a less-reactive metal from one of its salts. You may come across this and other similar reactions as examples used in the measurement of temperature changes.

**fig A** The photo shows what happens when a copper wire is placed in silver nitrate solution for some time. You can see the results of the displacement reaction – the 'growth' on the wire is silver metal and the blue solution contains the copper sulfate that is formed.

## Metal displacement reactions in the solid state

Reaction 2 is used in the railway industry to join rails together. You might imagine that a good way to join rails together would be by welding, but the metal rails are good conductors of heat and it is very difficult to get the ends of two rails hot enough for them to melt and join together.

So, the thermite method is used. A mixture of aluminium and iron(III) oxide is put just above the place where the two rails are to be joined. A magnesium fuse is lit, and Reaction 2 occurs. It is so exothermic that the iron is formed as a molten metal, which is then poured into the gap between the two rails. After cooling, the rails are now joined together.

**fig B** The flame comes from the highly exothermic reaction forming molten iron. This will be used to fill the gap between two rails and form a strong join between them.

As in Reaction 1, the equation can be rewritten ionically and simplified:

$$2Al(l) + 2Fe^{3+}(l) + 3O^{2-}(l) \rightarrow 2Fe(l) + 2Al^{3+}(l) + 3O^{2-}(l)$$

and:

$$2Al(l) + 2Fe^{3+}(l) \rightarrow 2Fe(l) + 2Al^{3+}(l)$$

As with Reaction 1, this can be seen as a redox reaction. Electrons are transferred from aluminium atoms to iron(III) ions, so aluminium atoms are oxidised (loss of electrons) and iron(III) ions are reduced (gain of electrons).

## Displacement reactions involving halogens

In **Section 4.2.2** we looked at how more-reactive halogens can displace less-reactive halogens from their compounds. For example, chlorine will displace bromine from a potassium bromide solution. The full, ionic and simplified ionic equations for this reaction are:

$$Cl_2(aq) + 2KBr(aq) \rightarrow Br_2(aq) + 2KCl(aq)$$

$$Cl_2(aq) + 2K^+(aq) + 2Br^-(aq) \rightarrow Br_2(aq) + 2K^+(aq) + 2Cl^-(aq)$$

$$Cl_2(aq) + 2Br^-(aq) \rightarrow Br_2(aq) + 2Cl^-(aq)$$

As with the metal displacement reactions, this can be seen as a redox reaction. Electrons are transferred from bromide ions to chlorine, so bromide ions are oxidised (loss of electrons) and chlorine is reduced (gain of electrons).

### Learning tip

When describing displacement reactions, be careful to refer to the correct species. For example, in the reaction between magnesium and copper(II) sulfate, magnesium *atoms* and copper(II) *ions* are involved.

## Questions

1. Iron metal reacts with silver nitrate in a displacement reaction to form silver and iron(II) nitrate. Write a full equation, an ionic equation and a simplified ionic equation for this reaction.

2. A mixture of zinc metal and copper(II) oxide is ignited, causing an exothermic reaction to occur. Write a full equation, and an ionic equation for this reaction.

### Key definition

A **displacement reaction** is a reaction in which one element replaces another element in a compound.

## 5.6 2 Precipitation reactions

**By the end of this section, you should be able to...**

- relate ionic and full equations, with state symbols, to observations from simple test tube reactions, for precipitation reactions

### Introduction

You have already come across examples of **precipitation reactions**. In this section, we focus on two aspects of these reactions:

- their use in chemical tests
- their use in working out chemical equations.

### Chemical tests

#### Carbon dioxide

This may have been your earliest memory of a precipitation reaction. When carbon dioxide gas is bubbled through calcium hydroxide solution (often described as limewater), a white precipitate of calcium carbonate forms. The relevant equation is:

$$Ca(OH)_2(aq) + CO_2(g) \rightarrow CaCO_3(s) + H_2O(l)$$

The formation of the white precipitate was probably described as the limewater going milky or cloudy.

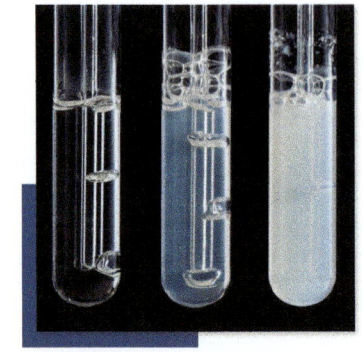

**fig A** Limewater is a colourless solution. As more carbon dioxide is bubbled through it, the amount of white precipitate increases.

#### Sulfates

The presence of sulfate ions in solution can be shown by the addition of barium ions (usually from barium chloride or barium nitrate). The white precipitate that forms is barium sulfate.

For example, when barium chloride solution is added to sodium sulfate solution, the relevant equations are:

$$Na_2SO_4(aq) + BaCl_2(aq) \rightarrow BaSO_4(s) + 2NaCl(aq)$$
$$SO_4^{2-}(aq) + Ba^{2+}(aq) \rightarrow Ba^{2+}SO_4^{2-}(s)$$

This test is covered in more detail in **Section 4.1.3**.

#### Halides

The presence of halide ions in solution can be shown by the addition of silver ions (from silver nitrate). The precipitates that form are silver halides.

For example, when silver nitrate solution is added to sodium chloride solution, the relevant equations are:

$$NaCl(aq) + AgNO_3(aq) \rightarrow AgCl(s) + NaNO_3(aq)$$
$$Cl^-(aq) + Ag^+(aq) \rightarrow Ag^+Cl^-(s)$$

This test is covered in more detail in **Section 4.2.4**.

### Working out equations

A good example of using a precipitation reaction to work out an equation is the reaction between aqueous solutions of lead nitrate and potassium iodide. Both reactants are colourless solutions. When they are mixed, a yellow precipitate of lead iodide forms.

The word equation for this reaction is:

lead nitrate + potassium iodide → lead iodide + potassium nitrate

## 5.6 Types of reaction

Here is an outline of the experiment.
- Place the same volume of a potassium iodide solution in a series of test tubes.
- Add different volumes of a lead nitrate solution to the tubes.
- Place each tube in a centrifuge and spin the tubes for the same length of time.
- Measure the depth of precipitate in each tube.

The concentration of both solutions is $1.0 \text{ mol dm}^{-3}$.

The depth of each precipitate indicates the mass of precipitate.

The table shows the results of one experiment.

| Tube | 1 | 2 | 3 | 4 | 5 | 6 | 7 |
|---|---|---|---|---|---|---|---|
| volume of potassium iodide solution/cm³ | 5.0 | 5.0 | 5.0 | 5.0 | 5.0 | 5.0 | 5.0 |
| volume of lead nitrate solution/cm³ | 0.5 | 1.0 | 1.5 | 2.0 | 2.5 | 3.0 | 3.5 |
| depth of precipitate/cm | 2.5 | 3 | 4 | 5 | 6 | 6 | 6 |

**table A** Results of the reaction between aqueous solutions of lead nitrate and potassium iodide in one experiment.

The diagram shows the tubes at the end of the experiment.

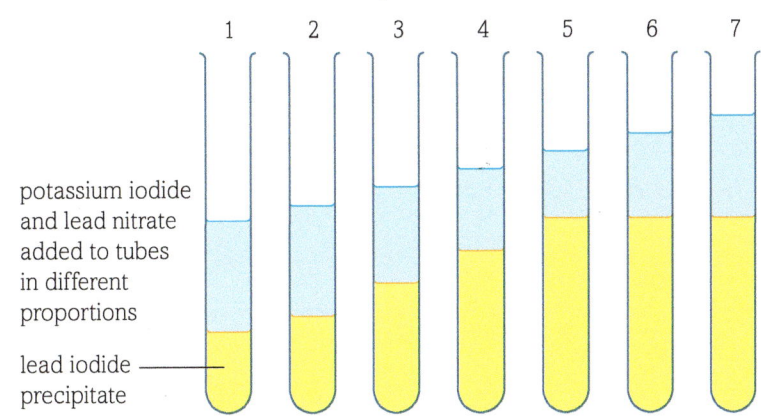

You can see that there is no increase in the amount of precipitate from tube 5 to tube 6. This shows that the reaction is incomplete in tubes 1, 2, 3 and 4, but is complete in tube 5. The amounts of reactants used in tube 5 are calculated as follows:

$n$(potassium iodide) = 0.005 × 1.0 = 0.005 mol

$n$(lead nitrate) = 0.0025 × 1.0 = 0.0025 mol

So, lead nitrate reacts with potassium iodide in the ratio 1 : 2.

The equations for the reaction are:

$Pb(NO_3)_2(aq) + 2KI(aq) \rightarrow PbI_2(s) + 2KNO_3(aq)$

$Pb^{2+}(aq) + 2I^-(aq) \rightarrow PbI_2(s)$

### Learning tip
Practise calculating the amounts of reactants and products in tubes 1–4.

## Questions

1. Write an ionic equation, including state symbols, for:
   (a) the test for a sulfate
   (b) the test for a chloride.

2. Calculate the amounts, in moles, of each reactant and product in tube 7.

### Key definition
**Precipitation reactions** are reactions in which an insoluble solid is one of the products.

# 5.6　3　Reactions of acids

By the end of this section, you should be able to...

- relate ionic and full equations, with state symbols, to observations from simple test tube reactions, for reactions of acids

## Introduction

Acids are common reagents in chemistry. In this section, we summarise some of their typical reactions, using hydrochloric, nitric, sulfuric and phosphoric acids.

In all of these reactions, a salt is formed. The reactions can be used to prepare samples of salts.

## Acids with metals

A general equation for these reactions is:

metal + acid → salt + hydrogen

The metal must be sufficiently reactive to react in this way. For example, magnesium reacts, but copper does not. Bubbles of hydrogen gas form, and if the salt formed is soluble, then a solution forms.

Typical equations for magnesium and hydrochloric acid are:

$Mg + 2HCl \rightarrow MgCl_2 + H_2$

$Mg(s) + 2H^+(aq) \rightarrow Mg^{2+}(aq) + H_2(g)$

These reactions can be classified as neutralisation reactions because the $H^+$ ions are removed from the solution (apart from the very small numbers present in water). They are also redox reactions because there is a transfer of electrons from the metal to the $H^+$ ions.

## Acids with metal oxides and insoluble metal hydroxides

A general equation for these reactions is:

metal oxide + acid → salt + water

metal hydroxide + acid → salt + water

The reactivity of the metal does not matter because in the reactant it is present as metal *ions*, not metal *atoms*. The only observation is likely to be the formation of a solution.

Typical equations for copper(II) oxide and zinc hydroxide reacting with sulfuric acid are:

$CuO + H_2SO_4 \rightarrow CuSO_4 + H_2O$

$CuO(s) + 2H^+(aq) \rightarrow Cu^{2+}(aq) + H_2O(l)$

$Zn(OH)_2 + H_2SO_4 \rightarrow ZnSO_4 + 2H_2O$

$Zn(OH)_2(s) + 2H^+(aq) \rightarrow Zn^{2+}(aq) + 2H_2O(l)$

These reactions can be classified as neutralisation reactions because the $H^+$ ions react with $O^{2-}$ or $OH^-$ ions. They are not redox reactions because there is no change in the oxidation number of any of the species.

## Acids with alkalis

A general equation for these reactions is:

metal hydroxide + acid → salt + water

You have already met these reactions in the sections on titrations. There are no visible changes during these reactions, although if a thermometer is used, a temperature rise can be noted.

Typical equations for sodium hydroxide reacting with phosphoric acid are:

$NaOH + H_3PO_4 \rightarrow NaH_2PO_4 + H_2O$

$2NaOH + H_3PO_4 \rightarrow Na_2HPO_4 + 2H_2O$

$3NaOH + H_3PO_4 \rightarrow Na_3PO_4 + 3H_2O$

There are three replaceable hydrogens in phosphoric acid, and the salt formed depends on the relative amounts of acid and alkali used. The ionic equation for all these reactions is:

$H^+(aq) + OH^-(aq) \rightarrow H_2O(l)$

These reactions can be classified as neutralisation reactions because the $H^+$ ions react with $OH^-$ ions. They are not redox reactions because there is no change in the oxidation number of any of the species.

## Acids with carbonates

A general equation for these reactions is:

metal carbonate + acid → salt + water + carbon dioxide

Bubbles of carbon dioxide gas form. If the salt formed is soluble, then a solution forms.

Typical equations for lithium carbonate reacting with hydrochloric acid are:

$Li_2CO_3 + 2HCl \rightarrow 2LiCl + H_2O + CO_2$

$CO_3^{2-}(aq) + 2H^+(aq) \rightarrow H_2O(l) + CO_2(g)$

These reactions can be classified as neutralisation reactions because the $H^+$ ions react with $CO_3^{2-}$ ions. They are not redox reactions because there is no change in the oxidation number of any of the species.

**fig A** Many public buildings are made from carbonates such as limestone. Centuries of reaction between limestone and acids in the atmosphere mean constant repair bills for York Minster.

## Hydrogencarbonates

Hydrogencarbonates are compounds containing the hydrogencarbonate ion ($HCO_3^-$), and they react with acids in the same way as carbonates do. The best-known example is sodium hydrogencarbonate ($NaHCO_3$), commonly known as bicarbonate of soda and baking soda. Baking soda is used in cooking at home and in the food industry. The 'lightness' of baked food such as cakes is due to the formation of bubbles of carbon dioxide in the cake mixture, which cause the cake to rise.

A word equation for the reaction between baking soda and the acid in lemon juice is:

sodium hydrogencarbonate + citric acid → sodium citrate + water + carbon dioxide

A suitable test for the presence of carbonate or hydrogencarbonate ions in a solid or solution is to add an aqueous acid and test the gas given off with limewater (see **Section 5.6.2**).

### Learning tip
Practise writing equations for different reactions of acids.

# Questions

1 Write full equations for the reactions between:
   (a) zinc and sulfuric acid
   (b) aluminium oxide and hydrochloric acid.

2 Write ionic equations for the reactions between:
   (a) zinc and hydrochloric acid
   (b) magnesium carbonate and nitric acid.

# THINKING BIGGER

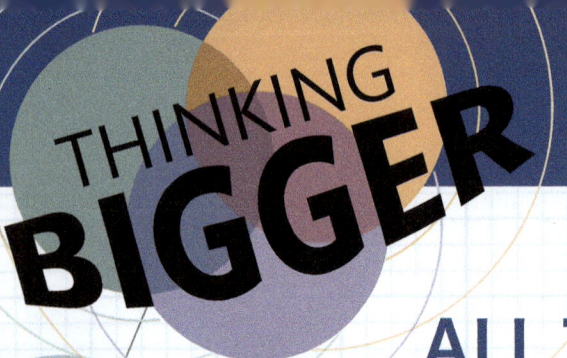

## ALL THAT GLISTERS

The chemical analysis of coinage from earlier periods in history can offer insights into the technology of metal extraction available at the time and give key insights into contemporary geopolitical questions. Read the following extract and answer the questions.

### 'ALL THAT GLISTERS IS NOT GOLD, OFTEN HAVE YOU HEARD THAT TOLD…'

**fig A** Trajan Decius, 249–251 BCE. Denomination: Silver Antoninianus. Mint: Rome.

Over the centuries, the base metals Fe, Cu, Ni, Zn, Al, Sn and Pb have been used as minor alloy constituents or as principal components in coins. Generally, a metal must be reasonably hard wearing to ensure the economic lifetime of a coin, and must retain an acceptable appearance on exposure to the atmosphere. And it must not be too expensive. As a soft and expensive metal, gold is now usually alloyed with copper to give a hard-wearing alloy. British gold sovereigns, which are still minted, are made from 2 carats of alloy and 22 carats of gold, and comprise 91% Au, 8.3% Cu. (A carat is a measure of the purity of gold, with pure gold being 24 carat.) Bronze (Cu–Sn alloy) farthings issued between 1897 and 1917 were darkened using $Na_2S_2O_3$, resulting in a surface layer of copper sulfide. This was done to avoid confusion between a newly minted farthing and a gold half-sovereign, the latter being worth 480 times the former. Bronze pennies issued between 1944 and 1946 were similarly treated to deter the hoarding of new pennies at the end of World War II.

Platinum made only a brief appearance as a coinage metal. A few high-value coins in Russia were made in the mid-nineteenth century, at a time when platinum had recently been found in the Ural Mountains. For a while, the value of platinum fell below that of gold, and this gave rise to the emergence of counterfeit sovereigns in the 1870s. These were made from platinum and had a thin layer of electrochemically deposited gold.

Silver coinage was the universal medium of trading for centuries. The silver content of the coins made in city mints was a reflection of the status of the city as a trading centre. Traders of good repute only dealt in high quality silver coins, as evidenced by the high silver content of those coins found along the old established trading routes, e.g. from the Mediterranean, through the Middle East and Central Asia, to China.

If military expenditure increased, or during times of declining prosperity, a government would put only enough silver in its coins for them to remain acceptable to the public. Sometimes, however, there was a more subtle reason for the decline in silver content. Up to circa 100 BCE, Roman silver coins contained more than 90% Ag. Thereafter, silver supplies grew scarcer, and Roman technology was unable to extract silver from low-grade ores, which by the third century BCE were the only primary sources of silver available. However, at around this time, deposits of AgCl (chlorargynite) were found in Cornwall, Brittany and Alsace. The Romans observed that if copper or bronze coins were dipped in molten AgCl (mp 455 °C), they became coated with silver:

$$Cu + 2AgCl \rightarrow 2Ag + CuCl_2$$

### Did you know?

In April 1940 during the Nazi occupation of Denmark, the Hungarian-born chemist George de Hevesy decided to hide the Nobel Prize medals of colleagues Max von Laue and James Franck by dissolving them in aqua regia (a mixture of concentrated nitric and hydrochloric acid). The resulting solution sat on a shelf for the next 5 years attracting no particular attention. At the end of the war, the gold was recovered from the solution and the two medals recast.

Where else will I encounter these themes?

1  2  3  4  5

# Thinking Bigger

Let us start by considering the nature of the writing in the article.

1. The article is about the development of a coin-based currency from the earliest times. In what ways is an article of this type different from a research paper? Which audience is this article aimed at? Is there any bias present in the report?

> See if you can find a similar article on the development of pigments and dyes over the ages or the development of textiles. How are the chemical ideas represented?

Now we will look at the chemistry in, or connected to, this article. Don't worry if you are not ready to give answers to these questions yet. You may like to return to the questions once you have covered other topics later in the book.

2. a. Modern gold sovereigns weigh 7.99 g (to two decimal places). Using the information in the article, calculate the number of moles of gold in a sovereign.
   b. Calculate the percentage of gold in an 18 carat ring.

> BCE stands for 'Before the Common Era' and means the same as the abbreviation BC.

3. Reread the third paragraph of the article. Using the information given, write half-equations for (a) The reduction of thiosulfate ions to sulfide in acidic solution, and (b) the oxidation of copper metal to copper(II) ions. Now write a redox equation for the reaction of copper metal with acidified sodium thiosulfate.

4. Explain why the reaction between copper and molten silver chloride can be described as a redox reaction. Illustrate your answer with appropriate half equations.

5. A gold sovereign is analysed using the following chemical procedure.
   1. The sovereign is reacted with excess concentrated nitric acid in a fume cupboard.
   2. The resultant mixture is filtered to remove the unreacted gold.
   3. The resultant solution is made up in distilled water to a volume of 250 cm$^3$
   4. A volume of 25 cm$^3$ of this solution is pipetted into a conical flask and 50 cm$^3$ of 0.1 mol dm$^{-3}$ KI(aq) is added (this is an excess). The resultant solution immediately turns an orange-brown colour.
   5. This solution is then titrated with 0.050 mol dm$^{-3}$ Na$_2$S$_2$O$_3$(aq) solution until the solution fades to a straw yellow colour.
   6. Starch indicator is then added to give a blue-black colour.
   7. Further Na$_2$S$_2$O$_3$(aq) solution is added until the solution becomes colourless.
   8. The first titration volume is recorded.
   9. The titration is repeated 3 times and the results recorded to the nearest 0.05 cm$^3$.

> At GCSE, you will have covered metals and their methods of extraction. This would be a good chance for you to review some of that work. How does metal reactivity relate to method of extraction?

| Titre number | 1 | 2 | 3 | 4 |
|---|---|---|---|---|
| Final volume/cm$^3$ | 26.00 | 26.35 | 25.90 | 25.85 |
| Initial volume/cm$^3$ | 0.00 | 0.55 | 0.00 | 0.00 |
| Titre/cm$^3$ | 26.00 | | | |

**Results table**

a. Suggest why step 1 must be done in a fume cupboard.
b. Complete the results table and calculate a suitable average titre.
c. Calculate the percentage uncertainty in the reading for titre number 3.
d. Match the equations below with the reactions taking place in the method described.
   Cu(s) + 4H$^+$(aq) + 2NO$_3^-$(aq) → Cu$^{2+}$(aq) + 2NO$_2$(g) + 2H$_2$O(l)
   2Cu$^{2+}$(aq) + 4I$^-$(aq) → 2CuI(s) + I$_2$(aq)
   2S$_2$O$_3^{2-}$(aq) + I$_2$(aq) → 2I$^-$(aq) + S$_4$O$_6^{2-}$(aq)
e. Using appropriate calculations, decide whether the sovereign is genuine or not.
f. What assumptions have been made in the procedure?

● From an article in *Education in Chemistry* magazine, published by the Royal Society of Chemistry

## Activity

The history of human technology is closely associated with metals.

Draw a timeline identifying the metals used and the cultures responsible for developing the technology.

- You should consider the following metals; gold, silver, copper, iron, aluminium, titanium and uranium.
- Consider also the methods of extraction and the key usages of the metals.

# 5 Exam-style questions

1. Magnesium chloride may be prepared by reacting magnesium with dilute hydrochloric acid.
   A sample of magnesium of mass 1.215 g was added to 60.0 cm³ dilute hydrochloric acid of concentration 2.00 mol dm⁻³.
   The equation for the reaction is
   Mg(s) + 2HCl(aq) → MgCl₂(aq) + H₂(g)
   (a) State two observations that are made when magnesium reacts with dilute hydrochloric acid. [2]
   (b) (i) Calculate the amount, in moles, of magnesium used.
           [$A_r$ of Mg = 24.3] [1]
       (ii) Calculate the amount, in moles, of hydrochloric acid used. [1]
       (iii) Use your answers to (b)(i) and (b)(ii) to show that the hydrochloric acid is in excess. [2]
   (c) Calculate the volume, measured at r.t.p. of hydrogen produced in this reaction.
       [The molar volume of hydrogen at r.t.p. is 24.0 dm³ mol⁻¹] [2]
   [Total: 8]

2. Ammonium nitrate (NH₄NO₃) can be prepared by the reaction of ammonia with nitric acid.
   The equation for the reaction is
   NH₃(aq) + HNO₃(aq) → NH₄NO₃(aq)
   (a) State the oxidation number of nitrogen in both ammonia and nitric acid. [2]
   (b) (i) Calculate the minimum volume of nitric acid, of concentration 0.500 mol dm⁻³, that is required to react completely with 25.0 cm³ of aqueous ammonia of concentration 2.00 mol dm⁻³. [2]
       (ii) How could a solid sample of ammonium nitrate be obtained from the reaction mixture? [1]
   (c) When ammonium nitrate is heated, it decomposes according to the following equation:
       NH₄NO₃(s) → N₂O(g) + 2H₂O(l)
       A 4.00 g sample of ammonium nitrate was carefully heated to produce only N₂O and water.
       (i) Calculate the amount, in moles, of ammonium nitrate used. [2]
           [$A_r$: N, 14.0; H, 1.00; O, 16.0]
       (ii) Calculate the volume of N₂O that was formed. [1]
           [Assume that the molar volume of N₂O is 24.0 dm³ mol⁻¹ under the experimental conditions.]
       (iii) N₂O can be decomposed into its elements by further heating. Write an equation for this reaction. Include state symbols. [2]
   [Total: 10]

3. Phosphorus forms three chlorides of molecular formula PCl₃, PCl₅ and P₂Cl₄.
   (a) State what is meant by the term **molecular formula**. [1]
   (b) PCl₅ can be prepared by reacting white phosphorus (P₄) with chlorine gas.
       Write an equation for this reaction. State symbols are not required. [1]
   (c) Solid PCl₅ reacts vigorously with water to form a solution containing a mixture of two acids, H₃PO₄ and HCl. Write an equation for this reaction. Include state symbols. [2]
   (d) A compound was found to have the following percentage composition by mass:
       P, 30.39%; Cl, 69.61%
       (i) Use this information to identify the chloride. [3]
       (ii) Give the systematic name for this chloride.
           [$A_r$: P, 31.0; Cl, 35.5] [1]
   [Total: 8]

4. The diagram shows the apparatus used to collect and measure the volume of hydrogen given off when a sample of a Group 2 metal reacts with dilute hydrochloric acid.

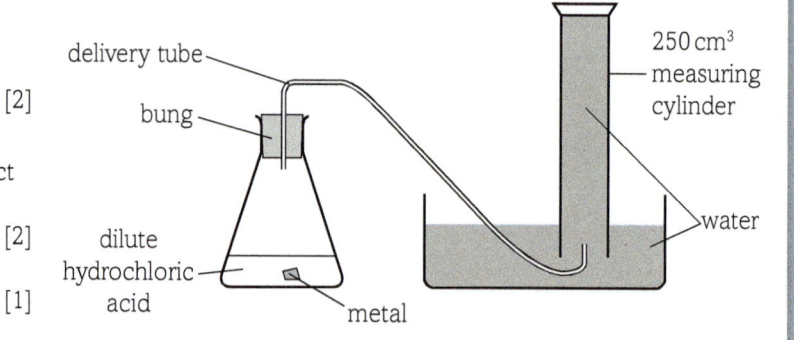

   The equation for the reaction is
   M(s) + 2HCl(aq) → MCl₂(aq) + H₂(g)
   where M is the symbol for the Group 2 metal.
   The apparatus was set up as shown in the diagram.
   A sample of the metal was weighed and then placed into the conical flask.
   The bung was removed and an excess of acid was then added to the metal. The bung was replaced.
   When the reaction was complete, and the gas collected had cooled to room temperature, the volume of gas collected in the measuring cylinder was measured.
   The results are shown in the table.

   | Mass of metal | 0.24 g |
   |---|---|
   | Volume of hydrogen collected | 230 cm³ |

(a) (i) Use the results to calculate the molar mass of the metal. Assume the molar volume of hydrogen is 24.0 dm³ mol⁻¹ under the experimental conditions [3]
   (ii) Give the most likely identity of the metal. Justify your answer. [2]
(b) (i) Identify the major procedural error in the experiment. [1]
   (ii) State a modification that would reduce this procedural error. [1]
(c) The measurement uncertainty in the use of the balance is ± 0.005 g.
   The measurement uncertainty in the use of the measuring cylinder is ± 1.0 cm³.
   (i) Calculate the percentage measurement uncertainty in the use of the balance and in the use of the measuring cylinder. [2]
   (ii) Explain how the major measurement uncertainty could be reduced. [2]
   **[Total: 11]**

5 Tungsten can be extracted from its ore, wolframite, in a multi-stage process. One of these processes involves a redox reaction between tungsten oxide ($WO_3$) and hydrogen:
   $WO_3(s) + 3H_2(g) \rightarrow W(s) + 3H_2O(g)$
   (a) Use oxidation numbers to show that the reaction between $WO_3$ and $H_2$ is a redox reaction. [3]
   (b) A sample of wolframite contains 2% by mass of $WO_3$. Calculate the maximum mass of tungsten formed from processing 100 tonnes of wolframite.
      [$A_r$: W, 184; O, 16.0; 1 tonne = $10^6$ g] [3]
   **[Total: 6]**

6 Azides are compounds of metals with nitrogen. Some are used as detonators in explosives. However, sodium azide ($NaN_3$) is used in air bags in cars.
   (a) Sodium azide decomposes into its elements when heated.
      $2NaN_3(s) \rightarrow 2Na(l) + 3N_2(g)$
      The volume of gas produced, measured at r.t.p. when one mole of sodium azide is decomposed is
      **A** 24 dm³   **B** 36 dm³   **C** 48 dm³   **D** 72 dm³
      [Assume one mole of gas occupies 24.0 dm³ at r.t.p.] [1]
   (b) (i) A student completely decomposed 3.25 g of sodium azide. Calculate the mass of sodium she obtained. [2]
      (ii) She then carefully reacted the sodium obtained with water to form 25.0 cm³ of aqueous sodium hydroxide.
         $2Na(s) + 2H_2O(l) \rightarrow 2NaOH(aq) + H_2(g)$
         Calculate the concentration, in mol dm⁻³, of the aqueous sodium hydroxide. [2]

(c) Liquid sodium is used as a coolant in the German designed sodium-cooled fast reactor (SFR).
   (i) State one property of sodium that makes it suitable as a coolant. [1]
   (ii) State a potential hazard in the use of sodium in the SFR. [1]
   **[Total: 7]**

7 When 30 cm³ of a gaseous hydrocarbon were completely burned in an excess of oxygen, 90 cm³ of carbon dioxide and 80 cm³ of water vapour were formed. All volumes were measured at the same temperature and pressure.
   The molecular formula of the hydrocarbon is
   **A** $C_3H_4$   **B** $C_3H_6$   **C** $C_3H_8$   **D** $C_6H_{14}$ [1]
   **[Total: 1]**

8 A student wanted to find the percentage purity of a sample of anhydrous sodium carbonate.
   He dissolved 1.00 g of the impure sample in water to make 250 cm³ of solution, using a volumetric flask.
   Using a volumetric pipette, he placed 25.0 cm³ of this solution into a conical flask and then added a few drops of methyl orange indicator.
   He then titrated this with 0.100 mol dm⁻³ hydrochloric acid. He repeated the titration as necessary to obtain concordant results.
   His mean titre value was 17.80 cm³.
   (a) State the colour of the methyl orange at the beginning and at the end of the titration. [1]
   (b) What is meant by the term **concordant results** in a titration experiment? [1]
   (c) Use the results to calculate the concentration, in g mol⁻¹, of the sodium carbonate solution. [3]
   (d) Calculate the percentage purity of the sample of sodium carbonate. [2]
   (e) The measurement uncertainty in the use of each piece of apparatus is:
      Volumetric flask ± 0.30 cm³
      Pipette ± 0.060 cm³
      Burette ± 0.10 cm³ (±0.05 cm³ for each reading)
      (i) Calculate the overall total percentage uncertainty in the experiment. [4]
      (ii) The manufacturers of the sodium carbonate claim that its degree of purity is 95% ± 2%.
         Are the manufacturers justified in their claims? [1]
         **[Total: 12]**

# TOPIC 6

# Organic chemistry

## Introduction

Along with physical chemistry and inorganic chemistry, organic chemistry is one of the traditional branches of chemistry. Students of biology will understand its importance because most of the compounds in this topic are found in, or are formed in, plants and animals, including the human body. Many aspects of our lives have been revolutionised by the production of new organic compounds, for example:

- New polymers with special properties
- More effective drugs to treat diseases
- The ongoing search for new antibiotics
- Sustainable fuels to replace fossil fuels.

Fertilisers and pesticides have increased crop yields to feed the world's growing population, but this is an example of where the application of the knowledge of chemistry has caused unforeseen problems. Some people prefer food grown naturally, without the use of man-made chemicals – ironically these foods are often described as 'organic'!

In this topic, you will learn about some of the reactions of organic compounds, starting with simple hydrocarbons, then moving on to compounds containing oxygen (alcohols) and halogens (halogenoalkanes). Alcohols have many uses, including as solvents and fuels as well as in drinks. Halogenoalkanes are used in making some polymers. Others have been used in refrigeration, but some were found to be responsible for damaging the atmospheric ozone layer. Since then, chemists have developed replacements that do not damage the ozone layer.

### All the maths you need

- Use ratios to construct and balance equations
- Represent chemical structures using angles and shapes in 2D and 3D structures

## What have I studied before?
- The names of simple organic compounds
- Homologous series and general formula
- The oxidation of ethanol
- Calculation of empirical and molecular formulae
- Representing organic compounds by structural formulae

## What will I study later?
- Optical isomerism (A level)
- Reactions of carbonyl and carboxyl compounds (A level)
- The type of bonding in benzene and other aromatic compounds (A level)
- Condensation polymers (A level)
- Nitrogen-containing compounds, including amino acids and proteins (A level)
- Planning reaction schemes to prepare organic compounds (A level)
- More practical techniques for preparing and purifying organic compounds (A level)

## What will I study in this topic?
- Using different types of formulae to represent organic compounds
- Structural isomerism and stereoisomerism
- Problems caused by the combustion of fuels, and solutions to these problems
- Using reaction mechanisms to understand how organic reactions occur
- The formation of polymers and dealing with polymer waste
- Practical techniques used to prepare and purify organic compounds

# 6.1 1 What is organic chemistry?

**By the end of this section, you should be able to...**

- understand that a hydrocarbon is a compound of hydrogen and carbon only

**fig A** Friedrich Wöhler was a German chemistry professor and an early pioneer in the field of organic chemistry. He is best remembered for making urea (an organic compound) starting only from inorganic compounds.

## Early days

Since the 1800s, our knowledge and understanding of chemistry has grown rapidly. To make sense of all of this knowledge, chemists divided chemistry into three main categories: inorganic, organic and physical chemistry. Each category is equally important. There are millions of different compounds in existence, and the vast majority are organic compounds. So, what is organic chemistry?

Today, the word 'organic' has a very different meaning in everyday life, and is often applied to farming and food. Back in the 1800s, people believed that there was something special about some substances – that they were only made in plants or animals. One example is the compound called urea, which is present in human urine. People used to believe that it could only be produced in the human body. In 1828, the German chemist Friedrich Wöhler discovered that urea could be made by heating a compound (ammonium cyanate) that was not organic. This meant that the idea of organic compounds only coming from living things was no longer correct.

## Hydrocarbons

The main feature of an organic compound is that it contains carbon. Almost all of these compounds also contain hydrogen. Some of the most important compounds contain elements such as nitrogen and oxygen. In this section, we look at some of the large numbers of compounds that contain only carbon and hydrogen. These are called **hydrocarbons**.

If an organic compound contains other elements as well as carbon and hydrogen, then it is not a hydrocarbon. For example, many foods contain a sugar called sucrose. Sucrose contains carbon and hydrogen, but also oxygen, so it is not a hydrocarbon. It is an example of a carbohydrate – the -ate ending shows that it contains oxygen.

**fig B** Work still goes on to make new organic compounds.

## Saturated or unsaturated?

Although there are many thousands of different hydrocarbons, most of them are classed as **saturated** or **unsaturated**. Like many chemical terms, these words have a very different meaning in everyday life. Someone who has been caught in a heavy rain shower may say that his or her clothes are saturated, which means that they have absorbed as much water as they possibly can.

# Introduction to organic chemistry 6.1

In organic chemistry, these terms have nothing to do with water, although there is a connection. A hydrocarbon that is saturated contains as much hydrogen as possible, which depends on the number of carbon atoms in the molecule. If a hydrocarbon has fewer hydrogen atoms than the maximum, then it is not saturated – we say it is unsaturated.

The formula of the simplest hydrocarbon, containing only one carbon atom, is:

$$\begin{array}{c} H \\ | \\ H-C-H \\ | \\ H \end{array}$$

When a hydrocarbon contains two carbon atoms, there is a maximum of six hydrogen atoms.

There is a hydrocarbon that contains two carbon atoms and six hydrogen atoms, but also one that contains two carbon atoms and only four hydrogen atoms. The formulae of these hydrocarbons are:

saturated        unsaturated

You can see that in all three examples, each carbon atom has four bonds to other atoms. This is a general rule for organic compounds – in most cases, every carbon atom has four bonds.

The difference between a saturated hydrocarbon and an unsaturated hydrocarbon is to do with whether there is room, or not, for more hydrogen atoms.

- If there is no room, then the hydrocarbon is saturated.
- If there is room, then the hydrocarbon is unsaturated.

One easy way to decide whether a hydrocarbon is saturated or unsaturated is to look at structures like the ones above.

- If you can see two bonds between the same carbon atoms (a double bond), then the hydrocarbon is unsaturated.
- If there are only single bonds, then the hydrocarbon is saturated.

## Any complications?

You are not likely to meet examples like these in this book:

The first hydrocarbon has three bonds (a triple bond) between the same two carbon atoms. It is unsaturated.

The term **multiple bonds** includes double bonds and triple bonds.

The second hydrocarbon has three carbon atoms joined in what looks like a triangular arrangement (it is called a cyclic hydrocarbon), and is sometimes described as having a ring structure. It is saturated (there are only single bonds).

### Learning tip

Many hydrocarbons are represented by molecular formulae – the first three in this section can be shown as $CH_4$, $C_2H_6$ and $C_2H_4$. Make sure that you can decide whether these formulae and other similar formulae are saturated or unsaturated.

## Questions

1  A saturated hydrocarbon molecule contains four carbon atoms. How many hydrogen atoms does it have in a molecule?

2  Consider the compound with the formula $C_2F_4$.
   (a) Is it an organic compound? Explain your answer.
   (b) Is it saturated or unsaturated? Explain your answer.

### Key definitions

A **hydrocarbon** is a compound that contains only carbon and hydrogen atoms.

**Saturated** refers to a compound containing only single bonds.

**Unsaturated** refers to a compound containing one or more multiple bonds.

A **multiple bond** is two or more covalent bonds between two atoms.

# 6.1 2 Different types of formulae

**By the end of this section, you should be able to...**

- represent organic molecules using displayed formulae, molecular formulae, skeletal formulae, empirical formulae and structural formulae

## Using diagrams to refer to organic compounds

There are many millions of organic compounds, so it can be challenging to make it clear which ones we are referring to.

There are two main ways to refer to organic compounds. We can use:

- formulae
- names.

In this section we will consider how to refer to organic compounds using formulae.

## Displayed formulae

The formulae you saw in the previous section are all **displayed formulae** – they show (display) every atom and every bond separately. In many situations, these are the best type of formulae to use, but sometimes it is better to simplify them.

Consider the hydrocarbon with this displayed formula (its name is butane):

$$\begin{array}{c} H \quad H \quad H \quad H \\ | \quad | \quad | \quad | \\ H-C-C-C-C-H \\ | \quad | \quad | \quad | \\ H \quad H \quad H \quad H \end{array}$$

## Structural formulae

One way to simplify this displayed formula is to group all the atoms joined to a particular carbon atom together. We can choose to show the bonds between the carbons, or we can leave them out. These are both **structural formulae** of butane:

$$CH_3-CH_2-CH_2-CH_3 \quad \text{and} \quad CH_3CH_2CH_2CH_3$$

## Skeletal formulae

Another way to represent a compound is by a **skeletal formula**. The word skeletal is connected with the word skeleton, which, as you know, shows only the bones in a human body.

A skeletal formula is a zig-zag line that shows only the bonds between the carbon atoms. Every change in direction and every ending means that there is a carbon atom (with as many hydrogen atoms as needed). Atoms other than carbon and hydrogen need to be shown.

This is the skeletal formula of butane:

The start and end both represent $CH_3$, and the two junctions between lines each represent $CH_2$.

## Molecular formulae

The displayed, structural and skeletal formulae above show the structures of the molecules unambiguously. In other words, each formula represents only one compound. With a displayed formula, this is very clear. With a structural formula, you have to imagine how the atoms are joined together in groups such as $CH_2$ and $CH_3$, but that is very straightforward. With a skeletal formula, once you know the rules, you can be sure how every atom is arranged in the molecule.

Now consider the formula $C_3H_7Cl$. This is clearly not a displayed formula or a skeletal formula, but it is also not a structural formula. This is because, with three carbon atoms, the chlorine atom could be attached to the middle carbon atom or to either of the end carbon atoms. The formula $C_3H_7Cl$ actually represents two different compounds.

Formulae like this are called **molecular formulae** – they only show the numbers of each type of atom in the molecule, and not its structure. Of course, in very simple molecules such as $CH_3Cl$, the molecular formula can be used to work out the displayed, structural and skeletal formulae because there is only one way in which these five atoms can be joined together.

## Empirical formulae

Another type of formula is an **empirical formula**. This shows the compound like a molecular formula, but the numbers of each atom are in their simplest possible whole-number ratio. So, butane (molecular formula $C_4H_{10}$) has an empirical formula of $C_2H_5$.

In chemistry, the word empirical usually means 'as found from practical evidence'. You would normally work out this type of formula mathematically from the results of an experiment.

## Different types of formula for chloroethane

So far, we have only considered the different types of formula using a hydrocarbon as the example.

Consider an example containing a third element – chloroethane. The table shows its different types of formula.

| Type of formula | Formula |
|---|---|
| displayed formula | H—C(H)(H)—C(H)(H)—Cl |
| structural formula | $CH_3$—$CH_2$—Cl or $CH_3CH_2Cl$ |
| skeletal formula | ⌒Cl |
| molecular formula | $C_2H_5Cl$ |
| empirical formula | $C_2H_5Cl$ |

**table A** Different types of formula for chloroethane.

## Questions

1. The displayed formula of a compound is:

   H—C(H)(H)—C(Cl)(H)—C(H)(H)—H (with Cl on middle carbon)

   Use a table (like the one for chloroethane) to show its structural, skeletal, molecular and empirical formulae.

2. The skeletal formula of a compound is:

   ⌒⌒ (with a vertical line on the middle)

   Use a table (like the one for chloroethane) to show its displayed, structural, molecular and empirical formulae.

### Key definitions

A **displayed formula** shows every atom and every bond.
A **structural formula** shows (unambiguously) how the atoms are joined together.
A **skeletal formula** shows all the bonds between carbon atoms.
A **molecular formula** shows the actual numbers of each atom in the molecule.
An **empirical formula** shows the numbers of each atom in the simplest whole-number ratio.

### Learning tip

For some compounds, the numbers in the molecular formula cannot be simplified – this means that the molecular formula and the empirical formula are identical.

# 6.1  3  Functional groups and homologous series

**By the end of this section, you should be able to...**
- know what is meant by the terms 'homologous series' and 'functional group'
- represent organic molecules using general formulae
- recall the general characteristics of a homologous series

## Functional group

A **functional group** in a molecule is an atom or group of atoms that gives the compound some distinctive and predictable properties. For example, the functional group of atoms shown as COOH gives molecules containing this group a sour, acidic taste.

There are many organic compounds containing this group. Here are some examples:

$$HCOOH \quad CH_3COOH \quad CH_3CH_2COOH \quad CH_3CH_2CH_2COOH$$

## Homologous series

If you look at the formulae above, you can see that each formula has one more carbon atom and two more hydrogen atoms than the previous one – they differ by $CH_2$. These compounds are the first four members of what is called a **homologous series**. This is a set of compounds with the same functional group, similar chemical properties and physical properties that show a gradation (a gradual change from one to the next).

## Alkanes

The organic compounds that are mainly used as fuels are the alkanes (you will learn more about alkanes later in this book). They are not considered to contain a functional group, but otherwise they form a homologous series. The displayed formulae of some of them are:

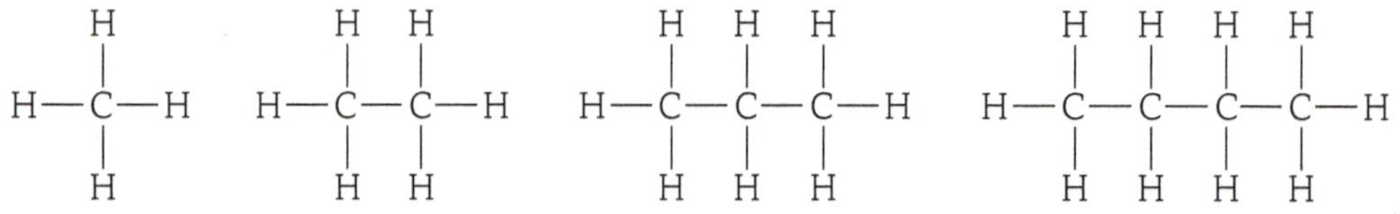

## General formulae

In the previous section we looked at five different types of formulae. Now we are going to look at another type of formula. For the compounds in a homologous series, we can use a general formula to represent all of them. This is done by using the letter n for the number of carbon atoms, excluding any in the functional group. For the compounds with formulae ending in COOH, the general formula is $C_nH_{2n+1}COOH$.

The table shows some of the homologous series in this book.

| Name | General formula | Example |
|---|---|---|
| alkane | $C_nH_{2n+2}$ | $CH_4$ |
| alkene | $C_nH_{2n}$ | $C_2H_4$ |
| halogenoalkane | $C_nH_{2n+1}X$ | $CH_3CH_2Br$ |
| alcohol | $C_nH_{2n+1}OH$ | $CH_3CH_2OH$ |

**table A** Examples of homologous series used in this book.

# Introduction to organic chemistry 6.1

## Properties of a homologous series

### Alkanes

We can use the alkanes to illustrate the similarity in chemical properties of a homologous series. For example, when alkanes are burned completely in air, they all form the same two products: carbon dioxide and water.

The commonest alkane is methane. The equation for its complete combustion is:

$$CH_4 + 2O_2 \rightarrow CO_2 + 2H_2O$$

### Alcohols

We can use the alcohols to illustrate the gradation in physical properties of a homologous series. For example, the boiling temperatures of the first four alcohols are shown in this table.

| Formula | Boiling temperature/°C |
|---|---|
| $CH_3OH$ | 65 |
| $CH_3CH_2OH$ | 79 |
| $CH_3CH_2CH_2OH$ | 97 |
| $CH_3CH_2CH_2CH_2OH$ | 117 |

You can see that as the number of carbon and hydrogen atoms increases, so does the boiling temperature.

### Learning tip

When looking at molecular models, remember that different elements are represented by different colours. The commonest are black for carbon, white for hydrogen and red for oxygen.

**fig A** Molecular models are very useful in organic chemistry. Both of these structures contain an oxygen atom (shown in red), but you can see that they belong to different homologous series.

## Questions

1. The equation for the complete combustion of propane is:

    $$C_3H_8 + 5O_2 \rightarrow 3CO_2 + 4H_2O$$

    What is the equation for the complete combustion of the alkane with five carbon atoms in a molecule?

2. The structural formula of a compound is $CH_3CH_2CHO$. What are the formulae of the two simpler compounds in the same homologous series?

### Key definitions

A **functional group** is an atom or group of atoms in a molecule that is responsible for its chemical reactions.

A **homologous series** is a family of compounds with the same functional group, which differ in formula by $CH_2$ from the next member.

171

# 6.1 4 Nomenclature

By the end of this section, you should be able to...

- name compounds using the rules of IUPAC nomenclature

## Why do we need rules for naming organic compounds?

As the number of known organic compounds has increased, it has become harder to continue to find new names for them. In the previous section, we referred to the simplest organic compound ($CH_4$) as methane, but it was originally known as marsh gas (because it was found in marshes, where it was formed by the decay of plants). Many other organic compounds were named in similar ways.

An organisation called the International Union of Pure and Applied Chemistry made some rules about how to name organic compounds. The organisation's name is usually abbreviated to IUPAC ('eye-you-pack'). These rules are often known as 'nomenclature'. The detailed rules needed for naming very complicated compounds are complex, but the simpler rules for the compounds described in this book are much easier to understand and apply.

## The simple rules of nomenclature

The table summarises the principles of naming organic compounds, including rules for **prefixes**, **suffixes** and **locants**.

| The part of the name | How to write it | Example |
|---|---|---|
| Number of carbon atoms | This is shown by using a letter code (usually 3 or 4 letters). | meth = one carbon atom |
| Prefixes Suffixes | The presence of atoms other than carbon and hydrogen is shown by adding other letters before or after the code for the number of carbon atoms. | bromo = an atom of bromine ol = a hydroxyl group (OH) |
| Multiplying prefixes | The presence of two or more identical groups is shown by using the prefixes di-, tri-, etc. | di = two |
| Locants | Where atoms and groups can have different positions in a molecule, numbers and hyphens are used to show their positions. The numbers represent the carbon atoms in the longest chain that the atoms and groups are attached to. | 2- = the atom or group is attached to the second carbon atom in the chain |

**table A** The principles of naming organic compounds.

The letter codes for the number of carbon atoms (up to five) are:

| Number | Code | Prefix |
|---|---|---|
| 1 | meth | methyl |
| 2 | eth | ethyl |
| 3 | prop | propyl |
| 4 | but | butyl |
| 5 | pent | pentyl |

# Introduction to organic chemistry 6.1

## Applying the rules to write names

### Alkanes

We can see how these rules work for some of the alkanes. The names of all the alkanes end in -ane.

| Structural formula | Name |
|---|---|
| $CH_3-CH_2-CH_3$ | propane |
| $CH_3-CH(CH_3)-CH_3$ | methylpropane<br>The locant 2- is not needed because if the methyl group below the horizontal chain were attached to one of the carbon atoms at either end of the chain, then there would be a sequence of four carbon atoms, and the compound would be named butane. |
| $CH_3-CH_2-CH(CH_3)-CH_2-CH_3$ | 3-methylpentane<br>The longest carbon chain contains five carbon atoms, and there is a methyl group attached to the third one. |
| $CH_3-CH_2-CH_2-CH(CH_3)-CH_3$ | 2-methylpentane<br>This is not 4-methylpentane because another rule is that the lowest locant numbers should be used. |
| $CH_3-CH(CH_3)-CH(CH_3)-CH_3$ | 2,3-dimethylbutane<br>This example shows the use of a comma between the locants when the attached groups are the same. |
| $CH_3-CH_2-CH(CH_2CH_3)-CH(CH_3)-CH_3$ | 3-ethyl-2-methylpentane<br>This example illustrates the rule about prefixes being in alphabetical order. Ethyl comes before methyl because e comes before m in the alphabet.<br>Notice also that it is not called 3-ethyl-4-methylpentane because these numbers (3 + 4) total more than the numbers 3 + 2 in the correct name. |

**table B** Naming alkanes from structural formulae using the rules of IUPAC nomenclature.

### Alcohols

Next, look at the alcohols. The rules for these are a bit different because the presence of the alcohol functional group is indicated by a suffix, not a prefix.

The names for all the alcohols end in -ol.

| Structural formula | Name |
|---|---|
| $CH_3-CH_2-OH$ | ethanol |
| $CH_3-CH_2-CH_2-OH$ | propan-1-ol<br>This time, the locant appears near the end of the name, but, as before, it appears directly before the letters representing the group. |
| $CH_3-CH_2-CH(OH)-CH(CH_3)$... with OH on C2 and CH_3 on C3 | 3-methylbutan-2-ol<br>This example illustrates the use of both a prefix and a suffix.<br>This is not called 2-methylbutan-3-ol the lowest number locant should be used for the suffix functional group (-2-ol not -3-ol). |
| $CH_3-C(CH_3)_2-CH_2-CH_2-OH$ | 3,3-dimethylbutan-1-ol<br>This example shows the use of prefixes, a suffix, locants and a comma!<br>As with alkanes, the name uses locants that add up to the smallest possible number. |

**table C** Naming alcohols from structural formulae using the rules of IUPAC nomenclature.

## Applying the rules to write formulae

Here are some examples of applying the rules the other way round, i.e. writing a structural formula for a compound from its IUPAC name.

| Name | Structural formula |
|---|---|
| dimethylpropane | prop indicates a chain of three carbon atoms<br>dimethyl indicates two methyl groups attached to the chain<br>No locants are needed, so the two methyl groups must be attached to the carbon chain in a way that does not make the longest carbon chain any longer than three carbon atoms.<br>So the structural formula is:<br><br>$$\begin{array}{c} CH_3 \\ | \\ CH_3-C-CH_3 \\ | \\ CH_3 \end{array}$$ |
| 3-methylbutan-1-ol | but indicates a chain of four carbon atoms<br>methyl indicates a $CH_3$ group<br>1- and 3- indicate attachments to the first and third carbon atoms in the chain.<br>So the structural formula is:<br><br>$$\begin{array}{cccc} CH_2-CH_2-CH-CH_3 \\ | \qquad\qquad\quad | \\ OH \qquad\qquad\ CH_3 \end{array}$$ |

**table D** Writing structural formulae from IUPAC names.

### Learning tip

When you practise writing names from structural formulae, always check that the code you have used for the longest carbon chain is for the longest chain. This may not be the one shown horizontally. When you practise writing structural formulae from names, always check that each carbon has only four bonds. Showing three or five bonds is a common error.

# Questions

1  Write IUPAC names for the compounds with these structural formulae.

$$\begin{array}{ccc} CH_2-CH-CH_3 & \quad & CH_2-CH-CH_3 \\ | \quad\ | & & | \quad\ \ | \\ Br \quad Br & & Br \quad OH \end{array}$$

2  Write structural formulae for the compounds with these IUPAC names.

2,2-dimethylpentane      2,3-dimethylbutan-2-ol

### Key definitions

A **prefix** is a set of letters written at the beginning of a name.
A **suffix** is a set of letters written at the end of a name.
A **locant** is a number used to indicate which carbon atom in the chain an atom or group is attached to.

# 6.1 5 Isomerism

By the end of this section, you should be able to...
- explain what isomerism is and how it arises
- explain the difference between structural isomerism and stereoisomerism

## Structural isomerism

Consider these two structures:

$$CH_3-CH_2-CH_2-CH_3 \qquad CH_3-CH-CH_3$$
$$\qquad\qquad\qquad\qquad\qquad\qquad\qquad |$$
$$\qquad\qquad\qquad\qquad\qquad\qquad\qquad CH_3$$

butane         methylpropane

You can see that they are different compounds because their names and structures are different. However, their molecular formulae are the same – they can both be represented by $C_4H_{10}$. These two compounds are simple examples of **structural isomers**. In other words, they have the same molecular formula but different structural formulae.

The molecular formula $C_4H_{10}$ represents only two possible structures, but more complicated molecular formulae can be represented by several possible structures. As the number of carbon atoms increases, the number of possible structures can be hundreds or thousands!

**fig A** By counting the atoms you can see that both structures have the molecular formula $C_4H_{10}$.

## Types of structural isomerism

There are two important types of structural isomerism.

### Chain isomerism

Chain isomerism refers to molecules with different carbon chains. Butane and methylpropane (shown above) are examples of chain isomers because their carbon chains are different.

### Position isomerism

Position isomerism refers to molecules with the same functional group attached in different positions on the same carbon chain. Propan-1-ol and propan-2-ol are simple examples of position isomerism:

$$CH_2-CH_2-CH_3 \qquad CH_3-CH-CH_3$$
$$|\qquad\qquad\qquad\qquad\qquad\qquad |$$
$$OH\qquad\qquad\qquad\qquad\qquad\; OH$$

propan-1-ol     propan-2-ol

They are examples of position isomers because the carbon chains are the same, but the two OH groups are attached to different carbon atoms in the chain.

You might see examples where both of these types of isomerism are present:

$$CH_2-CH_2-CH-CH_3 \atop \phantom{xx}| \phantom{xxxxxxx} | \atop \phantom{xx}OH \phantom{xxxxxx} CH_3$$

$$CH_3-CH_2-CH-CH_2-CH_3 \atop \phantom{xxxxxxxxx} | \atop \phantom{xxxxxxxxx} OH$$

## Stereoisomerism

You will recognise the 'stereo' part of this word from its use in music systems – it means that the sound comes from more than one speaker. In organic chemistry, the term refers to two different ways to arrange atoms or groups of atoms.

**Stereoisomers** have the same molecular formula and the same structural formula, but they still have a different arrangement! A good example is the alkene but-2-ene:

$$CH_3-CH=CH-CH_3$$

The name comes from four carbon atoms in a chain (but-), a carbon-carbon double bond (-ene), with 2- showing that the double bond comes after the second carbon atom in the chain.

### Geometric isomers

Unfortunately, showing the structure like this (with the atoms in a straight line) does not help us understand what stereoisomerism is. We need to show the bonds at angles of 120° to each other. Then you can see that there are two different arrangements:

The two $CH_3$ groups are further apart from each other in the first structure – they are across the molecule. We add the abbreviation *trans-* (Latin for 'across') to the beginning of the name to indicate this. Think of the word transatlantic, which means at opposite ends of the Atlantic Ocean.

In the second structure, the two $CH_3$ groups are still on opposite sides of the C=C bond, but they are both shown above the double bond and not at opposite ends of the molecule. We add the term *cis-* (Latin for 'on this side') to the name of this compound.

The two compounds are known as **geometric isomers**. This type of stereoisomerism is described as *cis-trans* isomerism or geometric isomerism.

This type of isomerism can exist in alkenes but not in alkanes. This is because there needs to be a C=C double bond for *cis-* and *trans-* isomers to occur. The presence of a C=C double bond leads to **restricted rotation**, so that the groups attached to the C=C carbon atoms cannot move around. In alkanes, which do not have any double bonds, carbon atoms and their attached hydrogen atoms can rotate freely, without restriction. This is much easier to understand if you have access to molecular models.

### How to tell if a compound is a stereoisomer

Look at this compound:

It is *not* a stereoisomer of either of the compounds above. When both $CH_3$ groups are on the same side of the double bond, then the carbon chain is different. There is no chain of four carbon atoms, and this compound has the name methylpropene: it is a structural isomer of the two types of but-2-ene.

## E-Z notation

There is a problem with the *cis-trans* notation – it only works with some compounds! Consider these two examples:

$$\begin{array}{c}H\\F\end{array}C=C\begin{array}{c}Cl\\Br\end{array} \qquad \begin{array}{c}H\\F\end{array}C=C\begin{array}{c}Br\\Cl\end{array}$$

Because there are four different groups, the idea of two identical groups being in a *cis-* or *trans-* arrangement cannot work because there are not two groups that are the same. Choosing names using the *E-Z* system is more complicated than using *cis-trans* notation, so break it down into steps.

| Step | What to do |
|---|---|
| 1 | Work out the part of the name that can be used for both isomers using the normal nomenclature rules. In this example, the name is 1-bromo-1-chloro-2-fluoroethene. |
| 2 | Use the priority rules to decide which of the two atoms on the left of the double bond has the higher priority. Priority is decided by which atom has the higher atomic number. You can check this if you are not sure by looking them up in the Periodic Table. In this case, H = 1 and F = 9, so fluorine has the higher priority. |
| 3 | Do the same as in Step 2 for the two atoms on the right of the double bond. In this case, Cl = 17 and Br = 35, so bromine has the higher priority. |
| 4 | Now decide whether the two atoms with the higher priorities from steps 2 and 3 are at opposite ends of the molecule (the equivalent of *trans-*). If they are, then you have the *E*-isomer. If not, then you have the *Z*-isomer. |

**table A** Step-by-step guide to *E-Z* notation.

In the examples above, F and Br are at opposite ends of the second molecule, so it is the *E*-isomer, and the first molecule is the *Z*-isomer.

### Remembering the difference between *E* and *Z*

There are several ways to remember the difference between *E* and *Z*, and you might be able to work out a memorable way yourself. For now, try thinking about enemies, a word that begins with E. Enemies are opposite sides of the molecule! You don't need a separate way to remember Z, as the *Z*-isomer is the one that isn't the *E*-isomer. If you understand German, you might not need to remember this, because the letters E and Z come from German words that have opposite meanings: *E* = entgegen, or opposite, and *Z* = zusammen, or together.

The rules are a bit more complicated when there are groups instead of single atoms attached to the C=C bond, but we'll tackle them later in the book.

### Learning tip

Try writing structures like the ones in this section for all the possible structures with the molecular formula $C_4H_9Br$. You will probably write structures that look different on paper but are actually the same compound. A good way to be sure whether they are different is to work out their names. If the names are the same, then the isomers are the same.

## Questions

1. Draw a structure for each of the five isomers with the molecular formula $C_6H_{14}$.

2. There are three ways to attach two $CH_3$ groups and two Cl atoms to C=C. Draw these structures and use your knowledge of nomenclature and *cis-trans* isomerism to name them.

### Key definitions

**Structural isomers** are compounds with the same molecular formula but with different structural formulae.

**Stereoisomers** are compounds with the same structural formula (and the same molecular formula), but with the atoms or groups arranged differently in three dimensions.

**Geometric isomers** are compounds containing a C=C bond with atoms or groups attached at different positions.

**Restricted rotation** around a C=C bond fixes the position of the atoms or groups attached to the C=C atoms.

## 6.2 1 Alkanes from crude oil

By the end of this section, you should be able to...
- understand that alkane fuels are obtained from the fractional distillation, cracking and reforming of crude oil

### The need for fuels

The worldwide demand for energy is huge – and steadily rising. Most of this energy comes from burning fossil fuels, in the form of coal, crude oil and natural gas. Most compounds in crude oil and natural gas are alkanes.

**fig A** Petrol is just one product of crude oil, but is perhaps the best known.

In this section we will look at the three main processes used to convert crude oil into fuels. They are:

- fractional distillation
- cracking
- reforming.

These processes are used in oil refineries that are located all over the world. You will already know something about fractional distillation, but you may be less familiar with cracking and reforming.

### Fractional distillation

Crude oil is a complex mixture of compounds, mostly hydrocarbons. The composition of the mixture varies quite a lot depending on which part of the world the crude oil comes from. The process is sometimes called 'fractionation' because it involves converting the crude oil into a small number of fractions. The number of fractions varies between different refineries but is typically six. Fractionation is done in a distillation column.

**fig B** Several different processes take place in oil refineries. This distillation plant is just one small part of a large refinery.

### The process

The crude oil is first heated in a furnace, which turns most of it into vapour, which is then passed into the column near the bottom. There is a temperature gradient in the column: it is hotter near the bottom and cooler near the top.

As the vapour passes up the column through a series of bubble caps, different fractions condense at different heights in the column, depending on the boiling temperature range of the molecules in the fraction.

- Near the bottom of the column, the fractions contain larger molecules with longer chains and higher boiling temperatures.

- Near the top of the column, the fractions contain smaller molecules with shorter chains and lower boiling temperatures.
- Some of the hydrocarbons in crude oil are dissolved gases, and they rise to the top of the column without condensing.

Some fractions still contain many different compounds, so they may undergo further fractional distillation separately.

## Cracking

The world has fewer uses for longer-chain hydrocarbons so there is a surplus of these. The demand for shorter-chain hydrocarbons is much higher because they are better fuels and can be used to make other substances such as polymers. Unfortunately, there are not enough of these to satisfy the demand. The solution is to convert the longer chains into shorter chains, which is what happens in cracking.

## The process

Cracking is done by passing the hydrocarbons in the heavier fractions through a heated catalyst, usually of zeolite, which is a compound of aluminium, silicon and oxygen. This causes larger molecules to break up into smaller ones. From one large molecule, at least two smaller molecules are formed.

A good example is the cracking of decane into octane and ethene:

decane $C_{10}H_{22}$

↓

octane $C_8H_{18}$

+

ethene $C_2H_4$

This is a good example because the two smaller molecules that are formed have familiar uses. Octane is one of the hydrocarbons in petrol and ethene is used to make polymers.

## Reforming

So far, we haven't mentioned one important point about the alkanes used as fuels. During the very rapid combustion that occurs in vehicle engines, not all hydrocarbons of the right size burn in the same way. Those with straight chains burn less efficiently than those with branched chains and those with rings (cyclic compounds). The process of reforming is used to convert straight-chain alkanes into branched-chain alkanes and cyclic hydrocarbons by heating them with a catalyst, usually platinum. This helps them to burn more smoothly in the engine.

## Examples

Here are some examples of reforming reactions, using skeletal formulae.

In the first one, pentane ($C_5H_{12}$) is converted into a cyclic alkane:

pentane → cyclopentane + $H_2$

In the second one, heptane ($C_7H_{16}$) is converted into methylbenzene, which is a cyclic hydrocarbon but not an alkane. You will learn the meaning of the circle inside the hexagon later.

heptane → methylbenzene + $4H_2$

In each example, hydrogen is formed. It is a useful by-product.

### Learning tip

Be careful not to confuse fractional distillation and cracking. Fractional distillation involves separating existing compounds, not making new ones in a chemical reaction. Cracking involves a chemical reaction in which new compounds are formed.

## Questions

1. One molecule of the alkane $C_{12}H_{26}$ is cracked to form two molecules of ethene and one molecule of a different alkane. What is the molecular formula of the alkane formed?

2. The products of a cracking reaction are two molecules of ethene and one molecule of pentane. What is the molecular formula of the alkane that is cracked?

### Key definitions

**Fractional distillation** is the process used to separate a liquid mixture into fractions by boiling and condensing.

**Cracking** is the breakdown of molecules into shorter ones by heating with a catalyst.

**Reforming** is the conversion of straight-chain hydrocarbons into branched-chain and cyclic hydrocarbons.

# 6.2  2  Alkanes as fuels

By the end of this section, you should be able to...
- know that pollutants, including carbon monoxide, oxides of nitrogen and sulfur, carbon particulates, and unburned hydrocarbons are formed during the combustion of alkane fuels
- understand some of the problems arising from pollutants from the combustion of fuels
- understand how catalytic converters solve some problems caused by pollutants

## The complete combustion of alkanes

As we mentioned earlier, alkanes can burn. They are burned in vast quantities to provide the world's energy. For example, propane is sold in containers at high pressure for use as a fuel in homes and when camping. The equation for its **complete combustion** is:

$$C_3H_8 + 5O_2 \rightarrow 3CO_2 + 4H_2O$$

## The products of combustion

One of the products of combustion is water, which is not a problem as it simply adds to the total quantity of global $H_2O$.

The other product is carbon dioxide. As you know, this is a greenhouse gas and its increasing production is considered to be responsible for global warming, climate change and other problems.

Unfortunately, other problems are caused by using alkanes as fuels. The water and carbon dioxide formed are not considered to be pollutants by most people, but some other compounds formed during the combustion of alkanes are definitely pollutants.

## Incomplete combustion

Sometimes the combustion of an alkane is incomplete because there is insufficient oxygen present, or because the combustion is very rapid. All of the hydrogen atoms in an alkane molecule are converted into water, but some of the carbon atoms can form gaseous carbon monoxide or solid carbon. These products can cause problems.

### Carbon

You can often see when **incomplete combustion** forms solid carbon – this can be seen as smoke in the air or soot on the burner. One example of an equation for a reaction in which carbon is formed is:

$$C_3H_8 + 4O_2 \rightarrow C + 2CO_2 + 4H_2O$$

Notice that in this reaction two of the carbon atoms in propane undergo complete combustion and one does not. Tiny particles of carbon in the atmosphere can be harmful, but there is another product of combustion that can be fatal.

### Carbon monoxide

Carbon monoxide is a toxic gas that causes the death of many people each year. It acts by preventing the transport of oxygen around the body. It is colourless and odourless, so people breathe it into their lungs without knowing, which is why it is sometimes described as 'the silent killer'. Here is an example of an equation for a reaction in which carbon monoxide is formed:

$$C_3H_8 + 4O_2 \rightarrow 2CO + CO_2 + 4H_2O$$

### Unburned hydrocarbons

The ultimate example of incomplete combustion is when the hydrocarbon does not burn at all! A small proportion of the hydrocarbons in a fuel are released into the atmosphere unchanged. They are known as unburned hydrocarbons (sometimes abbreviated to HC).

## Oxides of sulfur

Some of the molecules in crude oil contain atoms of sulfur, and these may not be removed by the fractional distillation, cracking or reforming processes. During the combustion of alkanes, these atoms of sulfur form sulfur dioxide, and then can react in the atmosphere to form sulfur trioxide. The equations for these reactions are:

$$S + O_2 \rightarrow SO_2 \quad \text{and} \quad 2SO_2 + O_2 \rightarrow 2SO_3$$

Both of these gases are acidic oxides. So, when they dissolve in water in the atmosphere, they form sulfurous acid and sulfuric acid:

$$SO_2 + H_2O \rightarrow H_2SO_3 \quad \text{and} \quad SO_3 + H_2O \rightarrow H_2SO_4$$

Both of these acids contribute to the formation of acid rain. This is responsible for a lot of environmental damage, including damage to aquatic life in lakes and rivers, and damage to crops and forests.

## Oxides of nitrogen

Although very few molecules used as alkane fuels contain atoms of nitrogen, their combustion occurs at very high temperatures. Under these conditions, especially around the spark plugs in cars, this very high temperature causes nitrogen molecules in the air to react with oxygen molecules. These reactions lead to the formation of what are collectively known as oxides of nitrogen. They are represented by the formula $NO_x$. There are several of these oxides, but the main ones are nitrogen monoxide (NO) and nitrogen dioxide ($NO_2$).

At very high temperatures, the main reaction is:

$$N_2 + O_2 \rightarrow 2NO$$

However, nitrogen monoxide can then react with more oxygen in the atmosphere as follows:

$$2NO + O_2 \rightarrow 2NO_2$$

Nitrogen dioxide is acidic and can dissolve in water in the atmosphere, forming nitrous acid and nitric acid:

$$2NO_2 + H_2O \rightarrow HNO_2 + HNO_3$$

Both of these acids contribute to the environmental damage in the same way as sulfurous acid and sulfuric acid.

**fig A** Air pollution is a growing problem in many cities.

## Catalytic converters to the rescue!

Cars and other road vehicles are responsible for a lot of air pollution. The widespread use of catalytic converters fitted to exhaust systems has made pollution less of a problem.

There are different types, but they all use small quantities of precious metals such as platinum, rhodium and palladium. These metals are spread thinly over a honeycomb mesh to increase the surface area for reaction (and also to save money!). One common type is known as a three-way catalyst because it can remove three different pollutants: carbon monoxide, unburned hydrocarbons and oxides of nitrogen.

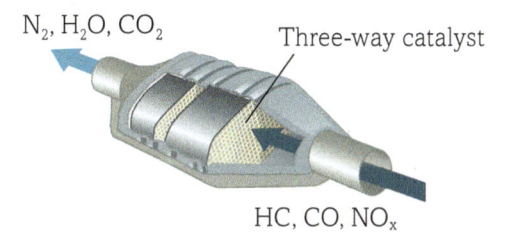

**fig B** A three-way catalytic converter.

As the exhaust gases from the engine pass through the catalytic converter, several reactions can occur.

Examples are the oxidation of the carbon monoxide and the oxidation of unburned hydrocarbons:

$$2CO + O_2 \rightarrow 2CO_2 \quad \text{and} \quad C_8H_{18} + 12.5O_2 \rightarrow 8CO_2 + 9H_2O$$

Here is another useful reaction that gets rid of two pollutants at the same time:

$$2NO + 2CO \rightarrow N_2 + 2CO_2$$

The catalysts currently used are not very good at removing sulfur compounds. The best way to prevent sulfur-based pollution is to remove the sulfur compounds from the fuel before the fuel is burned. This is done in some countries, where the resulting fuel is described as low sulfur or ultra-low sulfur fuel.

### Learning tip

Remember that even in the incomplete combustion of an alkane, all of the hydrogen atoms are completely oxidised to water.

## Questions

1. Summarise information about the products of combustion of alkanes, including names and whether complete or incomplete combustion was involved.

2. In a table, summarise the substances that react in a catalytic converter, and the products formed from them.

### Key definitions

**Complete combustion** means that all of the atoms in the fuel are fully oxidised.

**Incomplete combustion** means that some of the atoms in the fuel are not fully oxidised.

# 6.2 3 Alternative fuels

By the end of this section, you should be able to...
- understand what a biofuel is
- understand the use of alternative fuels in terms of a comparison with non-renewable fossil fuels
- outline the differences between biodiesel and bioalcohols

## The need for alternative fuels

There are serious concerns about relying on the combustion of fossil fuels to produce energy. We have already considered the pollution caused by the combustion of alkanes.

The other concerns are:
- the depletion of natural resources
- global warming and climate change.

Recently, there have been attempts to produce new fuels as alternatives to fossil fuels. Most of these fuels are described as **biofuels**, which means that they are obtained from living matter that has died recently, rather than having died many millions of years ago. A wide definition of biofuels would include wood, which has been used as a fuel for many centuries and is still important in some countries today.

The terms **renewable** and **non-renewable** are often used when discussing energy sources. Non-renewable usually refers to coal, oil and natural gas. Renewable sources include biofuels, but also sunlight, wind, waves and tides, and geothermal energy.

## Carbon neutrality

Fuels can be considered in terms of their carbon neutrality. Ideally, a fuel should be completely carbon neutral, although few are. The closer a fuel is to being carbon neutral, the better.

### What does carbon neutral mean?

'Carbon neutral' is a term used to represent the idea of carbon dioxide neutrality. For example, when a tree grows, it absorbs carbon dioxide from the atmosphere, and the carbon atoms become part of the structure of the tree. If the tree is cut down and the wood is burned, then carbon dioxide is formed during its combustion. If the amount of carbon dioxide formed in the combustion is the same as the amount absorbed during the tree's growth, then the wood used is described as carbon neutral. This is because, over the time period between the tree starting to grow and the use of its wood as a fuel, the amount of carbon dioxide in the atmosphere has not been altered by its combustion.

You might imagine that fossil fuels, such as those formed from trees, could be described as carbon neutral because they too absorbed carbon dioxide during their growth and form the same amount of carbon dioxide when they are burned. The reason that they are not considered to be carbon neutral is that the carbon dioxide was absorbed from the atmosphere millions of years ago, when the amount in the atmosphere was much higher. So, when fossil fuels are burned, this increases the amount of carbon dioxide in the atmosphere.

## Biofuels

In this section we will focus on two examples of biofuels that have been developed in recent decades. You might think that they are carbon neutral fuels because they are made from recently grown plants that have absorbed carbon dioxide from the atmosphere. However, this overlooks the fact that the plants have to be harvested, transported to a factory and processed in the factory, and the products transported to a point of sale. All of these stages involve the use of energy, much of which involves the formation of carbon dioxide. So, overall, the use of a biofuel involves forming more carbon dioxide than is absorbed. Even so, biofuels are closer to being carbon neutral than are fossil fuels.

### Biodiesel

Biodiesel is steadily growing in importance. The starting materials for making **biodiesel** are present in vegetable oils such as those obtained from rapeseed and sunflowers. Their chemical nature and processing is outside the scope of this book, but biodiesel is proving a very effective alternative to ordinary diesel. It can also be mixed with ordinary diesel.

**fig A** Biodiesel is increasingly available in filling stations in the USA.

### Bioalcohols

Currently, the commonest **bioalcohol** is bioethanol. For centuries, ethanol has been produced by the fermentation of sugars to produce alcoholic drinks. This involves the use of yeasts that contain enzymes, but there is an upper limit to the concentration of the ethanol in the solution. The ethanol has then to be separated from the much larger amount of water before it can be used as a fuel, and this separation requires energy.

Bioethanol now often refers to ethanol produced in a different way, involving bacteria rather than enzymes. It is now possible to use a much wider range of plants, and also plant waste. The upper limit to the amount of ethanol that can be obtained from a given amount of starting material is increasing, and is much higher than in traditional fermentation. In the United States, corn is the main source of ethanol used in cars.

**fig B** Not everyone agrees that it is a good idea for corn to be used as a fuel when it could be used to feed people.

## Comparing fuels
### Biofuels

The choice of alternative fuels is continually changing, as new sources of starting material and new processing methods are investigated. There are many factors to consider in any comparison, but for biofuels these include the following.

- Land use – how much land is used to grow the crop? Should the land be used for other purposes, especially to grow food to feed people?
- Yield – how much of a crop can be grown on a given piece of land, and how quickly does it grow? What percentage of the carbon and hydrogen atoms in the crop ends up in the fuel?
- Manufacture and transport – how much energy is used in growing (including any fertilisers), processing and transporting the crop?
- Carbon neutrality – how close to being carbon neutral is the fuel?

### Fossil fuels and biofuels

You should also be able to compare fossil fuels with biofuels. Here is one example of a comparison between fuels.

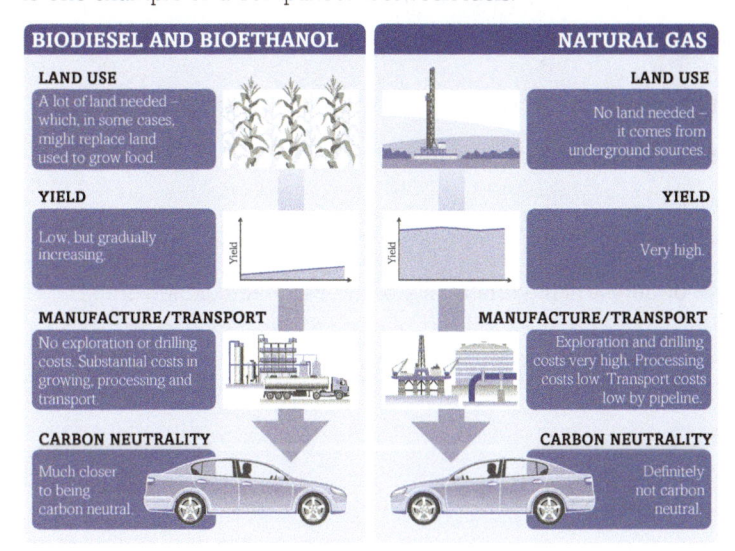

| BIODIESEL AND BIOETHANOL | NATURAL GAS |
|---|---|
| **LAND USE** A lot of land needed – which, in some cases, might replace land used to grow food. | **LAND USE** No land needed – it comes from underground sources. |
| **YIELD** Low, but gradually increasing. | **YIELD** Very high. |
| **MANUFACTURE/TRANSPORT** No exploration or drilling costs. Substantial costs in growing, processing and transport. | **MANUFACTURE/TRANSPORT** Exploration and drilling costs very high. Processing costs low. Transport costs low by pipeline. |
| **CARBON NEUTRALITY** Much closer to being carbon neutral. | **CARBON NEUTRALITY** Definitely not carbon neutral. |

**fig C** Comparing natural gas with biodiesel and bioethanol.

### Learning tip
Focus on the bigger picture of comparing fuels, not on the chemical reactions that occur in their manufacture.

## Questions

1. No carbon dioxide is formed when hydrogen is burned. Suggest why hydrogen is not a carbon neutral fuel.

2. Summarise reasons why a biofuel may not be carbon neutral.

### Key definitions

**Biofuels** are fuels obtained from living matter that has died recently.
**Renewable** energy sources use sources that can be continuously replaced.
**Non-renewable** energy sources are not being replenished, except over geological timescales.
**Biodiesel** is a fuel made from vegetable oils obtained from plants.
**Bioalcohols** are fuels made from plant matter, often using enzymes or bacteria.

# 6.2  4  Substitution reactions of alkanes

By the end of this section, you should be able to...

- know what a radical is and understand that homolytic fission of a covalent bond results in the formation of radicals
- understand the reactions of alkanes with halogens
- understand the limitations of the use of radical substitution reactions in the synthesis of organic molecules, in terms of further substitution reactions and the formation of a mixture of products

## What is a substitution reaction?

You already know that the most common use of alkanes is as fuels. Combustion reactions are very important in producing energy, but are not very interesting from a chemist's point of view. This section will give you a chance to increase your understanding of other reactions in organic chemistry.

Alkanes, apart from readily undergoing combustion, are fairly unreactive because they contain only carbon and hydrogen atoms and only single bonds. These bonds are also not very polar and so do not undergo reactions with substances that are considered to be very reactive, such as acids and alkalis and reactive metals.

There is a type of reaction that alkanes undergo – called a **substitution reaction** – which we will now look at in detail. Here is the equation for a reaction of the simplest alkane, methane:

$$CH_4 + Cl_2 \rightarrow CH_3Cl + HCl$$

You can see from the equation that one of the hydrogen atoms in methane has been replaced (or substituted) by an atom of chlorine. The reaction can be described as chlorination or, in general (if another halogen were used), halogenation.

## Mechanisms

As you study organic chemistry more thoroughly, you will come across reactions that have been carefully studied, and for which there are explanations of exactly how they occur. The equation shown above for the reaction of methane with chlorine only shows the formulae of the reactants and products. It does not show how or why the reaction occurs.

A **mechanism** tries to explain the actual changes that occur during a reaction, especially in the bonding between the atoms. A mechanism is a sequence of two or more steps, each one represented by an equation, that shows how a reaction takes place. Some mechanisms that you will meet later in this book involve the use of curly arrows, but we will not use them in this section.

## The chlorination of methane

When methane is just mixed with chlorine, no reaction occurs. If the temperature is increased, a reaction eventually occurs. However, the reaction will occur at room temperature if the mixture is exposed to ultraviolet radiation (or sunlight). We know that alkanes are not affected by ultraviolet radiation, but that ultraviolet radiation can affect chlorine. So, what happens in this reaction?

### Step 1

Ultraviolet radiation breaks the chlorine molecule into chlorine atoms. As the bond in the chlorine molecule consists of a shared pair of electrons, which are equally shared between the two atoms, then each chlorine atom takes one electron from the shared pair. This kind of bond breaking is called **homolytic fission**. 'Homo' indicates 'the same', 'lytic' indicates 'breaking down' and 'fission' is just another name for breaking.

This can be represented in an equation:

$$Cl_2 \rightarrow Cl\bullet + Cl\bullet$$

The dots on the products each represent an unpaired electron. The formula $Cl\bullet$ does not represent an ion or a molecule, and the term **radical** (sometimes free radical) is used for it. A radical is a species with an unpaired electron. 'Species' is a term used for any substance that can be represented by a formula, and includes atoms, molecules, ions and radicals.

Notice that the equation involves one molecule forming two radicals. This type of reaction is called **initiation**, which means it starts the sequence of steps that forms the overall reaction.

### Step 2

Chlorine radicals are very reactive species and when they collide with methane molecules they react by removing a hydrogen atom. An equation for this process is:

$$Cl\bullet + CH_4 \rightarrow HCl + CH_3\bullet$$

Notice that $CH_3\bullet$ is formed. This is a methyl radical and, like $Cl\bullet$, it is also very reactive. It can then react with chlorine molecules as follows:

$$CH_3\bullet + Cl_2 \rightarrow CH_3Cl + Cl\bullet$$

Notice that these two equations both involve one radical reacting with one molecule, and that the products are also one radical and one molecule. This type of reaction is called **propagation**, which means that the two steps considered together result in the conversion of $CH_4$ into the product $CH_3Cl$.

## Step 3

With all these radicals being formed, it is likely that two radicals will collide with each other. When this happens, they react to form a molecule, as the two unpaired electrons are shared to form a covalent bond. As there are two different radicals available, this means that there are three possibilities:

$Cl\bullet + Cl\bullet \rightarrow Cl_2$

$Cl\bullet + CH_3\bullet \rightarrow CH_3Cl$

$CH_3\bullet + CH_3\bullet \rightarrow C_2H_6$

These three equations all involve two radicals reacting with each other to form one molecule. This type of reaction is called **termination**, which means that the sequence of reactions comes to an end because two reactive species are converted into unreactive species.

## Further substitution reactions

You now know that a hydrogen atom in methane can be replaced by a chlorine atom in a substitution reaction. You can also see that the product chloromethane ($CH_3Cl$) still contains hydrogen atoms. These three hydrogen atoms can also be replaced, one by one, by chlorine atoms in similar substitution reactions.

It is not easy to prevent these further substitution reactions from occurring. As well as the formation of chloromethane, these other reactions occur and other products are formed:

the formation of dichloromethane     $CH_3Cl + Cl_2 \rightarrow CH_2Cl_2 + HCl$

the formation of trichloromethane     $CH_2Cl_2 + Cl_2 \rightarrow CHCl_3 + HCl$

the formation of tetrachloromethane   $CHCl_3 + Cl_2 \rightarrow CCl_4 + HCl$

Each of these overall reactions can be represented by the same sequence of initiation, propagation and termination steps as for the formation of chloromethane.

The likelihood of these further reactions occurring means that this is not a good method for the preparation of chloromethane. The yield will be low because of these further reactions, and because the products have to be separated.

**fig A** Free radicals are important for understanding many chemical reactions, but free radicals in the human body can be harmful.

## Questions

1. Write the six equations for the mechanism of the reaction between methane and bromine.

2. Classify each of these reactions as initiation, propagation or termination. Explain your choice in each case.
   (a) $C_2H_5\bullet + CH_3\bullet \rightarrow C_3H_8$
   (b) $2O\bullet \rightarrow O_2$
   (c) $F\bullet + CH_4 \rightarrow HF + CH_3\bullet$

### Learning tip

It helps to understand radical mechanisms if you focus on the three types of reaction. These are:

- inititation – one molecule becomes two radicals
- propagation – a molecule and a radical become a different radical and molecule, and there are two reactions in this step
- termination – two radicals become one molecule.

### Key definitions

A **substitution reaction** is one in which an atom or group is replaced by another atom or group.

A **mechanism** is the sequence of steps in an overall reaction. Each step shows what happens to the electrons involved in bond breaking or bond formation.

**Homolytic fission** is the breaking of a covalent bond where each of the bonding electrons leaves with one species, forming a radical.

A **radical** is a species that contains an unpaired electron.

An **initiation** step involves the formation of radicals, usually as a result of bond breaking caused by ultraviolet radiation.

The **propagation** steps are the two steps that, when repeated many times, convert the starting materials into the products of a reaction.

A **termination** step involves the formation of a molecule from two radicals.

# 6.2　5　Alkenes and their bonding

By the end of this section, you should be able to...

- know the general formula for alkenes
- know that alkenes and cycloalkenes are unsaturated hydrocarbons
- understand the bonding in alkenes in terms of sigma and pi bonds

## What are alkenes?

You have already discovered a lot of information about alkanes, but so far you have only come across alkenes as examples used to illustrate nomenclature and isomerism.

The main difference between alkanes and alkenes is that alkanes contain only single bonds, but alkenes contain at least one C=C double bond, so they are unsaturated. Alkenes are much less common than alkanes, but they can be made from alkanes in cracking reactions.

## General formula for alkenes

The structures and names of some common alkenes are shown in the table.

| Structure | Name |
|---|---|
| $CH_2=CH_2$ | ethene |
| $CH_2=CH-CH_3$ | propene |
| $CH_2=CH-CH_2-CH_3$ | but-1-ene |
| H₃C\C=C/H with CH₃ below (cis) | cis-but-2-ene |
| H\C=C/CH₃ with CH₃ and H (trans) | trans-but-2-ene |
| $CH_2=C-CH_3$ with $CH_3$ branch | methylpropene |

**table A** The structures and names of some common alkenes.

If you count the number of carbon and hydrogen atoms in each structure, you can see that there are twice as many hydrogen atoms as carbon atoms. This means that the general formula for the alkene homologous series is $C_nH_{2n}$.

When drawing the structures of alkenes, you normally show the bonds at angles of 120°. For example, ethene would be shown as:

$$\begin{array}{c} H \\ \phantom{H}\diagdown \\ \phantom{HH}C=C \\ \phantom{H}\diagup \\ H \end{array} \begin{array}{c} H \\ \diagup\phantom{H} \\ \phantom{HH} \\ \diagdown\phantom{H} \\ H \end{array}$$

Just as some alkanes are cyclic, so are some alkenes. A common cyclic alkene is cyclohexene, whose structure is:

(structure of cyclohexene)

Note that cyclohexene (and other cyclic alkenes) does not have the same general formula as non-cyclic alkenes. The molecular formula of cyclohexene is $C_6H_{10}$. So compared to hexene, it has two fewer hydrogen atoms.

What makes alkenes more interesting than alkanes is the C=C double bond. This makes them more reactive than alkanes and so they can be used in many more useful reactions. Before we look at these reactions, consider exactly what a C=C double bond is.

## What is a C=C double bond?

In some ways, the use of the symbol C=C to represent a double bond is very useful – it makes writing the structures of alkenes straightforward. However, it is somewhat misleading, because it implies that the two bonds between the carbon atoms are the same. They are not.

In an alkene molecule, both carbon atoms in the C=C double bond are joined to only three other atoms (in alkanes, it is four other atoms). You may remember from the sections about bonding that:

- electrons can exist in s orbitals and p orbitals
- pairs of electrons around an atom can be represented by a balloon shape.

## Sigma bonds

All of the single covalent bonds you have met so far involve the merging or overlapping of the orbitals of two different atoms. They may involve the overlapping of two s orbitals, or one s orbital and one p orbital, or two p orbitals. All of these bonds are represented by a line. So the covalent bond between the two hydrogen atoms in a hydrogen molecule can be shown as H—H. All of these types of bond can be referred to as **sigma bonds** ($\sigma$-bonds). When these sigma bonds are formed between carbon atoms in an alkene, their formation is sometimes described as formation by axial overlap, or end-on overlap.

## Pi bonds

Now, consider what happens when the three sigma bonds around each carbon atom are formed in ethene. After they are formed, each carbon atom has one electron in a p orbital that has not been used in bond formation. These p orbitals are parallel to each other and are not able to overlap in the same way as in the formation of sigma bonds. When they do overlap, this results in the formation of two regions of negative charge above and below the C—C sigma bond. This type of bond is formed by the sideways overlap of orbitals, and bonds of this type are referred to as **pi bonds** ($\pi$-bonds). Each of these regions of negative charge contains one electron, so together they make up a second shared pair of electrons between the two carbon atoms.

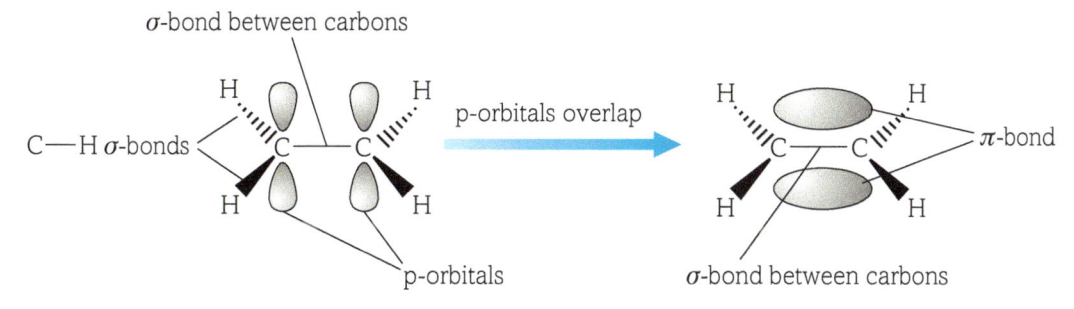

The diagram shows that these electrons seem to be further away from the carbon atoms than the electrons in the sigma bonds are. This means that they can be thought of as less under the control of the carbon atoms and so more available for reactions. We will look at these reactions in the next section.

### Learning tip
You can use the C=C symbol to represent double bonds in molecules when showing their structures. If you are explaining their reactions, it is better to refer to the separate sigma and pi bonds in C=C.

# Questions

1. The general formula for the alkenes is $C_nH_{2n}$.

   Why does cyclohexene (a cyclic hydrocarbon with five C—C single bonds and one C=C double bond) not have this general formula?

2. Alkenes are not used as fuels because they are more valuable for other purposes, but they do burn very well. Write an equation for the complete combustion of propene.

### Key definitions
**Sigma bonds** are covalent bonds formed when electron orbitals overlap axially (end-on).
**Pi bonds** are covalent bonds formed when electron orbitals overlap sideways.

# 6.2 6 Addition reactions of alkenes

By the end of this section, you should be able to...

- know why addition reactions of alkenes occur
- know the qualitative test for a C=C double bond using bromine water
- understand the addition reactions of alkenes with hydrogen, halogens, hydrogen halides, steam and potassium manganate(VII)

## Why do addition reactions occur?

In the previous section we learned that the C=C double bond in an alkene is made up of two different single bonds: a sigma bond and a pi bond. Because the sigma bond electrons are more tightly held between the two carbon atoms, a sigma bond is stronger than a pi bond. So a double bond is stronger than a single bond, but it is not twice as strong.

Most reactions of alkenes involve the double bond becoming a single bond. In these reactions, the sigma bond remains unchanged, but the pi-bond electrons are used to form new bonds with an attacking molecule. This reaction forms a product that is saturated. As the product contains sigma bonds, and not pi bonds, the bonds in the product are stronger, and so the product is more stable.

The equation for a typical addition reaction of an alkene, between ethene and bromine, is:

$$C_2H_4 + Br_2 \rightarrow C_2H_4Br_2$$

This reaction is used as a chemical test for the presence of C=C in a compound, because the products of these reactions are colourless. So, when this reaction occurs, the colour of bromine disappears. We often say that the bromine is decolorised. The bromine is normally used as an aqueous solution (often called bromine water), but the result is still decolorisation.

**fig A** The test for a C=C double bond. The tube on the left shows coloured bromine water with a layer of an organic compound on top. The tube on the right shows the mixture after shaking and leaving to settle. The bromine has been decolorised by the C=C bond.

You can see why this is called an **addition reaction**. Two molecules become one molecule. However, this equation does not show the mechanism of the reaction, i.e. how it occurs, in terms of the movement of electrons.

## Hydrogenation

Hydrogenation is an addition reaction in which hydrogen is added to an alkene. The simplest example is the **hydrogenation** of ethene:

$$\text{CH}_2=\text{CH}_2 + \text{H}_2 \xrightarrow{\text{Ni}} \text{CH}_3-\text{CH}_3$$

This reaction forms ethane, which is an alkane, and is done using heat and a nickel catalyst. You might be wondering why we would want to convert a useful alkene into an alkane, which has few uses except as a fuel. In fact, this particular reaction would never be done.

### Manufacture of margarine

This type of reaction is extensively used in industry to manufacture margarine. Naturally occurring vegetable oils are unsaturated and so contain C=C double bonds. When these react with hydrogen, some of the C=C double bonds become C—C single bonds. This process changes the properties of the vegetable oil and converts it into a solid: this is margarine.

There is much concern about fats in the human diet, and many people consider that monounsaturated fats (one C=C double bond per molecule) and polyunsaturated fats (two or more C=C double bonds per molecule) are better than saturated fats (no C=C double bonds):

$$\text{H}-\text{C}_3\text{H}_6-\text{CH}=\text{CH}-\text{C}_5\text{H}_{10}-\text{CH}=\text{CH}-\text{C}_{11}\text{H}_{22}-\text{COOH}$$

## Halogenation

Reactions between alkenes and bromine are examples of **halogenation**. The products of these reactions are dihalogenoalkanes. Reactions with chlorine are examples of chlorination (forming dichloroalkanes), and so on. Here are some examples of halogenation reactions:

$$\text{CH}_2=\text{CH}_2 + \text{Br}_2 \longrightarrow \text{CH}_2\text{Br}-\text{CH}_2\text{Br}$$
1,2-dibromoethane

$$\text{CH}_2=\text{CH}_2 + \text{Cl}_2 \longrightarrow \text{CH}_2\text{Cl}-\text{CH}_2\text{Cl}$$
1,2-dichloroethane

## Hydration

Hydration should not be confused with hydrogenation! **Hydration** means adding water, but you should consider it as adding H and OH to the two atoms in a C=C double bond. This reaction is usually done by heating the alkene with steam and passing the mixture over a catalyst of phosphoric acid. The reaction with ethene can be represented as:

$$\underset{\text{ethene}}{\text{CH}_2=\text{CH}_2} + \underset{\text{steam}}{\text{H}_2\text{O}} \xrightarrow{\text{H}_3\text{PO}_4} \underset{\text{ethanol}}{\text{CH}_3-\text{CH}_2\text{OH}}$$

Unlike the hydrogenation of ethene, the hydration of ethene forms ethanol, which is a useful product. This reaction and other similar ones used to make propanol are extensively used in industry.

## Addition of hydrogen halides

Another example of an addition reaction is the addition of a hydrogen halide (often hydrogen bromide or hydrogen chloride) to form a halogenoalkane (more specifically, a bromoalkane or a chloroalkane). Here is one example:

$$CH_2=CH_2 + H-Br \rightarrow CH_3-CH_2Br$$

The product is bromoethane. This reaction looks a lot like the one between ethene and bromine, but it cannot be used as a test for C=C because the reactants and the product are all colourless, so there is no colour change to observe.

ethene + HBr → bromoethane

## Oxidation to diols

This heading suggests a different type of reaction to the previous ones. In fact, the reaction involves both addition and oxidation. A **diol** is a compound containing two OH (alcohol) groups.

The oxidising agent is potassium manganate(VII) in acid conditions (usually dilute sulfuric acid). Although you do not need to know the full details of how this reaction occurs, you can think of the reaction as oxidation followed by addition. The potassium manganate(VII) provides an oxygen atom (oxidation) and the water in the solution provides another oxygen atom and two hydrogen atoms, so there is the addition of two OH groups across the double bond.

The equation for the reaction of ethene can be represented like this:

$$CH_2=CH_2 + [O] + H_2O \rightarrow CH_2OH-CH_2OH$$

The symbol [O] represents the oxygen supplied by the oxidising agent. You do not need to show the potassium manganate(VII) in the equation or know how it supplies the oxygen for the oxidation. The product is ethane-1,2-diol.

During the reaction the colour of the potassium manganate(VII) solution changes from purple to colourless. This colour change means that this reaction can be used like bromine to distinguish alkenes from alkanes (alkanes do not have double bonds and so are not oxidised in this way).

### Learning tip
Make sure you do not confuse hydrogenation with hydration.

## Questions

1. Write equations for the reactions between but-2-ene and:
   (a) hydrogen
   (b) bromine
   (c) hydrogen chloride.

2. Use IUPAC rules to write the names of the products of each of these reactions.

### Key definitions

An **addition reaction** is a reaction in which two molecules combine to form one molecule.
**Hydrogenation** involves the addition of hydrogen.
**Halogenation** involves the addition of a halogen.
**Hydration** involves the addition of water (or steam).
A **diol** is a compound containing two OH (alcohol) groups.

## 6.2 / 7 The mechanisms of addition reactions

**By the end of this section, you should be able to...**

- know what is meant by the term electrophile
- understand that heterolytic bond fission of a covalent bond results in the formation of ions
- understand the mechanism of the electrophilic addition reactions between alkenes and halogens, hydrogen halides and other binary compounds

## Background

We have already looked at reaction mechanisms with alkanes (initiation, propagation and termination). Now we can look in some detail at how addition reactions occur, which will involve the use of **curly arrows** to represent the movement of a pair of electrons.

Start with the reaction between ethene and hydrogen bromide, which we looked at in the previous section.

You already know that an alkene such as ethene has a pi bond, which is a region of high electron density (we could say that the molecule is electron rich around the C=C double bond). This makes it attractive to other species that are electron deficient, including molecules with polar bonds. Hydrogen bromide is a polar molecule because bromine is more electronegative than hydrogen, and can be shown with partial charges as:

$$\overset{\delta+}{H}-\overset{\delta-}{Br}$$

## Electrophile

When a hydrogen bromide molecule approaches an ethene molecule, the slightly positive end of the HBr molecule is attracted to the electrons in the pi bond in C=C. The HBr molecule is described as an **electrophile** when it does this. The 'phile' part of the term means 'liking' (a bibliophile likes books, a Francophile likes French things) and the 'electro' part refers to negative charge. So, an electrophile is a species that is attracted to negative charge.

The curly arrows used in reactions of this type must either:

- start from a bond and move to an atom, or
- start from a lone pair of electrons and move to an atom.

## Electrophilic addition of hydrogen halides

The complete name of this reaction is **electrophilic addition**. It involves addition and it involves attack by an electrophile. This is the mechanism of the reaction between ethene and hydrogen bromide. Notice that the hydrogen bromide molecule breaks so that both electrons in the H—Br bond go to one atom (in this case, bromine, because it is more electronegative than hydrogen).

This kind of bond breaking is called **heterolytic fission** (compare this with homolytic fission in the substitution reactions of alkanes).

### Step 1

a carbocation is formed

In this step, two ions are formed. The positive ion has its charge on a carbon atom, so is known as a **carbocation**.

In these diagrams, the charges are sometimes shown in circles to avoid possible confusion between + signs used to separate reactants and products, and − signs that could be confused with covalent bonds. Using these circles is a good idea but is not necessary.

The bromide ion is shown with four lone pairs, which is quite correct, but often only the lone pair that moves (where the arrow starts from) is shown.

### Step 2

The two oppositely charged ions attract each other and react to form a new covalent bond.

bromoethane

## Electrophilic addition of halogens

This reaction is very similar to that of hydrogen halides. The only difference is that the attacking bromine molecule does not have a polar bond. However, as it approaches the C=C bond, the electrons in the pi bond repel the electrons in the Br—Br bond and induce (cause) the molecule to become polar. After that happens, the mechanism is just the same as for hydrogen bromide.

**Step 1**

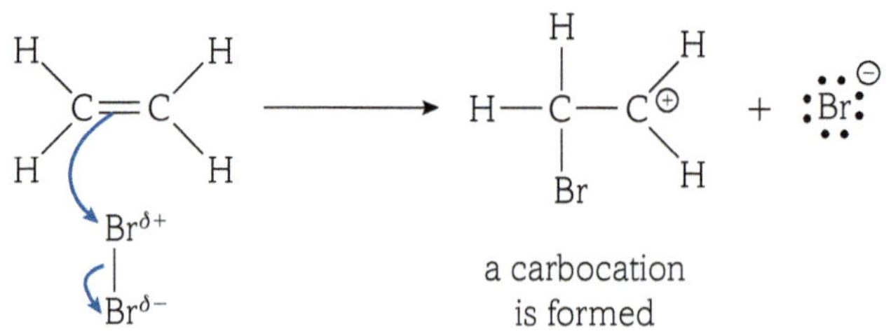

[Diagram: ethene with Br–Br attacking; curly arrow from C=C to Brδ+; forms H–C(H)(Br)–C⁺(H)(H) + :Br:⁻, labelled "a carbocation is formed"]

**Step 2**

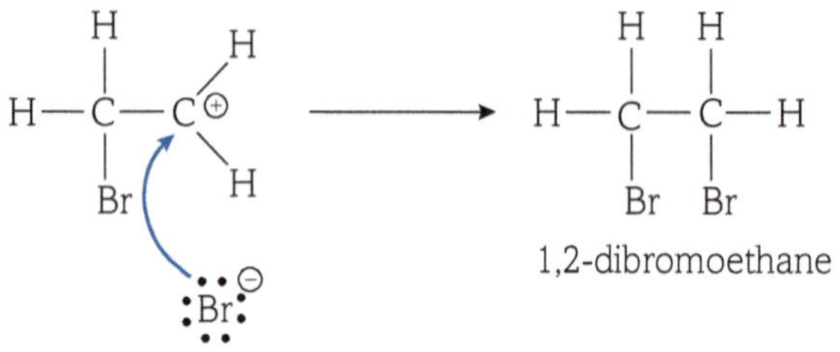

[Diagram: H–C(H)(Br)–C⁺(H)(H) attacked by :Br:⁻ → H–C(H)(Br)–C(H)(Br)–H, labelled "1,2-dibromoethane"]

## Unsymmetrical molecules

There is one more point to consider before leaving this section. When a molecule such as H—Br or Br—Br reacts with ethene in an addition reaction, there can only be one product. If both the alkene and the attacking molecule are unsymmetrical, then there are two possible products because the atoms in the attacking molecule can be added in two different places.

An unsymmetrical alkene is one in which the atoms on either side of the C=C bond are not the same. An unsymmetrical attacking molecule is one in which the atoms are different.

A good example is the reaction between propene and hydrogen bromide.

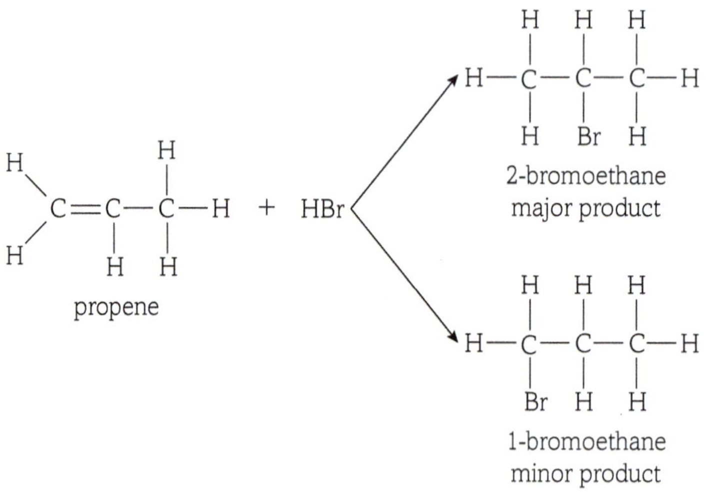

[Diagram: propene + HBr → 2-bromoethane (major product) and 1-bromoethane (minor product)]

In reactions of this type, one product is formed in greater amounts than the other – there is a major product and a minor product. Now we need to explain why, by considering the two possible carbocations formed in Step 1 of the reaction.

$$CH_3\overset{\oplus}{C}H-CH_3 \qquad \overset{\oplus}{C}H_2-CH_2-CH_3$$
structure A  structure B

Structure A shows the carbon atom with the positive charge joined to two alkyl groups, and is called a secondary carbocation.

In structure B, there is only one alkyl group joined to the carbon atom with the positive charge, so B is a primary carbocation.

A general principle to consider is that a carbocation in which the positive charge can be spread over more atoms is more stable than one in which there are fewer atoms available to spread the charge. Alkyl groups are **electron-releasing**, so when there are two of them, the positive charge is spread more than when there is only one.

In some reactions, there might be a tertiary carbocation (with three alkyl groups joined to the carbon atoms with the positive charge), and this would be more stable than a secondary carbocation.

## Summary

Reactions involving unsymmetrical molecules are complex, so here is a quick summary.

- Electrophilic addition reactions proceed via carbocations.
- Carbocations can be primary, secondary or tertiary.
- The stability of a carbocation is greatest for tertiary carbocations and least for primary carbocations.
- Carbocations are more stable when there are more electron-releasing alkyl groups attached to the carbon with the positive charge.
- The major product is formed from the more stable carbocation.

### Learning tip

When drawing organic structures, you already know that each carbon atom has four bonds to other atoms. This does not apply to carbocations – they only have three bonds around the carbon atom with the positive charge.

## Questions

1. Name the types of reactions that occur when:
   (a) alkanes react with halogens
   (b) alkenes react with halogens.

2. What is the name of the major product formed when but-1-ene reacts with hydrogen chloride?

### Key definitions

**Curly arrows** represent the movement of electron pairs.
An **electrophile** is a species that is attracted to a region of high electron density.
**Electrophilic addition** is a reaction in which two molecules form one molecule and the attacking molecule is an electrophile.
**Heterolytic fission** is the breaking of a covalent bond so that both bonding electrons are taken by one atom.
A **carbocation** is a positive ion in which the charge is shown on a carbon atom.
An **electron-releasing** group is one that pushes electrons towards the atom it is joined to.

# 6.2 8 Polymerisation reactions

By the end of this section, you should be able to...
- know that alkenes form polymers through addition polymerisation
- identify the repeat unit of an addition polymer given the monomer, and vice versa

## Alkenes used in addition polymerisation

Many compounds containing the C=C double bond can be polymerised. There is no need to know the mechanisms of these reactions, but you should describe them as addition reactions because the alkene molecules add together in vast numbers to form the polymer. You also do not need to know the exact conditions used in polymerisation reactions, but they mainly use a combination of high pressure and high temperature, which varies depending on the polymer.

You know that alkenes are hydrocarbons with the general formula $C_nH_{2n}$. When considering polymers, the term 'alkenes' is often widened to include other compounds containing C=C attached to other hydrocarbon groups and to halogens.

## Naming polymers

When alkene molecules are used in polymerisation, they are often referred to as **monomers**. The standard way to name a polymer is by writing 'poly', followed by the name of the monomer in brackets. The obvious example is the use of ethene to form poly(ethene). Most people abbreviate this name to polythene. Even though the polymer formed is saturated, the 'ene' ending is still used.

The table shows information about some common polymers.

| Monomer | Polymer | Common name |
|---|---|---|
| ethene | poly(ethene) | polythene |
| propene | poly(propene) | polypropene or polypropylene |
| chloroethene | poly(chloroethene) | polyvinyl chloride or PVC |
| tetrafluoroethene | poly(tetrafluoroethene) | PTFE or Teflon® |
| phenylethene | poly(phenylethene) | polystyrene |

**table A** Information about polymers, including their common name.

## Equations for polymerisation reactions

Because the polymers formed do not have a fixed molecular formula (their molecular masses can be anything from many tens of thousands to millions), we need to find a different way to show what happens in the reaction. The usual way to do this is to use the letter n to represent the number of monomer molecules reacting, then to show the **repeat unit** of the polymer inside a bracket (curved or square). The letter n is shown as a subscript after the bracket, and there are covalent bonds shown passing through the brackets to indicate that there is another repeat unit joined on to each side.

A general equation that you can modify for use with all addition polymerisation reactions is:

$$n \begin{array}{c} W \\ \phantom{x} \\ Y \end{array}\!\!\!C\!=\!C\!\!\!\begin{array}{c} X \\ \phantom{x} \\ Z \end{array} \longrightarrow \left[ \begin{array}{cc} W & X \\ | & | \\ -C-C- \\ | & | \\ Y & Z \end{array} \right]_n$$

monomer – any alkene    polymer (repeat unit)

## Examples of equations

Here is the equation showing the formation of poly(ethene):

$$n \; \underset{H}{\overset{H}{>}}C=C\underset{H}{\overset{H}{<}} \longrightarrow \left[ \begin{array}{c} H \; \; H \\ | \; \; \; | \\ -C-C- \\ | \; \; \; | \\ H \; \; H \end{array} \right]_n$$

ethene → poly(ethene) – polythene

Here is the equation showing the formation of poly(propene):

$$n \; \underset{H}{\overset{H}{>}}C=C\underset{H}{\overset{CH_3}{<}} \longrightarrow \left[ \begin{array}{c} H \; \; CH_3 \\ | \; \; \; | \\ -C-C- \\ | \; \; \; | \\ H \; \; H \end{array} \right]_n$$

propene → poly(propene) – polypropylene

This equation shows the formation of poly(chloroethene), better known as PVC:

$$n \; \underset{H}{\overset{H}{>}}C=C\underset{H}{\overset{Cl}{<}} \longrightarrow \left[ \begin{array}{c} H \; \; Cl \\ | \; \; \; | \\ -C-C- \\ | \; \; \; | \\ H \; \; H \end{array} \right]_n$$

chloroethene → poly(chloroethene) – PVC

Finally, this equation shows the formation of poly(phenylethene), better known as polystyrene:

$$n \; \underset{H}{\overset{H}{>}}C=C\underset{H}{\overset{C_6H_5}{<}} \longrightarrow \left[ \begin{array}{c} H \; \; C_6H_5 \\ | \; \; \; | \\ -C-C- \\ | \; \; \; | \\ H \; \; H \end{array} \right]_n$$

phenylethene → poly(phenylethene) – polystyrene

## Identifying the monomer

If you are given the repeat unit of a polymer, or a section of the polymer that contains several repeat units, you can work out the structure of the corresponding monomer. You need to identify the part of the structure that is repeated – this will be two carbon atoms in the chain and the four atoms or groups joined to them. The monomer structure is all of these atoms, but with a double bond between the two carbon atoms.

This is part of the structure of poly(methyl methylacrylate), better known as Perspex:

$$-\underset{\underset{H}{|}}{\overset{\overset{H}{|}}{C}}-\underset{\underset{COOCH_3}{|}}{\overset{\overset{CH_3}{|}}{C}}-\underset{\underset{H}{|}}{\overset{\overset{H}{|}}{C}}-\underset{\underset{COOCH_3}{|}}{\overset{\overset{CH_3}{|}}{C}}-\underset{\underset{H}{|}}{\overset{\overset{H}{|}}{C}}-\underset{\underset{COOCH_3}{|}}{\overset{\overset{CH_3}{|}}{C}}-$$

There are no brackets and no subscript n, because this shows part of the structure and not just the repeat unit. You can see that on alternate carbon atoms in the chain, there are first two hydrogen atoms, then one methyl group and one $COOCH_3$ group. It doesn't matter that you don't recognise the $COOCH_3$ group – you can still work out the monomer structure. The monomer structure is:

$$\underset{H}{\overset{H}{>}}C=C\underset{COOCH_3}{\overset{CH_3}{<}}$$

### Learning tip

In previous sections, you have seen alkene molecules drawn with angles of 120° between the bonds. When drawing the structures of polymers, use angles of 90°.

## Questions

1. The formula of a monomer used to make a polymer called PVA is:

$$\underset{H}{\overset{H}{>}}C=C\underset{OH}{\overset{H}{<}}$$

Draw the structure of the repeat unit of PVA.

2. Part of the structure of a polymer is:

$$-\underset{\underset{H}{|}}{\overset{\overset{H}{|}}{C}}-\underset{\underset{CN}{|}}{\overset{\overset{H}{|}}{C}}-\underset{\underset{H}{|}}{\overset{\overset{H}{|}}{C}}-\underset{\underset{CN}{|}}{\overset{\overset{H}{|}}{C}}-\underset{\underset{H}{|}}{\overset{\overset{H}{|}}{C}}-\underset{\underset{CN}{|}}{\overset{\overset{H}{|}}{C}}-$$

Draw the structure of the monomer used to make this polymer.

### Key definitions

**Monomers** are the small molecules that combine together to form a polymer.

The **repeat unit** of a polymer is the set of atoms that are joined together in large numbers to produce the polymer structure.

# 6.2 9 Dealing with polymer waste

By the end of this section, you should be able to...

- know that waste polymers can be separated into specific types of polymer for recycling, incineration (to release energy) and use as a feedstock for cracking
- understand how chemists limit the problems caused by polymer disposal
- understand, in terms of the use of energy and resources over the life cycle of polymer products, that chemists can contribute to the more sustainable use of materials

## Background

A hundred years ago, traditional materials used to make everyday objects were substances such as wood, metal, glass, wool and paper. Although these materials are still used today, to an increasing extent we have become more reliant on polymers (plastics) for many everyday objects. Reasons for the increasing use of polymers include the following.

- They can be manufactured on a large scale in a variety of complex shapes and with a wide range of physical properties. Think of plastic bottles that are rigid when used to hold bleach and squeezy when used to hold washing-up liquid.
- They are often lighter in weight than traditional alternatives. Think of milk in a glass bottle compared to a plastic bottle.
- They are unreactive and so can be used to contain many substances safely for long periods. Think of how metals corrode and wood rots.

Polymers are also relatively cheap to make when mass produced and have become, in the eyes of many people, disposable. Many years ago, 'disposable' meant that after use, the only thing to do was to throw the object away, perhaps in a bin, to be forgotten and taken away by the refuse collection service. Until a few years ago, most polymer waste ended up in landfill, but this method of disposal is now used much less. This is partly because of the limited space available in landfill sites, but also because there are now rules (backed by financial penalties) to decrease the use of landfill. More recently, public attitudes have changed, and people are becoming used to putting polymer waste in a designated wheelie bin or making a trip to the local recycling centre.

## Solutions to polymer waste

One way to reduce the problem of polymer waste is not to use polymers unnecessarily. The obvious example of this is some supermarkets discouraging the use of 'free' plastic bags and encouraging the use of 'bags for life' that can be used many times.

Another approach is not to throw away polymer waste, but to put it to other uses. Currently, the three main ways of putting polymer waste to other uses are:

- recycling
- incineration
- use as a chemical feedstock.

### Recycling

This word is often misused, but as far as polymer waste is concerned, **recycling** means converting it into other materials. For example, poly(ethylene terephthalate), better known as PET or PETE, is widely used in plastic bottles, and this polymer is now used on a large scale to make carpets.

The first stage in recycling is sorting. This is necessary because there are many types of polymer in use, and mixtures of these types cannot be effectively processed together. A lot of sorting is still done by hand, which is tedious and inefficient, but ways of automating the sorting process are being developed. The polymer used to make most plastic objects can be identified by a code, which helps in the sorting process.

The second stage is processing, which involves chopping the waste into small pieces and washing it. This material is then used to make new materials, using methods such as melting, moulding and fibre production.

|  |  |
|---|---|
| 1 PETE | Poly(ethene) terephthalate is used for plastic soft drink bottles and pre-packed meals. |
| 2 HDPE | High density poly(ethene) is used for milk bottles, detergent bottles and butter tubs. |
| 3 V | Polyvinyl chloride is used for medical tubing, cable insulation and drainpipes. |
| 4 LDPE | Low density poly(ethene) is used in squeezy bottles and frozen food bags. |
| 5 PP | Poly(propene) is used in yoghurt tubs and medicine bottles. |
| 6 PS | Polystyrene is used for cups, plates and cutlery, for CD cases and to serve hot food for its insulating properties. |
| 7 OTHER | Made from none of the above or made from a mixture of resin. |

**fig A** Polymer identification codes.

## Incineration

The elements present in polymer waste are mostly hydrogen and carbon, so they can be used as fuels, similar to other hydrocarbons. An **incinerator** takes in polymer waste and converts it into heat energy that can be used to heat homes and factories, or used to generate electricity. There is very little solid waste left after incineration, but that is because most of the atoms in the polymers end up in gaseous products which pass into the atmosphere via a chimney.

There is often local opposition when there is a proposal to build an incinerator because of concerns about air pollution. This is because, as well as hydrogen and carbon, there are other elements in the polymer waste – PVC contains chlorine, and there are small amounts of toxic heavy metals from the pigments used to colour plastics. These pollutants are difficult to remove from the waste gases released into the atmosphere.

## Use as a chemical feedstock

This is a process, similar to cracking, used to break down the polymer waste into gases (mainly hydrogen and carbon monoxide). This produces a **feedstock** that can be used in other chemical reactions, often to make new polymers.

## Biodegradable polymers

Traditional plastics put into landfill do not break down. The idea of using **biodegradable** polymers (sometimes described as biopolymers) is to allow them to be broken down by microbes in the environment. This sounds like a good idea, and some of them are used on a small scale in medicine (for sutures, often called stitches, and in drug delivery). However, there are some disadvantages.

- They are often made from plant material, so there is the same issue to consider as with biofuels – land is needed to grow the plants.
- They are designed to break down in the environment, so when they do, the hydrogen and carbon atoms they contain cannot be directly used. No recycling, incineration or use as a chemical feedstock is possible.

## Life cycle analysis

All polymers have to be manufactured, and then disposed of in the ways shown earlier in this section. The different stages in the life of a polymer have some environmental impact, especially in terms of resource use, energy use and disposal.

To help make decisions about the management of polymer waste in the future, to reduce the long-term impact on the environment, and to compare different polymer uses, a life cycle analysis is carried out. This is illustrated in the following diagram.

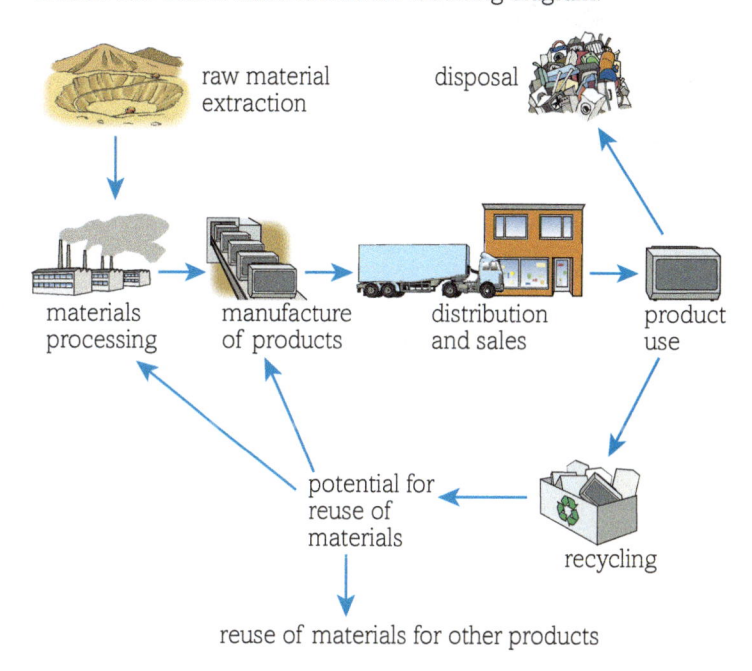

### Learning tip
Try to use the terms 'disposal' and 'recycling' correctly.

## Questions

1. Summarise the advantages of polymers over traditional materials.
2. Summarise the advantages and main disadvantage of incineration as a method of disposing of polymer waste.

### Key definitions

**Recycling** involves converting polymer waste into other materials.

An **incinerator** converts polymer waste into energy.

Use as a **feedstock** involves converting polymer waste into chemicals that can be used to make new polymers.

A **biodegradable** polymer is one that can be broken down by microbes.

# 6.3 1 Halogenoalkanes and hydrolysis reactions

By the end of this section, you should be able to...

- know that halogenoalkanes can be classified as primary, secondary or tertiary, and understand their trend in reactivity
- understand what is meant by the term 'nucleophile'

## What are halogenoalkanes?

The halogenoalkanes are a homologous series of compounds with the general formula $C_nH_{2n+1}X$. Think of them as the result of replacing a hydrogen atom in a hydrocarbon by a halogen atom. X represents a halogen atom – usually bromine or chlorine, but it could also be fluorine or iodine. You have already met halogenoalkanes in a previous chapter – chloromethane is the product of the reaction between methane and chlorine.

The symbol R (also used in a completely different context for the gas constant in the equation $pV=nRT$) is often used in organic chemistry to represent any alkyl group (such as methyl or ethyl). So, the formula of a halogenoalkane could be simplified to RX.

The number of halogen atoms in a halogenoalkane molecule can be more than one, so the general formula is different for these ($C_nH_{2n}X_2$ and $C_nH_{2n-1}X_3$ are other possible general formulae).

**fig A** The most notorious halogenoalkanes are those containing both chlorine and fluorine. They are known as chlorofluorocarbons (CFCs for short) and have been used in aerosol cans for decades. They are now known to damage the ozone layer in the upper atmosphere and, through international agreement, they have been mostly phased out.

### Naming halogenoalkanes

We have already discussed how to name halogenoalkanes, but here are some examples to remind you.

| Structure | Name |
|---|---|
| $CH_2Cl–CHCl–CH_3$ | 1,2-dichloropropane |
| $CH_2Br–CH_2–CH_2Cl$ | 1-bromo-3-chloropropane |
| $CCl_4$ | tetrachloromethane |
| $CH_3–CHF–CH_2–CH_3$ | 2-fluorobutane |

**table A** Examples of how to name halogenoalkanes.

If there are two or more different halogens, then their prefixes appear in alphabetical order. As for naming compounds in other homologous series, the longest carbon chain is the basis of the name, and when the prefix numbers are added together the total should be as small as possible.

## Classifying halogenoalkanes

You have met the terms 'primary', 'secondary' and 'tertiary' in the section referring to carbocations (see **Section 6.2.7**).

Halogenoalkanes are classified in a similar way, depending on the number of alkyl groups joined to the C atom bonded to the halogen atom. The table shows some examples.

| Structure | Abbreviated formula | Number of alkyl groups | Classification |
|---|---|---|---|
| $CH_3-CH_2-CH_2F$ | RX | 1 | primary |
| $CH_3-CHBr-CH_3$ | $R_2CHX$ | 2 | secondary |
| $(CH_3)_2CCl-CH_2-CH_3$ | $R_3CX$ | 3 | tertiary |

**table B** Examples of how to classify halogenoalkanes.

## What makes halogenoalkanes reactive?

Hydrocarbons contain only hydrogen and carbon atoms, which have similar electronegativities, so their bonds are almost non-polar.

Halogenoalkanes contain a halogen atom with an electronegativity higher than that of carbon, so the C—X bond is polar. This bond polarity can be indicated using the partial charges $\delta+$ and $\delta-$.

Down Group 7 of the Periodic Table, the electronegativities of the halogens decrease from fluorine to iodine, so the polarity of the C—X bond also decreases.

The carbon atom joined to the halogen is always the slightly positive ($\delta+$) or electron-deficient part of the molecule, and this is what makes halogenoalkanes react as they do.

These carbon atoms attract other species called **nucleophiles**. 'Nucleo' indicates positive and 'phile' indicates liking, so nucleophiles are species that are attracted to slightly positive or electron-deficient parts of a molecule. Nucleophiles are either negative ions or molecules with a slightly negative atom, but they always use a lone pair of electrons when attacking another species.

We will look at the mechanisms of halogenoalkane reactions in **Section 6.3.3**, but first take a look at a reaction that can be used to compare the reactivities of different halogenoalkanes.

## Hydrolysis reactions

When a halogenoalkane is added to water a reaction begins, but it may take a period of time to complete. A water molecule contains polar bonds, and the $\delta-$ oxygen atom in water is attracted to the $\delta+$ carbon atom in the halogenoalkane. The reaction that occurs can be represented by the equation:

$$RX + H_2O \rightarrow ROH + HX \quad \text{or} \quad RX + H_2O \rightarrow ROH + H^+ + X^-$$

The product ROH is an alcohol and, as both organic substances are usually colourless liquids, no colour change can be seen. You can see from this simple equation that the C—X bond breaks, so the RX molecule breaks into two parts (R and X), although the R group then combines with the OH group of water.

This type of reaction is known as **hydrolysis**. 'Hydro' refers to water and 'lysis' refers to splitting, so it means splitting with water.

# Questions

1 Write the IUPAC name for each of these halogenoalkanes and classify it as primary, secondary or tertiary.

(a) $CH_3-\underset{\underset{Cl}{|}}{\overset{\overset{CH_3}{|}}{C}}-CH_2-CH_3$

(b) $CH_3-\underset{\underset{Br}{|}}{CH}-CH_2-CH_3$

(c) $CH_3-\underset{\underset{CH_3}{|}}{CH}-CH_2-I$

2 Why are nucleophiles more strongly attracted to fluoroalkanes than to chloroalkanes?

### Key definitions

A **nucleophile** is a species that donates a lone pair of electrons to form a covalent bond with an electron-deficient atom.

A **hydrolysis** reaction is one in which water or hydroxide ions replace an atom in a molecule with an —OH group.

# 6.3 2 Comparing the rates of hydrolysis reactions

**By the end of this section, you should be able to...**

- understand that experimental observations and data can be used to compare the relative rates of hydrolysis of primary, secondary and tertiary halogenoalkanes, including chloro-, bromo- and iodoalkanes, using aqueous silver nitrate in ethanol
- understand, in terms of bond enthalpy, the trend in reactivity of chloro-, bromo- and iodoalkanes

## Practical aspects

If instead of adding water, silver nitrate solution is added, then the progress of the reaction can be followed. You may remember that silver nitrate can be used in a test for halide ions because the silver ions in silver nitrate react with the halide ions formed in the hydrolysis to give a precipitate:

$$Ag^+ + X^- \rightarrow AgX$$

This means that we can tell how quickly the hydrolysis reaction occurs by observing how quickly the precipitate of AgX forms.

Without going into full practical details, a comparison of the rates of these reactions involves:

- using ethanol as a solvent for the mixture (halogenoalkanes and aqueous silver nitrate do not mix, but form separate layers)
- controlling variables such as temperature and the concentration and quantity of halogenoalkanes
- timing the appearance of the precipitate (although this is difficult to do accurately because a precipitate may first appear faint, but become thicker with time).

In an experiment, three tubes containing silver nitrate dissolved in ethanol are left at the same temperature for several minutes. Each tube contains a different halogenoalkane.

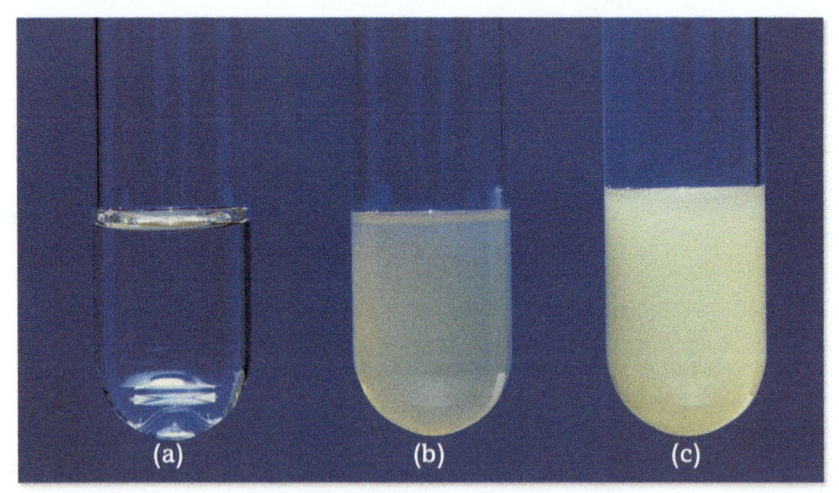

**fig A** Three tubes containing silver nitrate dissolved in ethanol, with a different halogenoalkane added to each one.

The figure shows that there is no precipitate in tube (a), a faint cream-coloured precipitate in tube (b) and a thicker pale yellow precipitate in tube (c). These observations are caused by a chloroalkane in (a), a bromoalkane in (b) and an iodoalkane in (c).

Two types of comparison can be made. You can:

- compare halogenoalkanes with the same structure but with different halogens
- compare halogenoalkanes with the same halogen but with different structures.

Fluoroalkanes are comparatively very unreactive, so reactions involving them are often omitted.

## Interpreting the results for different halogens

The table shows the trend when the halogen is different.

| Same structure but different halogen | Result |
|---|---|
| 1-iodobutane | fastest |
| 1-bromobutane | |
| 1-chlorobutane | slowest |

**table A** Examples of comparing halogenoalkanes with the same structure but with different halogens.

You might suppose that the halogenoalkane with the most polar bond would be the fastest to be hydrolysed (in this case, 1-chlorobutane) because the $\delta+$ charge on the carbon atom is greatest, so the attacking nucleophile should be attracted more strongly.

This is true, but there is another, more important factor to consider. Bond breaking requires energy, and weaker bonds break more easily than stronger bonds. The table of (mean) bond enthalpies shows that the C—I bond is the weakest and the C—Cl bond is the strongest.

| Bond | Bond enthalpy/kJ mol$^{-1}$ |
|---|---|
| C—Cl | +346 |
| C—Br | +290 |
| C—I | +228 |

**table B** Mean bond enthalpies.

So, under the same conditions, the C—I bond breaks most easily, forming I$^-$ ions, and so a precipitate of AgI forms more quickly.

The C—F bond is much stronger (+467 kJ mol$^{-1}$) than any of the others, which explains why fluoroalkanes are not often used in these hydrolysis experiments.

## Interpreting the results for different structures

The table shows the trend when the structure is different.

| Same halogen but different structure | Result |
|---|---|
| 2-bromo-2-methylpropane (tertiary) | fastest |
| 2-bromobutane (secondary) | |
| 1-bromobutane (primary) | slowest |

**table C** Examples of comparing halogenoalkanes with the same halogen but with different structures.

To explain why tertiary halogenoalkanes are more rapidly hydrolysed than secondary and primary compounds requires a detailed understanding of two different reaction mechanisms, and is beyond the scope of this book, but will be dealt with in **Book 2**.

#### Learning tip
Molecules with partial charges ($\delta+$ and $\delta-$) are not ions. They are molecules with no overall charge but with electron-rich and electron-deficient parts.

# Questions

1. 2-bromopropane and 2-iodopropane are put into separate test tubes and warmed with water. Explain which one is hydrolysed more quickly.

2. Why are fluoroalkanes not readily hydrolysed compared with chloroalkanes?

# 6.3   3   Halogenoalkane reactions and mechanisms

By the end of this section, you should be able to...

- understand the reactions of halogenoalkanes with aqueous potassium hydroxide, aqueous silver nitrate in ethanol, potassium cyanide, ammonia and ethanolic potassium hydroxide
- understand the mechanisms of the nucleophilic substitution reactions between primary halogenoalkanes and aqueous potassium hydroxide and ammonia

## Substitution reactions

The hydrolysis reactions in the previous section involved replacing a halogen atom (X) with a hydroxyl group (OH). These are substitution reactions, and here is a summary of the four reactions you need to know.

$$RX \xrightarrow[\text{Reaction 1}]{H_2O/\text{warm}} ROH$$

$$RX \xrightarrow[\text{Reaction 2}]{KOH/\text{heat under reflux}} ROH$$

$$RX \xrightarrow[\text{Reaction 3}]{KCN/\text{heat under reflux}} RCN$$

$$RX \xrightarrow[\text{Reaction 4}]{NH_3/\text{heat in sealed tube}} RNH_2$$

### Reaction 1

This is the hydrolysis reaction you met in the previous section.

### Reaction 2

Heating a halogenoalkane with aqueous potassium hydroxide under reflux is one way of making alcohols. The attacking nucleophile is the OH$^-$ ion. An example of an equation for this reaction is the conversion of 1-chloropropane into propan-1-ol:

$$CH_3CH_2CH_2Cl + KOH \rightarrow CH_3CH_2CH_2OH + KCl$$

or

$$CH_3CH_2CH_2Cl + OH^- \rightarrow CH_3CH_2CH_2OH + Cl^-$$

The advantage of an ionic equation is that nucleophile is clearly shown.

### Reaction 3

Heating a halogenoalkane with potassium cyanide dissolved in ethanol under reflux is one way of making **nitriles**. The attacking nucleophile is the CN$^-$ ion. An example of an equation for this reaction is the conversion of bromoethane into propanenitrile:

$$CH_3CH_2Br + KCN \rightarrow CH_3CH_2CN + KBr$$

or

$$CH_3CH_2Br + CN^- \rightarrow CH_3CH_2CN + Br^-$$

Note that the organic product contains one more carbon atom than the starting material. This reaction is a useful way of extending the carbon chain and is an important way to synthesise more complex compounds.

### Reaction 4

Heating a halogenoalkane with ammonia solution in a sealed tube is one way of making **primary amines**. The sealed tube is needed because ammonia is a gas and would otherwise escape from the apparatus before it could react. The attacking nucleophile is the NH$_3$ molecule. An example of an equation for this reaction is the conversion of 1-iodobutane into butylamine:

$$CH_3CH_2CH_2CH_2I + NH_3 \rightarrow CH_3CH_2CH_2CH_2NH_2 + HI$$

This equation looks similar to those for Reactions 1–3, but it isn't quite the full story. Like NH$_3$, the organic product is a base, so it would react with the inorganic product, the acid HI, to form a salt. So a better equation is:

$$CH_3CH_2CH_2CH_2I + NH_3 \rightarrow CH_3CH_2CH_2CH_2NH_3^+ + I^-$$

However, this is only the first step because the product is a salt, not a primary amine. To produce a high yield of the amine, the ammonia is used in excess, and some of this excess ammonia reacts in a second step to produce the amine:

$$CH_3CH_2CH_2CH_2NH_3^+I^- + NH_3 \rightarrow CH_3CH_2CH_2CH_2NH_2 + NH_4^+I^-$$

So, the final products are butylamine and ammonium iodide. These two steps are often combined as:

$$CH_3CH_2CH_2CH_2I + 2NH_3 \rightarrow CH_3CH_2CH_2CH_2NH_2 + NH_4^+I^-$$

## Nucleophilic substitution mechanisms

In previous sections you have met radical substitution and electrophilic addition mechanisms; here is a different one to learn about.

In each reaction the attacking species is a nucleophile, so the reaction type is better described as **nucleophilic substitution**. You only need to know about mechanisms for primary halogenoalkanes undergoing Reactions 2 and 4.

### Mechanism of Reaction 2

One example is the reaction between bromoethane and aqueous potassium hydroxide. The reaction starts with the donation of a lone pair of electrons from the oxygen of a hydroxide ion to the electron-deficient carbon atom and the formation of a C—O bond. At the same time, the electrons in the C—Br bond move to the Br atom, resulting in the breaking of the C—Br bond. This type of bond breaking is known as heterolytic fission. 'Hetero' indicates different, and you already know that 'lysis' indicates breaking.

$$CH_3-\overset{H}{\underset{H}{C}}-Br^{\delta-} \xrightarrow{\phantom{xx}} CH_3-\overset{H}{\underset{H}{C}}-OH + Br^-$$
$$\phantom{CH_3-C-}OH^-$$

## Mechanism of Reaction 4

One example is the reaction between chloroethane and ammonia. The first step of the reaction involves the donation of a lone pair of electrons from the nitrogen of an ammonia molecule to the electron-deficient carbon atom and the formation of a C—N bond. At the same time, the electrons in the C—Cl bond move to the Cl atom, resulting in the breaking of the C—Cl bond.

$$CH_3-\overset{H}{\underset{H}{C}}^{\delta+}-Cl^{\delta-} \xrightarrow{\phantom{xx}} CH_3-\overset{H}{\underset{H}{C}}-\overset{H}{\underset{H}{N^+}}-H + Cl^-$$
$$:NH_3$$

The second step of the reaction involves another ammonia molecule acting as a base and removing a hydrogen ion from the ion formed in the first step.

$$CH_3-\overset{H}{\underset{H}{C}}-\overset{H}{\underset{H}{N^+}}-H \xrightarrow{\phantom{xx}} CH_3-\overset{H}{\underset{H}{C}}-N\overset{H}{\underset{H}{\diagdown}} + NH_4^+$$
$$\phantom{CH_3-C-N^+-}:NH_3$$

## Elimination reactions

Reaction 2 above is a nucleophilic substitution reaction, but using a different solvent (ethanol instead of water) causes a different reaction to occur. When a halogenoalkane is heated with **ethanolic** potassium hydroxide, the OH⁻ ion acts as a base and not as a nucleophile. You know that a base reacts with a hydrogen ion (H⁺), but in the case of a halogenoalkane, the hydrogen that reacts with the OH⁻ ion is the one attached to a carbon atom next to the C in the C—Br bond.

For example, the equation for the reaction between 2-bromopropane and ethanolic potassium hydroxide is:

$$CH_3-CHBr-CH_3 + KOH \rightarrow CH_2=CH-CH_3 + H_2O + KBr$$

The organic product is propene (an alkene), and water and potassium bromide are the other products. You can see why the reaction is referred to as **elimination**. H and Br are removed from the halogenoalkane but are not replaced by any other atoms.

You do not need to know the mechanism for this elimination reaction.

### Learning tip

Both the hydroxide ion and the ammonia molecule can act as nucleophiles and as bases.

## Questions

1. Chloromethane is heated with:
   (a) ammonia
   (b) potassium cyanide dissolved in ethanol.
   Give the name of the organic product formed in each case.

2. 2-chlorobutane is heated with:
   (a) aqueous potassium hydroxide
   (b) ethanolic potassium hydroxide.
   State the mechanism of each reaction and explain why there are two different organic products in (b).

### Key definitions

**Nitriles** are organic compounds containing the C–CN group.
**Primary amines** are compounds containing the C–NH₂ group.
A **nucleophilic substitution** reaction is one in which an attacking nucleophile replaces an existing atom or group in a molecule.
An **ethanolic** solution is one in which ethanol is the solvent.
An **elimination** reaction is one in which a molecule loses atoms attached to adjacent carbon atoms, forming a C=C double bond.

# 6.4 1 Alcohols and some of their reactions

By the end of this section, you should be able to...

- know that alcohols can be classified as primary, secondary or tertiary
- understand the combustion, halogenation and dehydration reactions of alcohols

## What are alcohols?

The alcohols are a homologous series of compounds with the general formula $C_nH_{2n+1}OH$. Think of them as the result of replacing a hydrogen atom in a hydrocarbon with a hydroxyl group. You have already met alcohols in **Section 6.3.1** as the products of the hydrolysis of halogenoalkanes.

The symbol R can be used to represent an alkyl group, as with halogenoalkanes. So, the formula of an alcohol could be simplified to ROH.

## Naming alcohols

We have already discussed how to name alcohols, but here are some examples to remind you.

| Structure | Name |
|---|---|
| $CH_3-CH(OH)-CH_3$ | propan-2-ol |
| $CH_3-CH_2-CH_2-CH_2OH$ | butan-1-ol |
| $(CH_3)_3C-CH_2OH$ | 2,2-dimethylpropan-1-ol |
| $CH_2(OH)-CH(OH)-CH_2OH$ | propane-1,2,3-triol |

**table A** Examples of how to name alcohols.

**fig A** Not all alcohols are used as drinks.

You will not meet the last example in this book, but you will have come across it in foods, medicines and personal care products. Its common names are glycerol and glycerine. You can see how the IUPAC system can be adapted to name compounds with more than one OH group.

## Classifying alcohols

You have met the terms primary, secondary and tertiary in the sections referring to carbocations and halogenoalkanes (**Section 6.2.7** and **Chapter 6.3**).

Alcohols are classified in a similar way, depending on the number of alkyl groups joined to the C atom bonded to the hydroxyl group. The table shows some examples.

| Structure | Abbreviated formula | Number of alkyl groups | Classification |
|---|---|---|---|
| $CH_3-CH_2-CH_2OH$ | ROH | 1 | primary |
| $CH_3-CH(OH)-CH_3$ | $R_2CHOH$ | 2 | secondary |
| $(CH_3)_2C(OH)-CH_2-CH_3$ | $R_3COH$ | 3 | tertiary |

**table B** Examples of how to classify alcohols.

## Reactions

In this section, the reactions of alcohols we cover are:

- combustion
- conversions to halogenoalkanes
- dehydration to alkenes.

Reaction mechanisms are not required for any of these reactions.

## Combustion

You have seen in a previous section (**Section 6.2.3**) that alcohols are used as biofuels. If combustion is complete, the products are carbon dioxide and water. This is the equation for the complete combustion of ethanol:

$$C_2H_5OH + 3O_2 \rightarrow 2CO_2 + 3H_2O$$

## Conversions to halogenoalkanes

These reactions involve replacing the hydroxyl group in an alcohol molecule with a halogen atom. The reaction is often known as **halogenation**. However, adding a halogen to an alcohol does not work. A different method is needed for each halogen.

Chlorination is carried out using phosphorus(V) chloride (a white solid, also known as phosphorus pentachloride). The reaction is very vigorous at room temperature, so the alcohol and phosphorus(V) chloride reaction mixture does not need heating. There are also two inorganic products: phosphorus oxychloride and hydrogen chloride. This is the equation for the reaction with propan-1-ol:

$$CH_3CH_2CH_2OH + PCl_5 \rightarrow CH_3CH_2CH_2Cl + POCl_3 + HCl$$

Chlorination of tertiary alcohols can be done in a different way, using a method that does not work well for primary and secondary alcohols. The alcohol needs only to be mixed (by shaking) with concentrated hydrochloric acid at room temperature. This is the equation for the reaction with 2-methylpropan-2-ol:

$$(CH_3)_3COH + HCl \rightarrow (CH_3)_3CCl + H_2O$$

Bromination is carried out using a mixture of potassium bromide and about 50% concentrated sulfuric acid. The reaction mixture is warmed with the alcohol. It is better to write two equations, rather than one, as the inorganic reagents react together to form hydrogen bromide:

$$KBr + H_2SO_4 \rightarrow KHSO_4 + HBr \quad \text{or} \quad 2KBr + H_2SO_4 \rightarrow K_2SO_4 + 2HBr$$

The other inorganic product is either potassium hydrogensulfate or potassium sulfate. This is the equation for the reaction with butan-1-ol:

$$CH_3CH_2CH_2CH_2OH + HBr \rightarrow CH_3CH_2CH_2CH_2Br + H_2O$$

Iodination is carried out using a mixture of red phosphorus and iodine. The reaction mixture, including the alcohol, is heated under reflux. As with bromination, it is better to write two equations, as the inorganic reagents first react to form phosphorus(III) iodide:

$$2P + 3I_2 \rightarrow 2PI_3$$

This is the equation for the reaction with ethanol:

$$3C_2H_5OH + PI_3 \rightarrow 3C_2H_5I + H_3PO_3$$

The inorganic product is phosphonic acid (often known as phosphorous acid).

## Dehydration to alkenes

This is done by heating the alcohol with concentrated phosphoric acid. The reaction is similar to the elimination reaction of a halogenoalkane, with the OH group and a hydrogen atom from an adjacent carbon atom being removed and a C=C double bond formed in the carbon chain.

You can see why the reaction is described as **dehydration**, as water is the only inorganic product. These are the equations for a reaction in which there are two possible products, starting with butan-2-ol:

$$CH_3CH(OH)CH_2CH_3 \rightarrow CH_2=CHCH_2CH_3 + H_2O$$
but-1-ene

and

$$CH_3CH(OH)CH_2CH_3 \rightarrow CH_3CH=CHCH_3 + H_2O$$
but-2-ene

Remember that but-2-ene actually exists as a pair of *E–Z* (*cis-trans*) isomers. It is therefore correct to state that there are three products – but-1-ene and the two isomeric forms of but-2-ene.

The formula for phosphoric acid does not appear in the equation. The water formed in the reaction mixes with the concentrated phosphoric acid to dilute the acid.

### Learning tip

When you balance an equation for the combustion of an alcohol, remember that there is already one atom of oxygen in an alcohol molecule.

## Questions

1. Write an equation for the complete combustion of butanol, $C_4H_9OH$.

2. When butan-2-ol is dehydrated, there is more than one organic product. Explain why there is only a pair of E/Z isomers when pentan-3-ol is dehydrated.

### Key definitions

A **halogenation** reaction results in the replacement of the hydroxyl group in an alcohol molecule by a halogen atom.

A **dehydration** reaction results in the removal of the hydroxyl group in an alcohol molecule, together with a hydrogen atom from an adjacent carbon atom, forming a C=C double bond.

# 6.4 2 Oxidation reactions of alcohols

**By the end of this section, you should be able to...**

- understand the techniques of distillation and heating under reflux
- understand the different oxidation reactions of primary and secondary alcohols, and the techniques used to maximise the yields of different products

## Background

This book considers oxidation and reduction in different ways. In this section, the relevant way to consider oxidation is as the loss of hydrogen from an alcohol molecule. Unlike dehydration, covered in the previous section, oxidation affects only one carbon atom. The atoms removed from an alcohol molecule are the hydrogen of the OH group and a hydrogen atom from the carbon atom joined to the OH group, as shown in the diagram.

The organic product contains a C=O group, known as a carbonyl group.

The diagrams below should help you to understand why primary and secondary alcohols, but not tertiary alcohols, can be oxidised in this way.

primary    secondary    tertiary

Only the primary and secondary structures have a hydrogen atom on the C of the C—OH group – the tertiary structure does not.

## The products of oxidation

### Ketones

When a secondary alcohol is oxidised, the organic product belongs to a homologous series called **ketones**. A simplified formula for a ketone is RCOR. The two alkyl groups can be the same or different.

### Aldehydes and carboxylic acids

When a primary alcohol is oxidised, the organic product belongs to a homologous series called **aldehydes**. A simplified formula for an aldehyde is RCHO. (This should not be written as RCOH, which would imply that the molecule contains an OH group.)

There is a complication with aldehydes that is not the case with ketones. This is that aldehydes are more easily oxidised than alcohols, so when a primary alcohol is oxidised, the aldehyde formed may be oxidised further. When an aldehyde is oxidised, the process involves the gain of an oxygen atom, not the loss of hydrogen. The oxygen atom gained goes between the C and H of the CHO group. The organic product belongs to a homologous series called **carboxylic acids**. A simplified formula for a carboxylic acid is RCOOH.

These reactions are summarised below.

secondary alcohol → ketone

primary alcohol → aldehyde → carboxylic acid

The usual reagent for these oxidation reactions is a mixture of potassium dichromate(VI) and dilute sulfuric acid. Unlike with the inorganic reagents used in previous sections, equations involving this mixture do not need to be written (because they are very complicated!). Instead, the oxidising agent is represented by [O]. This symbol simply represents an oxygen atom provided by the oxidising agent. Whenever this oxidising agent is used, there is a colour change from orange to green.

So now we can write some equations for examples of oxidation reactions. The names of the organic products are given, but a fuller study of these products is outside the scope of this book.

Propan-1-ol to propanal:

$CH_3CH_2CH_2OH + [O] \rightarrow CH_3CH_2CHO + H_2O$

Propanal to propanoic acid:

$CH_3CH_2CHO + [O] \rightarrow CH_3CH_2COOH$

Propan-2-ol to propanone:

$CH_3CH(OH)CH_3 + [O] \rightarrow CH_3COCH_3 + H_2O$

You do not need to know the mechanisms of any of these reactions.

## Different practical techniques

Because of the easier oxidation of aldehydes compared with alcohols, two different techniques are used. These are:

- heating under reflux
- distillation with addition.

### Heating under reflux

When the oxidation is intended to be complete (to obtain a ketone or a carboxylic acid), the apparatus used is **heating under reflux**, as shown in the diagram.

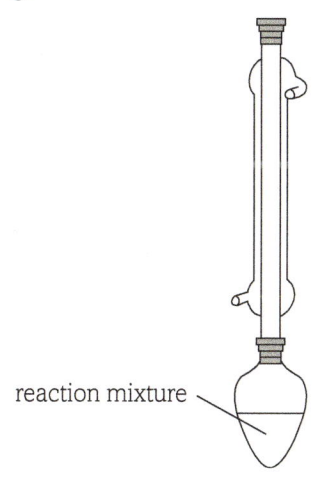

In this apparatus, the products of oxidation stay in the reaction mixture, because if they do boil off, they condense in the vertical condenser and return to the heating flask.

### Distillation with addition

When the oxidation is intended to be incomplete (to obtain an aldehyde, and not a carboxylic acid), the apparatus used is **distillation with addition**, as shown in the diagram.

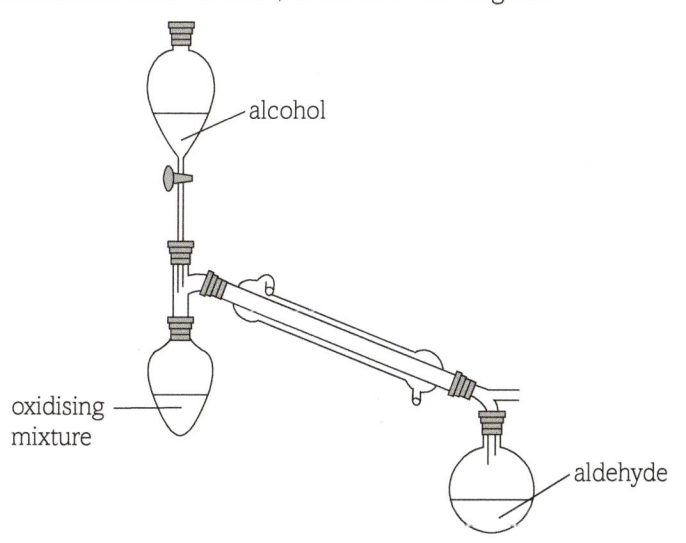

In this apparatus, only the oxidising agent is heated, and the alcohol is slowly added to the oxidising agent. When the aldehyde is formed, it immediately distils off (it has a much lower boiling temperature than the alcohol used to make it), and is collected in the receiver.

### Learning tip

Practise writing equations to show how a given alcohol can be separately dehydrated and oxidised. This will help you to understand the difference between the two reactions.

## Questions

1. Butan-1-ol can be dehydrated to form one organic product, but can be oxidised to form two organic products. Give the names and structures of these three products.

2. Pentan-2-ol can be dehydrated to form more than one organic product, but can be oxidised to form only one organic product. Give the names and structures of these products.

### Key definitions

**Ketones** are a homologous series of organic compounds formed by oxidation of secondary alcohols.

**Aldehydes** are a homologous series of organic compounds formed by the partial oxidation of primary alcohols.

**Carboxylic acids** are a homologous series of organic compounds formed by the complete oxidation of primary alcohols.

**Heating under reflux** involves heating a reaction mixture with a condenser fitted vertically.

**Distillation with addition** involves heating a reaction mixture, but adding another liquid and distilling off the product as it forms.

# 6.4 / 3 — Purifying an organic liquid

**By the end of this section, you should be able to...**

- know when and how to use techniques to purify organic liquids, including simple distillation, fractional distillation, solvent extraction and drying
- understand how measuring the boiling temperature of an organic liquid can be used to assess its purity

## Background

So far, you have met many different reactions that can be used in the preparation of organic compounds. In most cases the intended organic product is far from pure. It could be contaminated with:

- unreacted starting materials
- other organic products
- the inorganic reagents used, or the inorganic products formed from them
- water.

This means that organic chemists need to use several techniques to separate the intended product from a reaction mixture. These techniques will be different, depending on whether the intended product is a gas, a liquid or a solid. In this section we consider only the techniques used to purify an organic liquid.

The techniques we consider are:

- simple distillation
- fractional distillation
- solvent extraction
- drying.

## Apparatus

In a chemistry textbook you will see many different diagrams of apparatus used in organic chemistry. You will have some of these pieces of apparatus in your laboratory, but you will probably not be able to set up every piece of apparatus you see in the book.

In simple laboratory experiments not involving organic compounds, you are likely to use pieces of glassware connected by tubing and corks or bungs. This kind of apparatus is relatively inexpensive to buy and easy to use. If one of the gases you are using, such as carbon dioxide, leaks from the apparatus, this is not a major problem.

Organic compounds are more of a problem because of their flammable and sometimes toxic nature. They may also attack corks and bungs, and so increase the risk of leaks and contamination. One solution to these problems is to use a type of apparatus made only (or mostly) of glass, and which can be fitted together tightly using ground-glass joints.

## Alcohols 6.4

A selection of this type of apparatus is shown below.

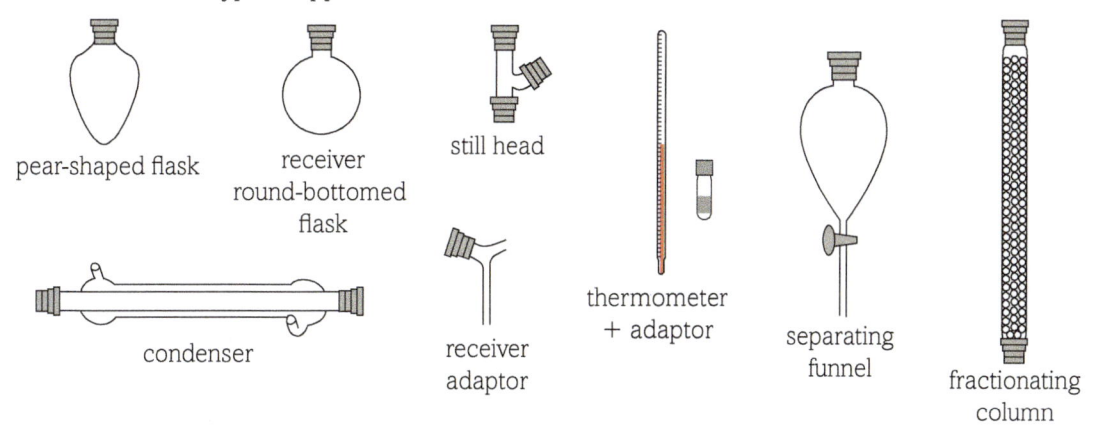

- pear-shaped flask
- receiver round-bottomed flask
- still head
- condenser
- receiver adaptor
- thermometer + adaptor
- separating funnel
- fractionating column

**fig A** A selection of apparatus made only (or mostly) of glass, ideal for experiments using organic compounds.

Using these eight pieces of apparatus, you can create a wide variety of experimental set-ups.

Heating is needed in many experiments. You can do this using a Bunsen burner, but often an electric heating mantle or a hot water bath or hot oil bath is preferred for safety reasons.

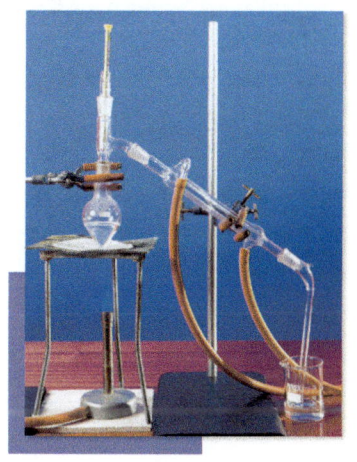

**fig B** This is just one of the many apparatus set-ups using ground-glass joints.

## Simple distillation

Here is a summary of the process of **simple distillation**.

- Distillation of an impure liquid involves heating it in a flask connected to a condenser.
- The liquid with the lowest boiling temperature evaporates or boils off first and passes into the condenser first. This means it can be collected in the receiver separately from any other liquid that evaporates later.
- The purpose of the thermometer is to monitor the temperature of the vapour as it passes into the condenser. If the temperature remains steady, this is an indication that one compound is distilling over. If, after a while, the temperature begins to rise, this indicates that a different compound is distilling over.

The diagram below illustrates the process.

- reaction mixture
- distillate

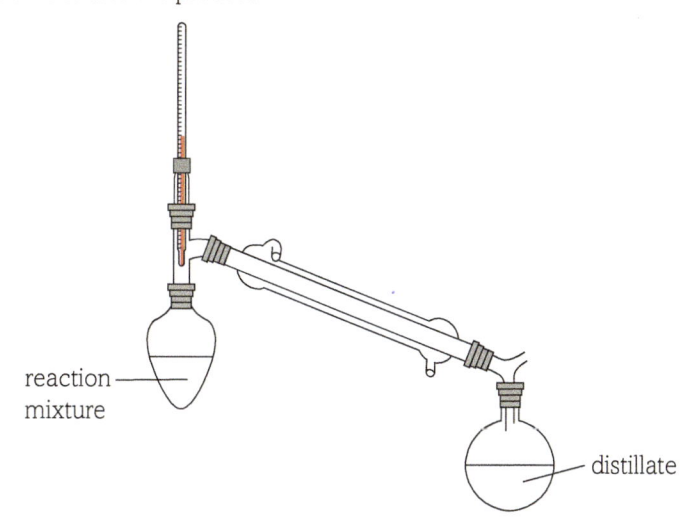

**fig C** Apparatus used in the process of simple distillation.

## Advantages and disadvantages

The advantages of using simple distillation, rather than fractional distillation, are that it is easier to set up and is quicker.

The disadvantage is that it does not separate the liquids as well as fractional distillation. It should only be used if the boiling temperature of the liquid being purified is very different from the other liquids in the mixture, ideally a difference of more than 25 °C.

## Fractional distillation

Here is a summary of the process of **fractional distillation**.

- Fractional distillation uses the same apparatus as simple distillation, but with a fractionating column between the heating flask and the still head.
- The column is usually filled with glass beads or pieces of broken glass, which act as surfaces on which the vapour leaving the column can condense, and then be evaporated again as more hot vapour passes up the column.
- Effectively, the vapour undergoes several repeated distillations as it passes up the column, which provides a better separation.

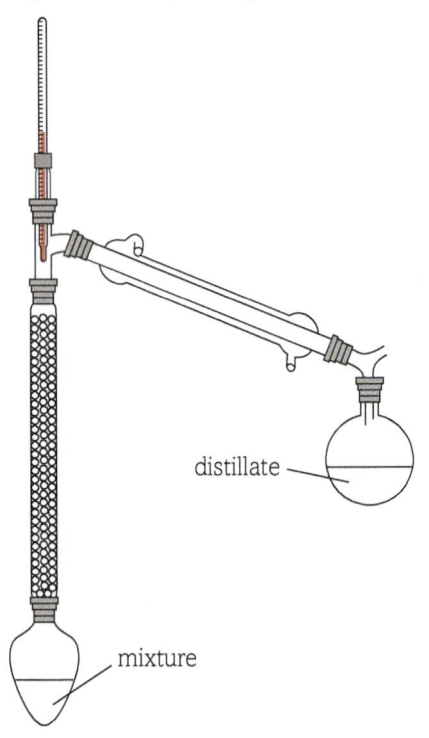

**fig D** Apparatus used in the process of fractional distillation.

Fractional distillation takes longer than simple distillation, and is best used when the difference in boiling temperatures is small, and when there are several compounds to be separated from a mixture.

## Solvent extraction

As the name suggests, this method involves using a solvent to remove the desired organic product from the other substances in the reaction mixture. There are several solvents that can be used, but the choice depends mainly on these features.

- The solvent added should be immiscible (i.e. not mix) with the solvent containing the desired organic product.
- The desired organic product should be much more soluble in the solvent added than in the reaction mixture.

Here is a summary of the process of **solvent extraction**.

- Place the reaction mixture in a separating funnel, and then add the chosen solvent – it should form a separate layer.
- Place the stopper in the neck of the funnel and gently shake the contents of the funnel for a while.
- Allow the contents to settle into two layers.
- Remove the stopper and open the tap to allow the lower layer to drain into a flask. Then pour the upper layer into a separate flask.

**fig E** Apparatus used in the process of solvent extraction.

If a suitable solvent is used and the method is followed correctly, most of the desired organic product will have moved into the added solvent. It is better to use the solvent in small portions than in a single large volume (for example, four portions of 25 cm$^3$ rather than one portion of 100 cm$^3$) because this is more efficient. Using more portions of solvent, but with the same total volume, removes more of the desired organic product.

The desired organic product has been removed from the reaction mixture, but is now mixed with the added solvent. So, simple distillation or fractional distillation now has to be used to separate the desired organic product from the solvent used.

## Drying

Many organic liquids are prepared using inorganic reagents, which are often used in aqueous solution. A liquid organic product may partially or even completely dissolve in water, so water may be an impurity that needs to be removed by a drying agent. One important feature of a drying agent is that it does not react with the organic liquid.

There are several drying agents available, but the most common ones are anhydrous metal salts, often calcium sulfate, magnesium sulfate and sodium sulfate. What these compounds have in common is that they form hydrated salts, so when they come into contact with water in an organic liquid they absorb the water as water of crystallisation. Anhydrous calcium chloride can also be used for some organic compounds, although it does react with others and is soluble in alcohols..

Here is a summary of the drying process.

- The drying agent is added to the organic liquid and the mixture is swirled or shaken, and then left for a period of time.
- Before use, a drying agent is powdery, but after absorbing water it looks more crystalline.
- If a bit more drying agent is added, and it remains powdery, this is an indication that the liquid is dry.
- The drying agent is removed either by decantation (pouring the organic liquid off the solid drying agent), or by filtration.

## A test for purity

So, using one or more of the purifying techniques, you have purified an organic compound. But is it pure?

For liquids, there is a simple way to test whether it is pure – measure its boiling temperature. Impurities raise the boiling temperature.

The boiling temperatures of pure organic compounds have been carefully measured and are widely available in data books and online. If you measure the boiling temperature of your organic compound, you can compare it with an accurate value and then make your decision about how pure it is.

The apparatus used depends on the volume of liquid available, and whether it is toxic or flammable. The apparatus used for simple distillation can be used.

A word of caution – this test may not be conclusive, because you may not be able to measure the boiling temperature of your organic compound accurately enough. Your thermometer might read too low or too high, so even if your measured boiling temperature exactly matches the one in the data book or online, you may wrongly assume that your compound is pure.

Also remember that different organic compounds can, by coincidence, have the same boiling temperature. For example, both 1-chloropentane and 2-methylpropan-1-ol boil at 108 °C.

### Learning tip
Give examples of when you would use fractional distillation instead of simple distillation.

## Questions

1. What is the purpose of using a thermometer during distillation?
2. What are the limitations of using the measurement of a boiling temperature as a way of assessing the purity of an organic liquid?

### Key definitions
**Simple distillation** is used to separate liquids with very different boiling temperatures.
**Fractional distillation** is used to separate liquids with similar boiling temperatures.
**Solvent extraction** is used to separate a liquid from a mixture by causing it to move from the mixture to the solvent.

# THINKING BIGGER

## TOWARDS A GREENER ENVIRONMENT

The extract below considers examples of the roles of catalysts in the modern petroleum industry.

# CATALYSTS FOR A GREEN INDUSTRY

### Important catalytic reactions
Today, the industrial world relies upon an enormous number of chemical reactions and an even greater number of catalysts. A selection of important reactions reveals the scope of modern catalysis and demonstrates how crucial it will be for chemists to achieve their environmental objectives.

### A sacrifice: worst case catalyst
A sacrificial, or stoichiometric, catalyst is used once and discarded. The amount of waste produced is not insignificant since these catalysts are used in stoichiometric amounts. For example, the catalyst may typically be in a 1 : 1 mole ratio with the main reactant.

In the manufacture of anthraquinone for the dyestuffs industry, for example, aluminium chloride is the sacrificial catalyst in the initial step, the acylation of benzene, see **fig A**. This is a type of Friedel–Crafts reaction in which the spent catalyst is discarded along with waste from the process. Fresh catalyst is required for the next batch of reactants. The problem is that the aluminium chloride complexes strongly with the products, i.e. Cl–, forming $[AlCl_4]^-$ and cannot be economically recycled, resulting in large quantities of corrosive waste.

**fig A** Dysprosium trifluoromethane sulfonate.

New catalysts, with better environmental credentials, are now being tried out. Compounds, such as the highly acidic dysprosium(III) triflate (trifluoromethane sulfonate, see **fig A**) offer the possibility of breaking away from the sacrificial catalyst by enabling the catalyst to be recycled.

### Low sulfur fuels: desulfurisation catalysis
Petroleum-derived fuels contain a small amount of sulfur. Unless removed this sulfur persists throughout the refining processes and ends up in the petrol or diesel. Pressure to decrease atmospheric sulfur has driven the development of catalytic desulfurisation. One of the problems was that much of the sulfur present was in compounds such as the thiophenes, which are stable and resistant to breakdown.

Aromatic hydrogenation

Hydrogenolysis

Elimination

Isomerisation

Double bond hydrogenation

**fig B** Desulfurisation of thiophenic compounds from petroleum

The catalyst molybdenum disulfide coated on an alumina support provided one solution. Cobalt is added as a promoter, suggesting that the active site is a molybdenum-cobalt sulfide arrangement. In the catalytic reaction (see **fig B**), which is essentially a hydrogenation sequence, the adsorbed thiophene molecule is hydrogenated and its aromatic stability destroyed. This enables the C–S bond to break and release the sulfur as hydrogen sulfide. This is an interesting example of a catalyst performing different types of reactions: hydrogenation, elimination and isomerisation.

Where else will I encounter these themes?

1  2  3  4  5

# Thinking Bigger 6

Let us start by considering the nature of the writing in the article.

> **1.** A good deal of scientific literacy is required to read this extract. Imagine that you are required to convince the general public that removing the sulfur from fuels is worth the extra cost. Design a pamphlet to present the argument as strongly as possible.

*Think about how illustrations can elicit very emotive responses. How might you use illustrations in your pamphlet?*

Now we will look at the chemistry in, or connected to, this article. Don't worry if you are not ready to give answers to these questions yet. You may like to return to the questions once you have covered other topics later in the book.

> **2. a.** Work out the molecular formula of thiophene (shown below).
>
> **b.** Calculate the percentage by mass of sulfur in the molecule.
>
> **c.** Write a balanced equation for the complete combustion of thiophene. (You can assume the oxidised product of sulfur is $SO_2$ only.)
>
> **3. a.** During the elimination (**fig B**) part of the reaction sequence, butan-1-thiol is converted into two products. Name them.
>
> **b.** The isomerisation process gives rise to two stereo isomers. Explain what is meant by a geometric isomer and name both.
>
> **c.** Why can the double bond hydrogenation reaction be considered to have 100% atom economy?

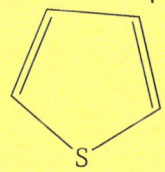

*In Chemistry, you will often need to make assumptions that allow simplification of calculations. In general these assumptions make little difference as they are chosen to ignore effects which have a very small impact on the real world answers.*

## Activity

You may wonder why sulfur appears in fossil fuels at all! The chemistry of sulfur gives it some special properties and, apart from carbon, hydrogen, oxygen and nitrogen, it is the only other element present in the building blocks of all proteins: amino acids. Prepare a 5 minute presentation to the class on the importance of sulfur in proteins. Your presentation should include:

- which amino acids contain sulfur
- what properties of sulfur make it so important in protein structure
- what the consequences of a diet low in sulfur can be

### Did you know?

Hydrogen sulfide ($H_2S$) is highly toxic but, luckily, most humans can detect it at concentrations of less than 0.5 parts per billion (or ppb; that's 1 molecule in $2 \times 10^9$ air molecules!). It also smells like rotten eggs so we get plenty of warning before the level of 800 000 ppb, which can be fatal, is reached!

- From an article in *Education in Chemistry* magazine, published by the Royal Society of Chemistry

# 6 Exam-style questions

1  The table lists the boiling temperatures of some alkanes.

| Alkane | Molecular formula | Boiling temperature/K |
|---|---|---|
| butane | $C_4H_{10}$ | 273 |
| pentane | $C_5H_{12}$ | 309 |
| hexane | $C_6H_{14}$ | 342 |
| heptane | $C_7H_{16}$ | 372 |
| octane |  | 399 |
| nonane | $C_9H_{20}$ |  |
| decane | $C_{10}H_{22}$ | 447 |

(a) Give the molecular formula of octane. [1]
(b) (i) Explain the trend in boiling temperature of the alkanes. [2]
   (ii) Predict a value for the boiling temperature of nonane. [1]
(c) Long chain alkanes, such as decane, can be cracked into shorter chain alkanes and alkenes.
   (i) Write an equation for the cracking of decane into octane and ethene. [1]
   (ii) The ethene produced can be converted into ethanol by direct hydration with steam.
       Write an equation for this reaction and state the conditions that are used in industry. [4]
(d) Reforming is a process used in the production of petrol. Unbranched-chain alkanes can be reformed to produce either branched-chain alkanes or cycloalkanes.
   The equation shows the reforming of decane into 2-methylnonane.

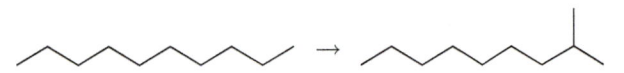

   (i) Using skeletal formulae, write an equation for the reforming of decane into 2,3-dimethyloctane. [1]
   (ii) Using skeletal formulae, write an equation for the reforming of heptane into methylcyclohexane. [2]
   (iii) State why reforming is used in the production of petrol. [1]
   [Total: 13]

2  Compound **Y** is a hydrocarbon containing 85.7% of carbon by mass.
(a) (i) Calculate the empirical formula of **Y**.
       [$A_r$: C, 12.0; H, 1.0] [2]
   (ii) The molar mass of **Y** is 56 g mol$^{-1}$. Show that the molecular formula of **Y** is $C_4H_8$. [1]
(b) There are six isomers for compound **Y**; four unsaturated molecules and two saturated molecules.
   Draw a displayed formula for each of the five isomers and name each compound. [6]
(c) Compound **Y** reacts with steam in the presence of an acid catalyst to form $CH_3CH_2CH_2CH_2OH$. Deduce the structural formula of **Y**. [2]
   [Total: 11]

3  Cyclohexane ($C_6H_{12}$) reacts with bromine to produce bromocyclohexane ($C_6H_{11}Br$).
   The mechanism for this reaction is radical substitution.
(a) (i) Write an equation for the initiation step in this reaction. [1]
   (ii) State the conditions necessary to bring about this initiation step. [1]
   (iii) The initiation step is followed by two propagation steps, resulting in the formation of $C_6H_{11}Br$. Write an equation for each of these two propagation steps. [2]
(b) Bromocyclohexane reacts with sodium hydroxide to produce either cyclohexanol or cyclohexene as the major product.

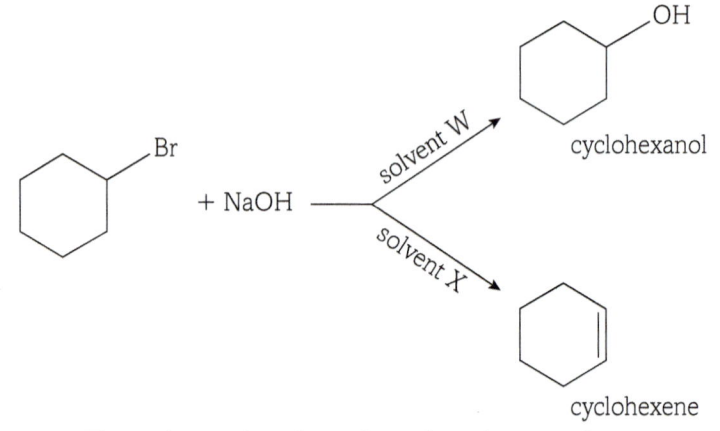

   The major product depends on the solvent used.
   (i) Identify solvent **W** and solvent **Y**. [2]
   (ii) Describe a simple test that could be performed to show that the product was cyclohexene and not cyclohexanol. [2]
(c) Cyclohexanol can be oxidised to cyclohexanone.
   (i) Name a suitable oxidising agent for this reaction. [1]
   (ii) Write an equation, using skeletal formulae, for the oxidation of cyclohexanol to cyclohexanone. Use [O] to represent the oxidising agent. [2]
   [Total: 11]

4 Halogenoalkanes undergo both substitution reactions with aqueous KOH and elimination reactions with ethanolic KOH.
  (a) Draw the mechanism for the substitution reaction between 1-chlorobutane and aqueous potassium hydroxide. [3]
  (b) Explain why substitution reactions are faster with 1-bromobutane than with 1-chlorobutane. [3]
  (c) Which pair of chlorobutanes would both give a hydrocarbon of formula $C_4H_6$ when separately treated with hot ethanolic KOH? [1]
    A $CH_3CH_2CH_2CH_2Cl$ and $CH_3CH_2CH_2CHCl_2$
    B $CH_3CHClCHClCH_3$ and $ClCH_2CH_2CH_2CH_2Cl$
    C $CH_3CH_2CH_2CH_2Cl$ and $ClCH_2CH_2CH_2CH_2Cl$
    D $CH_3CH_2CH_2CH_2Cl$ and $CH_3CH_2CHClCH_3$
  (d) Which chlorobutane is classified as a tertiary halogenoalkane? [1]
    A $H_3CH_2CH_2CH_2Cl$   B $CH_3CH_2CHClCH_3$
    C $(CH_3)_3CCl$   D $(CH_3)_2CHCH_2Cl$
[Total: 8]

5 Lavandulol is a compound that is found in lavender oil. The skeletal formula of lavandulol is

  (a) (i) Give the molecular formula of lavandulol. [1]
      (ii) Give the names of the two functional groups that are present in lavandulol. [2]
  (b) Lavandulol can be oxidised by acidified potassium dichromate(VI) solution to produce either compound **X** or compound **Y**.

    lavandulol        compound **X**        compound **Y**

    (i) State the conditions required to produce compound **X**. [2]
    (ii) State the conditions required to produce compound **Y**. [2]
    (iii) Describe a chemical test that could be performed to show that the compound formed was **X** and not **Y**. [2]
[Total: 9]

6 The following is a description of how to prepare a dry, pure sample of 2-chloro-2-methylpropane starting from methylpropan-2-ol.
The equation for the reaction is

$$CH_3-\underset{\underset{OH}{|}}{\overset{\overset{CH_3}{|}}{C}}-CH_3 + HCl \longrightarrow CH_3-\underset{\underset{Cl}{|}}{\overset{\overset{CH_3}{|}}{C}}-CH_3 + H_2O$$

methylpropan-2-ol        2-methyl-2-chloropropane

Step 1: Place about 9 cm³ of methylpropan-2-ol into a separating funnel and carefully add 20 cm³ of concentrated hydrochloric acid, about 3 cm³ at a time. After each addition, hold the stopper and tap firmly in place and invert the funnel a few times. Then, with the funnel in the upright position, loosen the stopper briefly to release any pressure.

Step 2: Leave the separating funnel and contents in a fume cupboard for about twenty minutes. Shake gently at intervals.

Step 3: Allow the layers in the separating funnel to separate and then run off, and discard, the lower aqueous layer.

Step 4: Add sodium hydrogencarbonate solution 2 cm³ at a time. Shake the funnel carefully after each addition and release the pressure of gas by loosening the stopper. Repeat until no more gas is evolved.

Step 5: Allow the layers to separate and then run off, and discard, the lower aqueous layer.

Step 6: Run the organic layer into a small, dry conical flask and add some anhydrous sodium sulfate. Swirl the flask occasionally for about five minutes.

Step 7: Carefully decant the organic liquid from the solid sodium sulfate into a pear-shaped flask.
Add a few anti-bumping granules and set up the flask for distillation. Collect the liquid that distils over between 47–53 °C.

  (a) Other than wearing safety spectacles and a protective coat, explain a safety precaution you should take when using concentrated hydrochloric acid. [2]
  (b) State the purpose of adding sodium hydrogencarbonate in step 4. [1]
  (c) State the purpose of adding anhydrous sodium sulfate in step 6. [1]
  (d) Draw a labelled diagram of the distillation of the organic liquid in step 7. [4]
  (e) A student started with 7.40 g of methylpropan-2-ol and obtained 7.82 g of 2-chloro-2-methylpropane. Calculate the percentage yield of the product. [3]
[Total: 11]

# TOPIC 7
# Modern analytical techniques

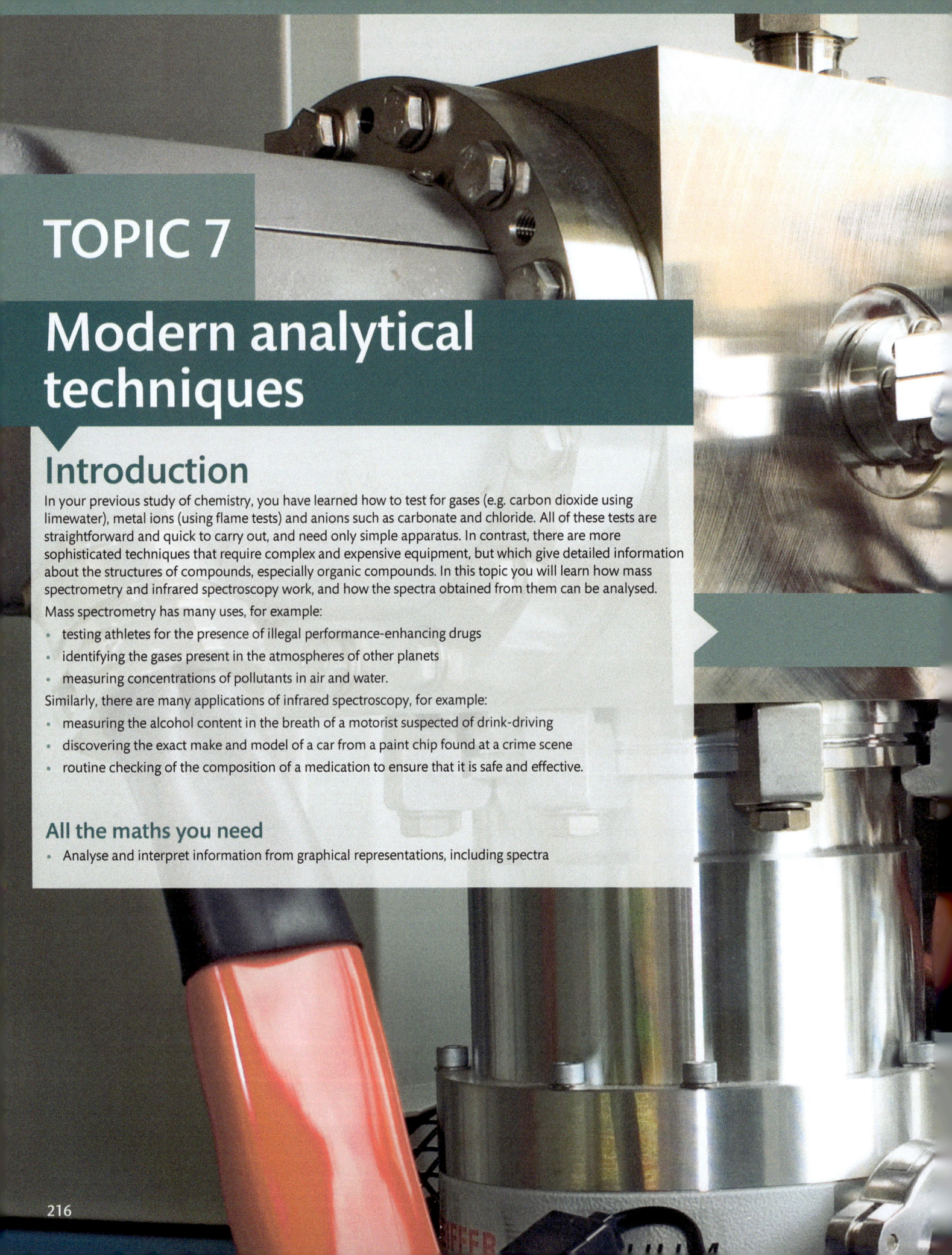

## Introduction

In your previous study of chemistry, you have learned how to test for gases (e.g. carbon dioxide using limewater), metal ions (using flame tests) and anions such as carbonate and chloride. All of these tests are straightforward and quick to carry out, and need only simple apparatus. In contrast, there are more sophisticated techniques that require complex and expensive equipment, but which give detailed information about the structures of compounds, especially organic compounds. In this topic you will learn how mass spectrometry and infrared spectroscopy work, and how the spectra obtained from them can be analysed.

Mass spectrometry has many uses, for example:
- testing athletes for the presence of illegal performance-enhancing drugs
- identifying the gases present in the atmospheres of other planets
- measuring concentrations of pollutants in air and water.

Similarly, there are many applications of infrared spectroscopy, for example:
- measuring the alcohol content in the breath of a motorist suspected of drink-driving
- discovering the exact make and model of a car from a paint chip found at a crime scene
- routine checking of the composition of a medication to ensure that it is safe and effective.

## All the maths you need
- Analyse and interpret information from graphical representations, including spectra

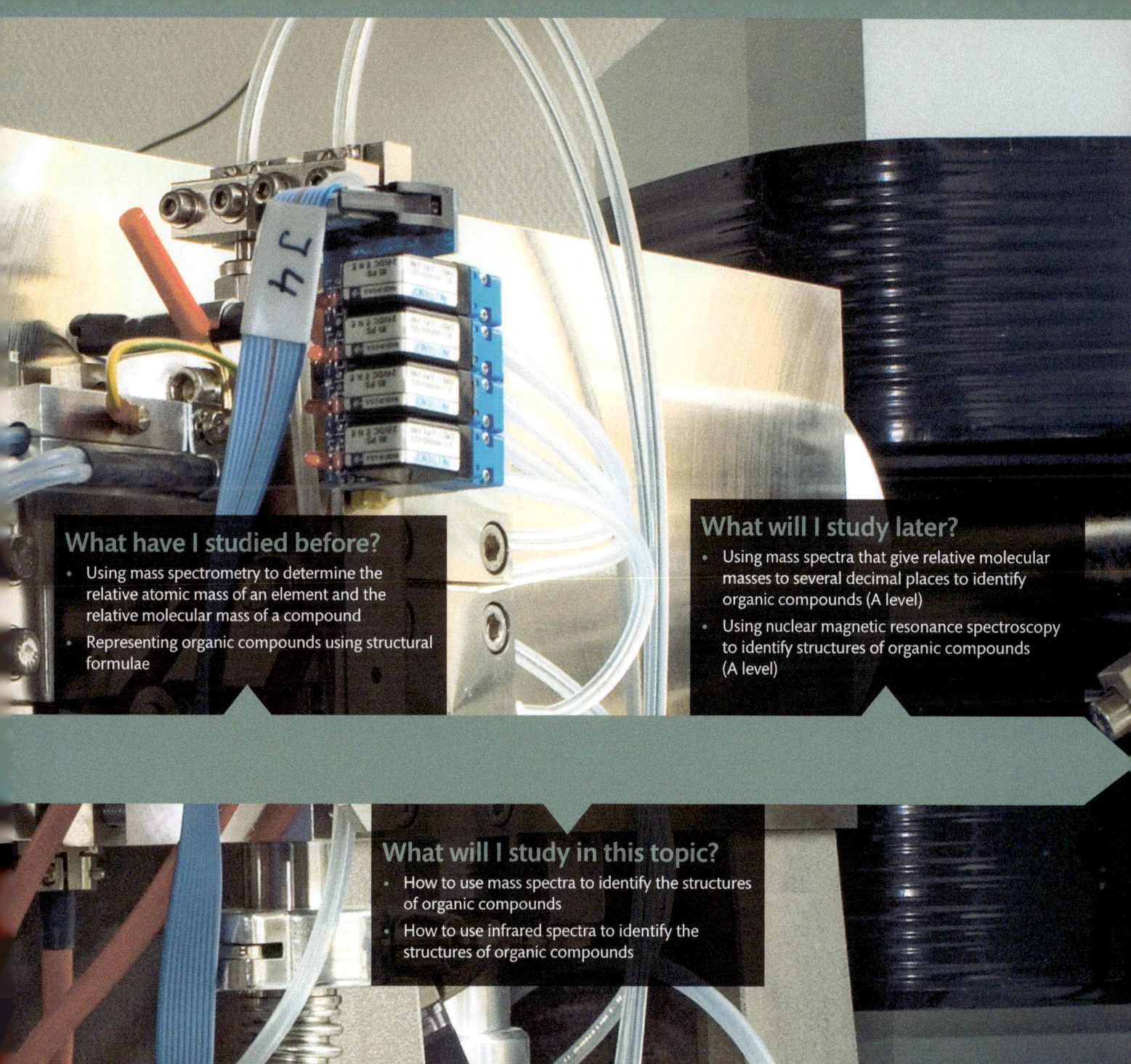

### What have I studied before?
- Using mass spectrometry to determine the relative atomic mass of an element and the relative molecular mass of a compound
- Representing organic compounds using structural formulae

### What will I study later?
- Using mass spectra that give relative molecular masses to several decimal places to identify organic compounds (A level)
- Using nuclear magnetic resonance spectroscopy to identify structures of organic compounds (A level)

### What will I study in this topic?
- How to use mass spectra to identify the structures of organic compounds
- How to use infrared spectra to identify the structures of organic compounds

# 7.1 1 Mass spectrometry in organic compounds

**By the end of this section, you should be able to...**

- use data from a mass spectrometer to determine the relative molecular mass of an organic compound from the molecular ion peak
- understand what happens during fragmentation

## Background to mass spectrometry

You have already come across the principles of mass spectrometry in **Section 1.1.2**. We will now look at how this technique can be used to determine the relative molecular mass of organic compounds, and also to determine the structures of some of them.

The mass spectrum of an element often appears very simple, with a very small number of vertical lines, called peaks, each one representing an isotope of the element.

In contrast, the mass spectrum of an organic compound often appears complex, with a large number of peaks. If all the peaks are included, there may be more peaks than there are atoms in a molecule of the compound, so something different is happening.

## The molecular ion peak

The obvious thing to do when first looking at the mass spectrum of an organic compound is to find the peak furthest to the right. This is the one with the greatest $m/z$ value (mass to charge ratio). This peak is the **molecular ion peak** – the result of the organic molecule losing an electron in the mass spectrometer. The equation for this process, using butane as an example, is:

$$C_4H_{10} + e^- \rightarrow C_4H_{10}^+ + 2e^-$$

An electron collides with a butane molecule and knocks out an electron, so forming a positive ion from the molecule. The $m/z$ value of this peak (58) indicates the relative molecular mass of butane.

## Other peaks

The spectrum may show a very small peak just to the right of the molecular ion peak (sometimes referred to as the $M+1$ peak). This is caused by the presence of a naturally occurring isotope of carbon ($^{13}C$, rather than the usual $^{12}C$ isotope) in the molecule. Approximately 99% of all carbon atoms are $^{12}C$, with most of the remaining 1% being $^{13}C$.

You may have met the $^{14}C$ isotope, which also occurs naturally in organic compounds. This isotope is radioactive and used in radiocarbon dating. The proportion of $^{14}C$ atoms in a sample of an organic compound is extremely small and can be ignored in mass spectrometry.

The peaks with smaller $m/z$ values result from fragmentation in the mass spectrometer, rearrangement reactions and the loss of more than one electron. Rearrangement is often unpredictable and will not be considered in this book, but it does help to explain the large number of peaks in some spectra. The breaking of a carbon–hydrogen bond can occur, but this is not usually described as fragmentation.

## Fragmentation in hydrocarbons

**Fragmentation** is very common and can often be used to work out the structure of an organic molecule.

Consider the breaking of a carbon–carbon bond in the molecular ion formed from a hydrocarbon. Two species are formed. They are:

- another positive ion
- a neutral species (usually a radical).

# Mass spectrometry 7.1

## Examples of fragmentation

A very simple example of fragmentation is the molecular ion of ethane, which can fragment to form a methyl cation and a methyl radical:

$$(CH_3-CH_3)^+ \rightarrow CH_3^+ + CH_3$$

The ethane molecule is symmetrical, and there is only one carbon–carbon bond in ethane, so you can imagine that the right-hand carbon is just as likely to become the positive ion. The equation for this fragmentation would be:

$$(CH_3-CH_3)^+ \rightarrow CH_3 + CH_3^+$$

You can see that the products are identical, so the spectrum does not depend on how the bond breaks – there will be a peak at $m/z = 15$.

Sometimes the radical formed is shown with a dot (representing the unpaired electron), and the molecular ion is shown with a dot as well as a positive charge, as in this example:

$$(CH_3-CH_3)^{+\bullet} \rightarrow CH_3^\bullet + CH_3^+$$

Now consider propane. There are two carbon–carbon bonds, but they are equivalent – they can both be described as the bond between the central carbon and one of the two terminal carbons. However, there are now two possible fragment ions that can form:

$$(CH_3-CH_2-CH_3)^+ \rightarrow CH_3^+ + CH_2-CH_3$$

$$(CH_3-CH_2-CH_3)^+ \rightarrow CH_3 + (CH_2-CH_3)^+$$

You would therefore expect to see peaks at $m/z = 15$ (the methyl cation) and $m/z = 29$ (the ethyl cation) in its spectrum. These peaks are present, but there are several others that are difficult to explain and are of no help in deducing the structure.

Radicals are not detected in a mass spectrometer, so all the peaks formed by fragmentation are caused by positive ions.

## Possible information

Here is the information you could be given in questions.

- A complete mass spectrum – the disadvantage of this is the possible large number of peaks that cannot be used to work out the structure and would be distracting. Another possibility is a complete mass spectrum, but with only the $m/z$ values of the useful peaks marked on the spectrum.
- A simplified mass spectrum showing only the peaks that will help you work out a structure.
- A list of the $m/z$ values of the useful peaks.

With practice, you will be able to work out the structure of an organic compound from this information.

## A typical mass spectrum

The traditional way to present a mass spectrum is to label the vertical axis as relative intensity (%), always from 0% to 100%. The horizontal axis is labelled $m/z$ (with no units). The horizontal axis usually, but not necessarily, starts from zero and continues to just beyond the molecular ion peak.

The tallest peak is sometimes referred to as the **base peak**.

This base peak represents the ion with the highest abundance, and is shown with a relative intensity of 100%. It represents the most stable fragment.

This is the mass spectrum of butane:

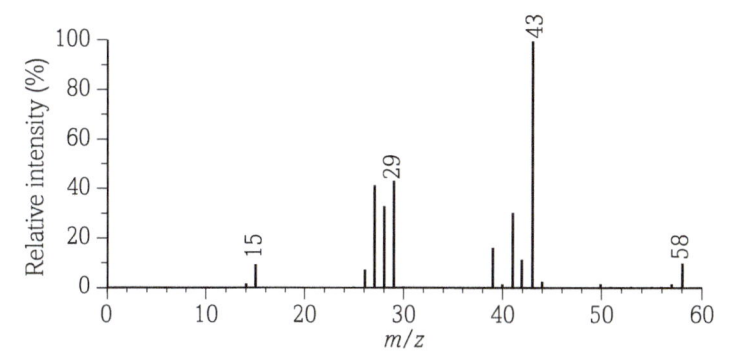

Among the visible peaks are those labelled 15, 29, 43 and 58. The table shows the origin of these peaks.

| $m/z$ | Ion | Notes |
|---|---|---|
| 15 | $CH_3^+$ | |
| 29 | $(CH_3-CH_2)^+$ | |
| 43 | $(CH_3-CH_2-CH_2)^+$ | This is the most abundant peak. |
| 58 | $(CH_3-CH_2-CH_2-CH_3)^+$ | This is the molecular ion peak. |

**table A** The origin of the peaks in the mass spectrum of butane.

### Learning tip

Be clear about the difference between a peak with a large $m/z$ value and a peak with a large abundance.

The $m/z$ value indicates which ion is present, but the height of the peak indicates its abundance.

## Questions

1. Explain the origin of the M+1 peak seen in some mass spectra.

2. Write an equation to show the formation of the base peak in the fragmentation of the molecular ion of butane.

### Key definitions

The **molecular ion peak** indicates the species formed from the molecule by the loss of one electron.

**Fragmentation** occurs when the molecular ion breaks into smaller pieces.

The **base peak** indicates the peak with the greatest abundance.

# 7.1 2 Deducing structures from mass spectra

**By the end of this section, you should be able to...**

- use data from a mass spectrometer to suggest possible structures of a simple organic compound from the m/z of the molecular ion and fragmentation patterns

| m/z value | Possible ions |
|---|---|
| 15 | $CH_3^+$ |
| 17 | $OH^+$ |
| 28 | $CO^+$ |
| 29 | $CH_3CH_2^+$ and $CHO^+$ |
| 31 | $CH_2OH^+$ |
| 43 | $CH_3CH_2CH_2^+$ and $CH_3CHCH_3^+$ and $CH_3CO^+$ |
| 45 | $COOH^+$ and $CH_3CHOH^+$ |
| 57 | $C_4H_9^+$ (this represents four possible structures) |

**table A** Common m/z values and the possible ions responsible for these peaks.

## Fragmentation in other organic compounds

So far we have looked only at mass spectra of alkanes. Now let's look at other organic compounds – those containing oxygen.

In this section we will look at two examples of simplified mass spectra and see how they can be used to work out the structures of the compounds responsible for them.

Oxygen is present in many organic compounds, so you need to be familiar with other m/z values. You may also come across compounds containing nitrogen (amines) and halogens (halogenoalkanes).

The table shows some common m/z values and possible ions responsible for these peaks.

### Example 1

Two compounds, A and B, have the molecular formula $C_3H_6O$. Their simplified mass spectra are shown below.

Can you deduce the structure of each one?

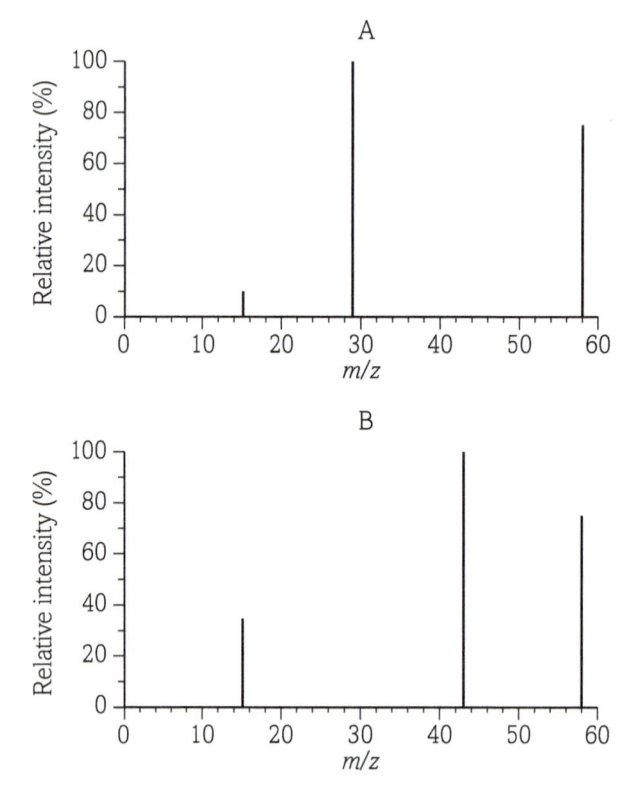

### Interpretation

Here are the main points that can be deduced from the simplified mass spectra.

- Both mass spectra show a peak at m/z = 58. This corresponds to the molecular ion $C_3H_6O^+$ (relative molecular mass of $C_3H_6O$ is 58.0), but this does not help in deducing the structure.

- Both A and B have a peak at $m/z = 15$, which is caused by the $CH_3^+$ ion. As this is present in both, it does not help to distinguish between the two structures.

- One obvious difference is a major peak at $m/z = 29$ in A, which is not present in B. This could be caused by either $CH_3CH_2^+$ or $CHO^+$, or by both of them. The structure $CH_3CH_2CHO$ fits perfectly – this is the aldehyde propanal.

- The other obvious difference is a major peak at $m/z = 43$ in B, which is not present in A. This could not be caused by either $CH_3CH_2CH_2^+$ or $CH_3CHCH_3^+$ because the radical produced at the same time would have a mass of 15 (they must add up to 58), and oxygen has a mass of 16. The other possibility is $CH_3CO^+$, which when considered with the peak at $m/z = 15$ ($CH_3^+$) suggests the structure $CH_3COCH_3$ – this is the ketone propanone.

You may be asked to write equations to show the formation of the ions you have used in your deduction.

In this example, they are:

A ($m/z = 29$)  $(CH_3CH_2CHO)^+ \rightarrow CH_3CH_2^+ + CHO\bullet$

and $(CH_3CH_2CHO)^+ \rightarrow CH_3CH_2\bullet + CHO^+$

B ($m/z = 43$)  $(CH_3COCH_3)^+ \rightarrow CH_3CO^+ + CH_3\bullet$

## Example 2

Two compounds, C and D, have the molecular formula $C_3H_8O$. Their simplified mass spectra are shown below.

Can you deduce the structure of each one?

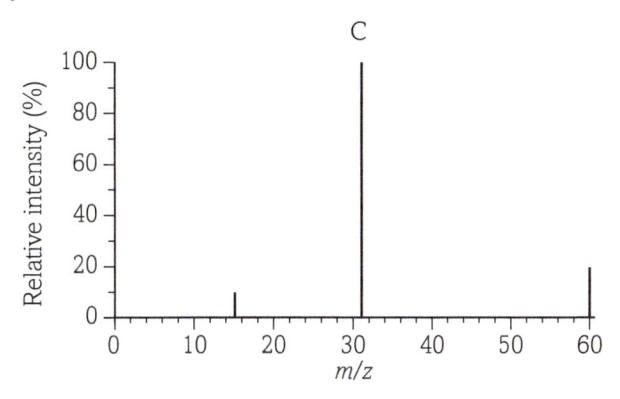

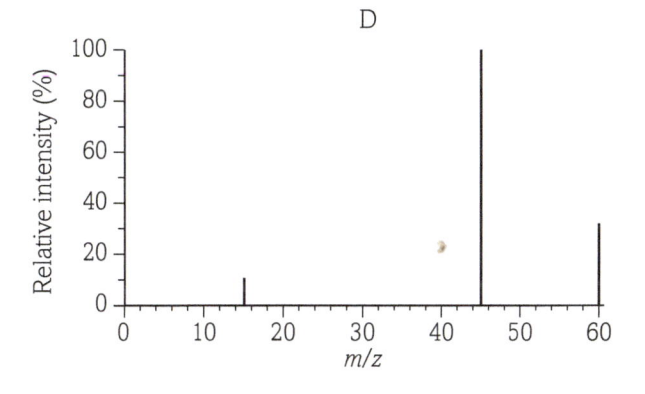

### Interpretation

Here are the main points that can be deduced from the simplified mass spectra.

- Both mass spectra show a peak at $m/z = 60$. This corresponds to the molecular ion $C_3H_8O^+$ (relative molecular mass of $C_3H_8O$ is 60.0), but this information does not help in deducing the structure.

- Both C and D have a peak at $m/z = 15$, which is caused by the $CH_3^+$ ion. As this is present in both, this information does not help to distinguish between the two structures.

- One obvious difference is a major peak at $m/z = 31$ in C, which is not present in D. This could be caused by $CH_2OH^+$. When considered with the peak at $m/z = 15$ ($CH_3^+$) and a $CH_2$ group, this suggests the structure $CH_3CH_2CH_2OH$, which is propan-1-ol.

- The other obvious difference is a major peak at $m/z = 45$ in D, which is not present in C. This could not be caused by $COOH^+$ because D contains only 1 oxygen atom. The other possibility is $CH_3CHOH^+$, which when considered with the peak at $m/z = 15$ ($CH_3^+$) suggests the structure $CH_3CH(OH)CH_3$, which is propan-2-ol.

### Learning tip

It is sometimes worth comparing the $m/z$ values of fragment ions with that of the molecular ion. The difference between the two indicates what has been lost during fragmentation.

For example, if the molecular ion peak is at $m/z = 60$ and there is a fragment ion with $m/z = 45$, the difference is 15, which suggests the loss of a $CH_3$ group.

## Questions

1. Write an equation to show the formation of the ethyl cation in the fragmentation of the molecular ion of pentane.

2. Write equations to show the formation of the ions used in the deduction of the structures of C and D (from Example 2 above).

# 7.2 1 Infrared spectroscopy

**By the end of this section, you should be able to…**

- understand what happens when some molecules absorb infrared radiation
- understand how infrared spectra can provide information about the bonds in a molecule

## Infrared radiation

The electromagnetic spectrum of radiation includes **infrared radiation**. The 'infra' part of the word comes from the Latin for 'below', so this radiation has a frequency below, or less than, that of red light.

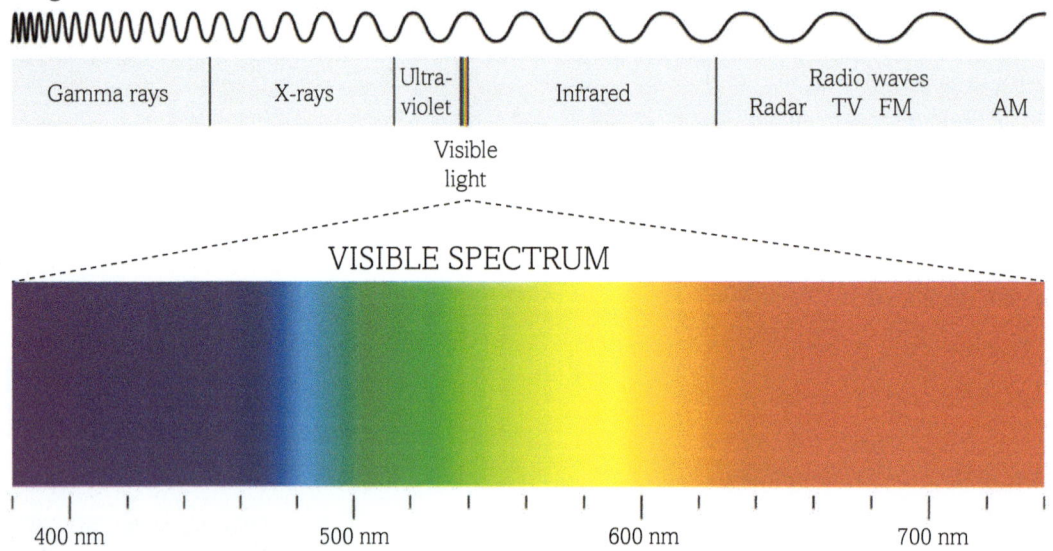

**fig A** The electromagnetic spectrum of radiation.

## What happens when some molecules absorb infrared radiation?

The importance of infrared radiation in chemistry is that it is absorbed by molecules and causes two possible effects, both described as vibrations. These effects are:

- **stretching** – where the bond length increases and decreases
- bending – where the bond angle increases and decreases.

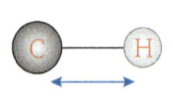

The C–H bond stretches when it absorbs infrared radiation.

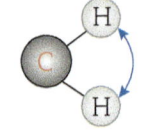

The C–H bond bends when it absorbs infrared radiation.

In this book, only stretching vibrations will be considered.

When a molecule absorbs infrared radiation, the amount of energy absorbed depends on:

- the length of the bond
- the strength of the bond
- the mass of each atom involved in the bond.

The absorption of infrared radiation is linked to changes in the polarity of the molecule, so simple non-polar molecules (such as $H_2$ and $Cl_2$) do not absorb infrared radiation.

## What does an infrared spectrum look like?

When a compound is irradiated by infrared radiation, the bonds in the molecules absorb radiation from some parts of the spectrum, but not from others.

### Axes

The spectrum is normally shown with the vertical axis labelled **transmittance**, shown as a percentage from 0 to 100. A value of 100% transmittance means that 100% of the radiation is transmitted and none is absorbed.

The horizontal axis could be labelled either as frequency or wavelength, but a different unit is used: **wavenumber**. This is the reciprocal of the wavelength, and so it represents frequency. It is usually quoted in the unit $cm^{-1}$. The numerical scale normally starts at $4000\,cm^{-1}$ and ends at $500\,cm^{-1}$. It may seem unusual for the numbers to decrease from left to right, but the left to right direction does represent increasing frequency. Another unusual feature is that the scale changes after $2000\,cm^{-1}$.

Here is an example.

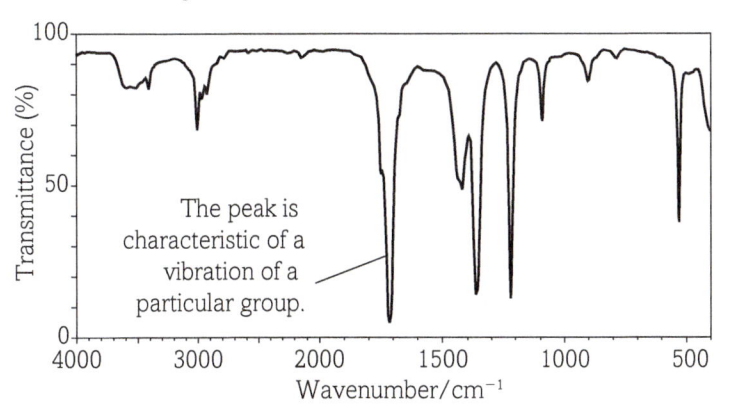

**fig B** A typical infrared spectrum.

### Absorptions and their intensities

You can see that much of the spectrum consists of an almost horizontal line close to 100%, but at specific wavenumbers there are dips or troughs. These are referred to as absorptions (and sometimes as peaks). The actual transmittance value of an absorption is not very important, but its **intensity** is.

Weak intensities refer to high transmittance values, and strong intensities to low transmittance values.

### Wavenumber values

The wavenumber values are also very important. The spectrum above shows that the absorption with the lowest transmittance occurs at about $1700\,cm^{-1}$. Sometimes, another important feature to note is whether the absorption is sharp (i.e. a narrow wavenumber range) or broad (a wide wavenumber range).

### Characteristic absorptions

You do not need to remember any of the information in the following table, as it will be provided in an examination. However, it is important to be familiar with how to use it when interpreting infrared spectra.

| Wavenumber/$cm^{-1}$ | Bond | Functional group |
|---|---|---|
| 3750–3200 | O–H | alcohol |
| 3500–3300 | N–H | amine |
| 3300–2500 | O–H | carboxylic acid |
| 3095–3010 | C–H | alkene |
| 2962–2853 | C–H | alkane |
| 2900–2820, 2775–2700 | C–H | aldehyde |
| 1740–1720 | C=O | aldehyde |
| 1725–1700 | C=O | carboxylic acid |
| 1720–1700 | C=O | ketone |
| 1669–1645 | C=C | alkene |

**table A** Information to help interpret an infrared spectrum.

In the example spectrum in this section, the strong absorption at $1700\,cm^{-1}$ is caused by C=O, but the compound responsible for this spectrum could be a carboxylic acid or a ketone.

Sometimes a missing absorption is just as useful. In the example, there is no absorption in the $3300$–$2500\,cm^{-1}$ region, so the compound does not contain an OH group. Therefore, the absorption at $1700\,cm^{-1}$ strongly suggests that the compound is a ketone and not a carboxylic acid.

### Learning tip

In an infrared spectrum, remember that the horizontal scale uses wavenumber to represent frequency, and that the scale changes at $2000\,cm^{-1}$.

## Questions

1. Explain why hydrogen fluoride absorbs infrared radiation but fluorine does not.

2. A compound has an infrared spectrum that shows a broad absorption centred on $2850\,cm^{-1}$ and a sharp absorption at $1710\,cm^{-1}$. Suggest what homologous series the compound belongs to.

### Key definitions

**Infrared radiation** is the part of the electromagnetic spectrum with frequencies below that of red light.

**Stretching** occurs when a bond absorbs infrared radiation and uses it to alter the length of the bond.

The **transmittance** value in an infrared spectrum represents the amount of radiation absorbed at a particular wavenumber.

The **wavenumber** of an infrared absorption represents the frequency of infrared radiation absorbed by a particular bond in a molecule.

The **intensity** of an infrared absorption describes the amount of infrared radiation absorbed.

# 7.2 2 Using infrared spectra

**By the end of this section, you should be able to...**

- use data from infrared spectra to deduce functional groups present in organic compounds and to predict infrared absorptions, given wavenumber data, caused by familiar functional groups

## Different ways of using infrared spectra

There are three main ways your understanding of infrared spectra could be tested, and you need to become familiar with all of them. They are:

- predicting the spectrum of an organic compound
- deducing the functional groups from a list of wavenumbers
- deducing the structure from wavenumbers and molecular formula.

They all require the use of wavenumber data, like the data from the table in the previous section. This will either be provided in the question, or you will use information from the data book provided in an examination. We have reproduced the table from the previous section here for convenience.

| Wavenumber/cm$^{-1}$ | Bond | Functional group |
|---|---|---|
| 3750–3200 | O–H | alcohol |
| 3500–3300 | N–H | amine |
| 3300–2500 | O–H | carboxylic acid |
| 3095–3010 | C–H | alkene |
| 2962–2853 | C–H | alkane |
| 2900–2820<br>2775–2700 | C–H | aldehyde |
| 1740–1720 | C=O | aldehyde |
| 1725–1700 | C=O | carboxylic acid |
| 1720–1700 | C=O | ketone |
| 1669–1645 | C=C | alkene |

**table A** Information to help interpret an infrared spectrum.

## Fingerprint region

You may come across the term 'fingerprint region', although you do not need to know anything about it.

The term is worth explaining. The table of wavenumbers starts at 3750 and ends at 1645 cm$^{-1}$, even though there are many absorptions in the infrared region between 1500 and 500 cm$^{-1}$. Most of the absorptions in this region result from bending vibrations (not considered in this book) or from absorptions by bonds not listed in the table of wavenumbers.

This region is sometimes referred to as the 'fingerprint region' because, although individual absorptions are not easily recognised, the whole pattern acts like a fingerprint that is slightly different for similar molecules. It is recommended that you ignore this region completely.

## Predicting the spectrum of an organic compound

Suppose you are given the identity of an organic compound – this might be a formula (displayed, skeletal or structural) or a name. You should then be able to predict the wavenumber ranges of the compound's infrared spectrum.

## Example 1

What absorptions would you expect to find in the infrared spectrum of propanal?

Once you recognise propanal as an aldehyde, you could predict:
- weak absorptions in the ranges 2900–2820 and 2775–2700 cm$^{-1}$ resulting from the C—H bond
- strong absorption in the range 1740–1720 cm$^{-1}$ resulting from the C=O bond.

## Example 2

What absorptions would you expect to find in the infrared spectrum of CH$_3$CH(OH)CH$_3$?

This compound is an alcohol, so you could predict a broad absorption in the range 3750–3200 cm$^{-1}$ resulting from the O–H bond.

# Deducing the functional groups from a list of wavenumbers

## Example 1

An organic compound absorbs in the infrared region at these wavenumbers: 3675, 2870 and 1735 cm$^{-1}$. Which functional groups does it contain?

**Answer**

O—H (alcohol), C—H and C=O (aldehyde).

## Example 2

An organic compound has absorptions in these wavenumber ranges in the infrared region: 3500–3300 and 3300–2500 cm$^{-1}$. Which functional groups does it contain?

**Answer**

O—H (carboxylic acid) and N—H (amine): it could be an amino acid.

# Deducing the structure from wavenumbers and molecular formula

This is a bit more complicated. Some molecular formulae could represent different combinations of functional groups, so to make the decision you need to consider the actual functional groups from the spectrum and how they could be used in conjunction with the molecular formula.

## Example 1

A compound has infrared absorptions at 1730 and 3450 cm$^{-1}$, and has a molecular formula of C$_2$H$_4$O$_2$. Deduce a possible structure for it.

**Answer**

The functional groups are O—H (alcohol) and C=O (aldehyde). The only structure with the molecular formula C$_2$H$_4$O$_2$ that fits is CH$_2$(OH)CHO. Note that it is not CH$_3$COOH (ethanoic acid) because this would have an absorption in the range 3300–2500 cm$^{-1}$, not at 3450 cm$^{-1}$.

## Example 2

An organic compound has a molecular formula of C$_3$H$_6$O$_2$. Its infrared spectrum is shown below.

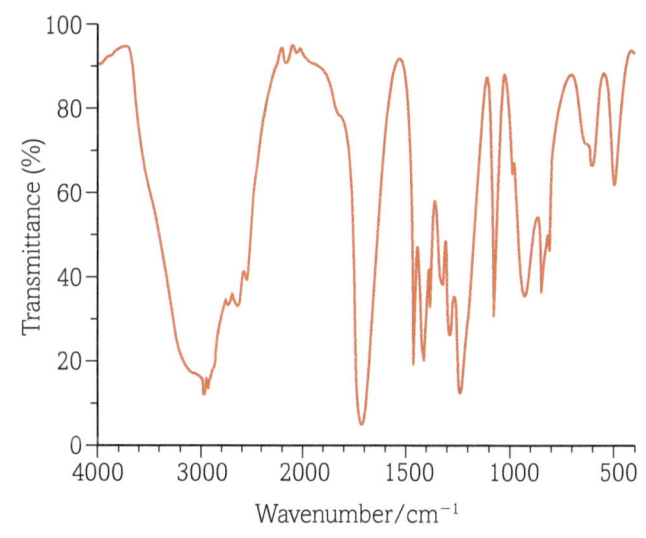

**fig A** The infrared spectrum of an unknown compound.

Deduce a possible structure for it.

**Answer**

There is a broad absorption at 3000 cm$^{-1}$ and a strong narrow absorption at 1700 cm$^{-1}$. These suggest the presence of O—H (carboxylic acid) and C=O (carboxylic acid). So the compound must contain the COOH group, which leaves C$_2$H$_5$ to make up the rest of the molecular formula. The only possible structure is CH$_3$CH$_2$COOH, which is propanoic acid.

### Learning tip

Concentrate on the region between 4000 and 1500 cm$^{-1}$. Practise making predictions of the spectra of molecules in the organic chemistry topic in this book.

# Questions

1. Why is it not easy to use infrared spectra to distinguish between propan-1-ol and propan-2-ol?

2. A compound has the molecular formula C$_3$H$_6$O. How can its infrared spectrum be used to show that it has the structure (CH$_3$)$_2$CO and not CH$_3$C(OH)=CH$_2$?

# THINKING BIGGER

## UNBEATABLE RECORDS?

The following extract comes from an article entitled 'Five rings good, four rings bad'. The article highlights the problem of drug misuse in sport and, in particular, the use of 'designer' steroids that are not detectable by routine drug testing. This extract focusses on how analytical chemists identified a sample of an unknown steroid as tetrahydrogestrinone (THG).

## HOW DON CATLIN'S TEAM CRACKED THG

On 13 June 2003, Don Catlin received a methanolic solution of an unknown steroid, recovered from a hypodermic syringe. He ran standard GC-MS tests on the solution, and synthesised several derivatives. Attempts to identify the steroid failed because the mass spectrum contained a large number of unidentifiable peaks. The only compound that they could identify at this stage was a small amount of another anabolic steroid, norbolethone, evidently present as an impurity.

Catlin suspected that the 'unknown' shared a common carbon skeleton with norbolethone. However, they noted a peak in its mass spectrum with $m/z = 312$, and thought this was the molecular ion. Accurate mass measurement gave 312.2080, from which they deduced the compound had the molecular formula $C_{21}H_{28}O_2$.

**fig A** Don Catlin.

When they compared the mass spectrum of the unknown with other steroids, it became clear that it shared features with gestrinone and trenbolone. All three compounds had the same fragments with $m/z$ values trenbolone.

All three compounds had the same fragments with $m/z$ values at 211 and below present, so Catlin deduced that they contained the same A, B and C rings.

Furthermore, when the MS of the unknown was compared with gestrinone, the fragments with $m/z$ above 240 occur 4 Da higher in the unknown, suggesting that it was gestrinone with four additional hydrogen atoms. A possibility was that the terminal alkyne group in gestrinone had been reduced to an ethyl group.

Having tentatively identified the unknown steroid as tetrahydrogestrinone, the team then prepared an authentic sample of THG by catalytic hydrogenation of gestrinone. This required careful control of conditions (0 °C) to prevent hydrogenation of C=C double bonds (see equation). The retention time and mass spectra of the synthetic THG matched the unknown material exactly.

gestrinone + 2H$_2$ $\xrightarrow{\text{0 °C} \atop \text{Pd/C catalyst}}$ tetrahydrogestrinone

**fig C** The mass spectrums of gestrinone (L) and tetrahydrogestrinone (R).

To study the metabolism of THG in mammals, the team gave intravenous doses of THG to a baboon, and collected urine samples from the animal over several days. Detectable amounts of THG were found in urine for many hours after administration.

THG was thus directly detectable in urine samples, though it defies detection by the standard procedure involving derivatisation into the Me$_3$Si derivatives.

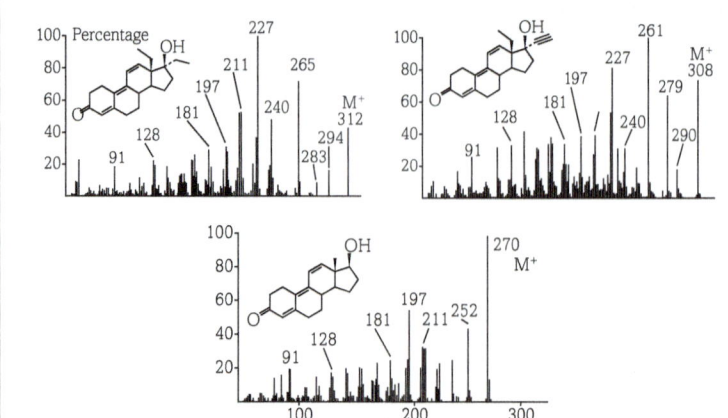

**fig B** The mass spectra of the unknown substance, gestrinone and trenbolone.

Where else will I encounter these themes?

1    2    3    4    5

# Thinking Bigger

The different groups of questions below will ask you to think about this article in different ways, including considering the way the author has presented the science for the audience, the writing itself, as well as the ways in which the article is linked to science you have studied in this topic and that you will go on to study later.

The first question is about the nature of the scientific writing itself rather than the science being communicated.

> **1. a.** Who do you think is the intended audience for this article? Evaluate the ways in which the author has written for them.
> **b.** Why do you think value judgements are avoided by the author even though the article considers a very emotive issue?

*Command word*
*An evaluation should review all the information to form a conclusion. You should think about the strengths and weaknesses of the evidence and information, and come to a supported judgement.*

Now we will look at the chemistry in, or connected to, this article. Don't worry if you are not ready to give answers to these questions yet. You may like to return to the questions once you have covered other topics later in the book.

> **2.** The analytical techniques of IR spectroscopy and mass spectrometry can be used to identify unknown molecules. Suggest which of these two techniques would be more useful in distinguishing samples of two steroids with similar structures. You should be prepared to justify your choice.
> **3.** The mass spectrum of trenbolone is shown in fig B. Suggest how the fragment at $m/z = 252$ is generated.

*The Dalton (Da) is the standard unit for indicating mass on an atomic or molecular scale.*

## Activity

The free-to-access database at NIST Chemistry WebBook (webbook.nist.gov/chemistry) allows you to search for a range of organic compounds and related data.

Use the database to find the mass spectrum of one of the following compounds:

(a) Chlorobenzene
(b) Bromoethane
(c) Ethylamine
(d) Cyclohexane

Prepare a 3–5 minute presentation showing the mass spectrum of your chosen molecule and identifying the most important peaks. Your presentation should include:

- a picture of the mass spectrum of your chosen molecule
- an identification of the main fragment and isotopic abundance peaks of your molecule explaining how each peak is formed.

### Did you know?

The women's 800 m record was set in 1983 by the Czechoslovakian Jarmila Kratochvilova, who ran the distance in 1:53.28. This was before the test for human growth hormone was in routine use at athletics competitions. Since then, only one athlete has managed to come within a second of her record. It has been speculated that many records in women's athletics may never be broken.

- From an article in *Education in Chemistry* magazine, published by the Royal Society of Chemistry

# 7 Exam-style questions

1 A primary alcohol can be oxidised by reaction with acidified potassium dichromate(VI). The major product obtained depends on the conditions used.

   If the oxidising agent is slowly added to the alcohol and then the product is distilled off as it forms, an aldehyde is collected.

   If the alcohol is heated under reflux with an excess of the oxidising agent, a carboxylic acid is formed.

   The infrared spectrum is that of a product formed by the oxidation of butan-1-ol.

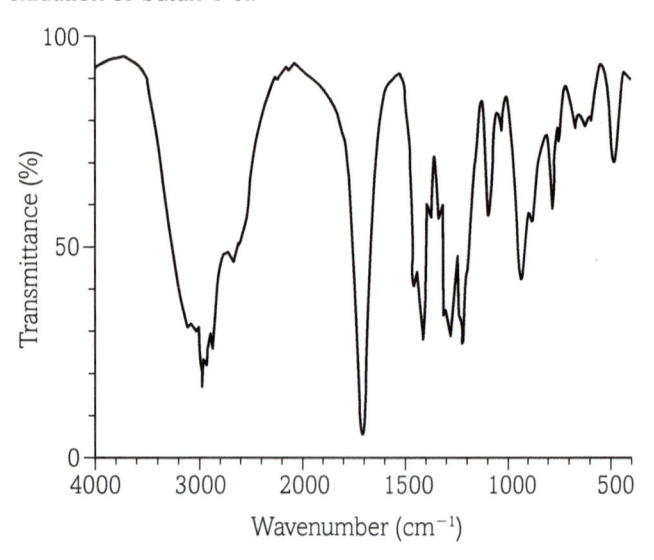

   (a) Identify the product and explain your reasoning. [3]
   (b) Write an equation for the oxidation of butan-1-ol to this product. Use [O] to represent the oxidising agent. [2]
   [Total: 5]

2 When halogenoalkanes are refluxed with a solution of sodium hydroxide, two products can be formed. One is an alcohol, the other is an alkene. The major product is determined by the solvent used.

   The skeletal formula of chlorocyclohexane is

   A student refluxed a solution of chlorocyclohexane with sodium hydroxide. The organic product was separated and analysed.

   The infrared spectrum of the organic product is shown.

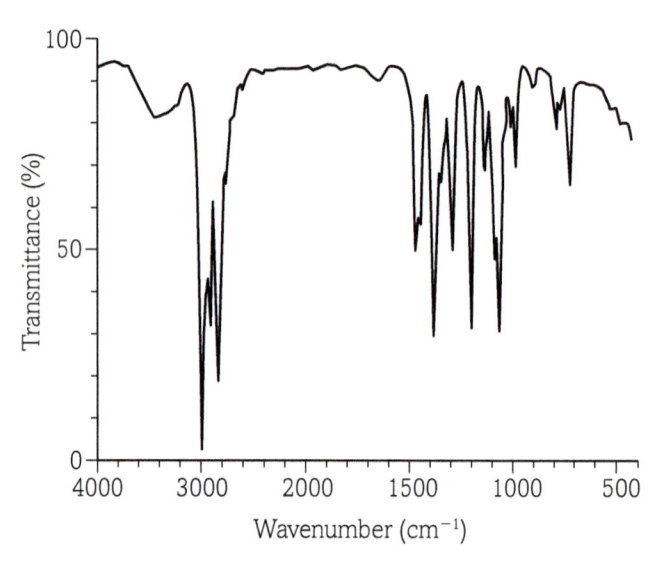

   The student concluded that cyclohexene had been produced.

   (a) State what is meant by the term **refluxed**. [1]
   (b) Explain why the student was justified in ruling out cyclohexanol as a product. [2]
   (c) Describe a simple chemical test to confirm that an alkene had been formed. [2]
   [Total: 5]

3 **Spectrum A** and **Spectrum B** are the mass spectra of pentan-2-one ($CH_3COCH_2CH_2CH_3$) and pentan-3-one, ($CH_3CH_2COCH_2CH_3$), but not necessarily in that order.

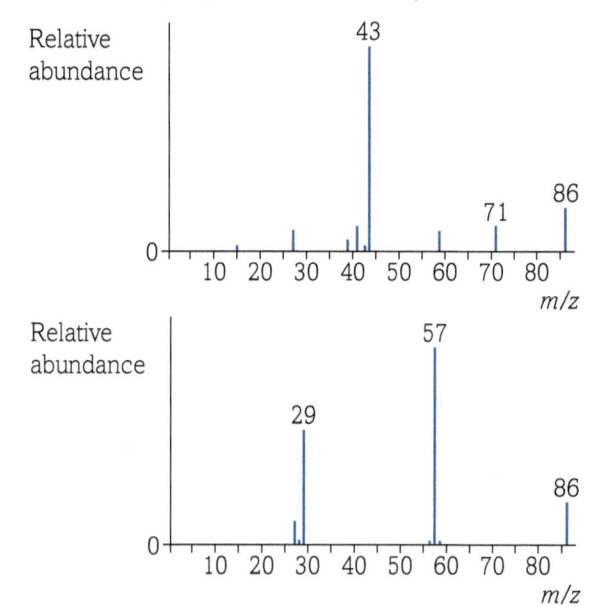

   Explain which spectrum belongs to each compound. [4]
   [Total: 4]

4   Propenal, CH₂=CHCHO, and propenoic acid, CH₂=CHCOOH, are used in industry for the manufacture of polymers and resins.

Both compounds can be made from prop-2-en-1-ol, CH₂=CHCH₂OH.

(a) (i) Draw a skeletal formula for each of the three compounds mentioned above. [3]
    (ii) Give the name of the functional group that is common to all three compounds. [1]
(b) Prop-2-en-1-ol can be oxidised to form either propenal or propenoic acid. Name a suitable oxidising mixture. [2]
(c) A sample of prop-2-en-1-ol was oxidised. The infrared spectrum of the product was

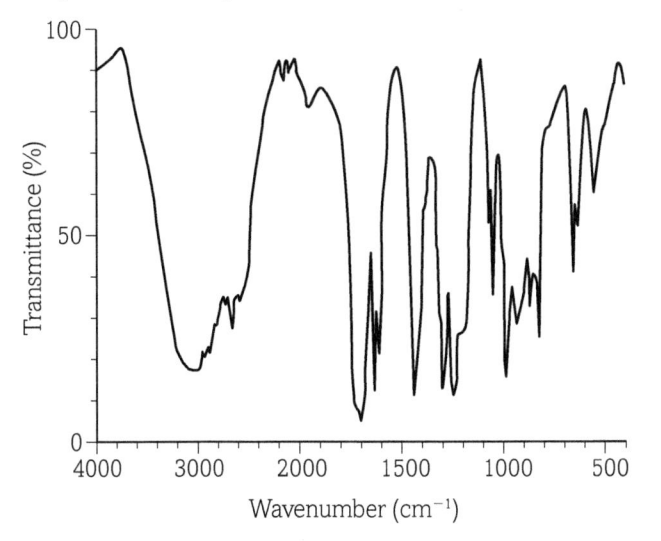

Explain whether the organic product was propenal or propenoic acid. [2]

[Total: 8]

5   Compound **X** has the following composition by mass: C, 62.07%; H, 10.34%; O, 27.59%.
(a) Calculate the empirical formula of compound **X**. [3]
(b) The mass spectrum of compound **X** is shown below.

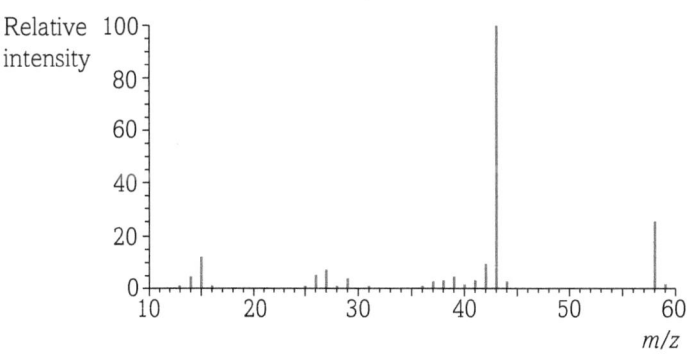

Deduce the molecular formula of compound **X**. Show how you obtained your answer. [3]

(c) Compound **X** is one of two structural isomers. One is an aldehyde, the other is a ketone.
The infrared spectrum of compound **X** is shown below.

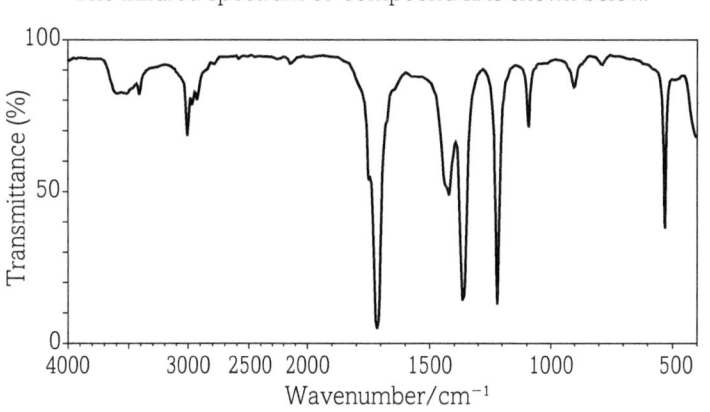

Compound **X** produced no observable change when heated with Fehling's solution.
   (i) Identify compound **X**. [1]
   (ii) Identify the species responsible for the peaks at 15 and 43 in the mass spectrum of compound **X**. [2]
   (iii) Identify the functional group responsible for the absorbance at 1700 cm⁻¹ in the infrared spectrum of compound **X**. [1]

[Total: 10]

# TOPIC 8

# Chemical energetics

## Introduction

You will be familiar with many exothermic reactions in everyday life even if you do not always realise it. Exothermic reactions can be easily identified because the reaction mixture gets hot. The heat energy generated can then be used for heating. Perhaps the most common example of this is burning natural gas: the heat energy generated can then be used to cook food.

If your hands have got very cold whilst outdoors you may have used a chemical hand warmer. One type of chemical hand warmer uses anhydrous calcium oxide (CaO) and water. These are kept in separate compartments and are then allowed to mix by breaking the seal. When the two chemicals mix, an exothermic reaction takes place and the mixture gets hot.

In contrast, endothermic reactions can often be recognised by the reaction mixture getting cold. They are less common, but you may be familiar with sherbet, which contains citric acid and sodium hydrogencarbonate. When you add water to this mixture an endothermic reaction takes place and the temperature of the mixture drops, which is why the inside of your mouth feels cold when you eat sherbet.

Interestingly, the chemical reactions that take place when an egg is cooked are also endothermic. Another way of recognising an endothermic reaction is that it needs a constant supply of energy for the reaction to continue. If you take the frying pan off the hob, then the egg will stop cooking.

### All the maths you need

- Recognise and make use of appropriate units in calculations
- Recognise and use expressions in decimal and ordinary form
- Use the appropriate number of significant figures
- Change the subject of an equation
- Substitute numerical values into algebraic equations using appropriate units for physical quantities
- Solve algebraic equations

## What have I studied before?
- Exothermic and endothermic reactions
- Energy level diagrams
- Simple experiments to determine temperature changes in chemical reactions such as dissolving and neutralisation

## What will I study later?
- The significance of bond enthalpy in determining the relative rates of reaction of different halogenoalkanes
- Lattice energy, electron affinity, enthalpy change of hydration and enthalpy change of solution (A level)
- Born–Haber cycles (A level)

## What will I study in this topic?
- Enthalpy change as the heat energy change measured at constant pressure
- The importance of standard conditions when comparing enthalpy changes
- Enthalpy changes of formation, combustion and neutralisation
- Experiments to obtain data required to calculate the enthalpy changes of reactions including combustion and neutralisation
- Calculations to determine the enthalpy changes of reactions from experimental data
- How Hess's Law can be used to determine enthalpy changes of reactions that cannot be determined directly
- Bond enthalpies and their use to calculate enthalpy changes of reaction, and mean bond enthalpies from enthalpy changes of reaction

# 8.1 **1** Introducing enthalpy and enthalpy change

**By the end of this section, you should be able to...**

- know that the enthalpy change is the heat energy change measured at constant pressure
- know that standard conditions for the measurement of enthalpy changes are 100 kPa and a specified temperature, usually 298 K
- define what is meant by standard enthalpy of reaction

## Chemical and heat energy

The first law of thermodynamics states that, during a chemical reaction, energy cannot be created or destroyed. However, one form of energy can be converted into another.

Various forms of energy are interesting to a chemist. Two of the most important ones are:

- chemical energy
- heat energy.

## Chemical energy

Chemical energy is made up of two components:

- Kinetic energy, which is a measure of the motion of the particles (atoms, molecules or ions) in a substance.
- Potential energy, which is a measure of how strongly these particles interact with one another (i.e. both attract and repel one another).

## Heat energy

Heat energy is the portion of the potential energy and the kinetic energy of a substance that is responsible for the temperature of the substance.

The heat energy of a substance is directly proportional to its absolute temperature (i.e. the temperature measured in Kelvin).

## Enthalpy and enthalpy changes

Enthalpy is a measure of the total energy of a system.

You cannot directly determine the enthalpy of a system, but you can measure the enthalpy change ($\Delta H$) that takes place during a physical or a chemical change.

The enthalpy change of a process is the heat energy that is transferred between the system and the surroundings at *constant pressure*.

## Exothermic and endothermic processes and reactions

Two types of process can take place. These are:

- **exothermic** – where heat energy is transferred from the system to the surroundings
- **endothermic** – where heat energy is transferred from the surroundings to the system.

### Learning tip

Do not confuse heat energy with heating. Heating is the result of a transfer of heat energy from one system to another, which in turn produces a change of temperature.

For example, when water at 60 °C is put in contact with air at 20 °C, heat energy will be transferred from the water to the air. This will result in an increase in temperature of the air, with a subsequent decrease in temperature of the water. This is why your cup of hot coffee gets cold when you forget to drink it!

### Additional reading

Enthalpy includes the internal energy, which is the energy required to create the system, and the amount of energy required to make room for it by displacing its environment and establishing its volume and pressure.

Enthalpy has the symbol $H$ and internal energy has the symbol $U$. This equation shows their relationship:

$$H = U + PV$$

where $P$ is the pressure of the system and $V$ is its volume.

Examples of exothermic and endothermic processes are given in the table.

| Exothermic | Endothermic |
|---|---|
| Freezing water | Melting ice |
| Condensing water vapour | Evaporating water |
| Dissolving sodium hydroxide in water | Dissolving ammonium nitrate in water |
| Reaction between dilute hydrochloric acid and aqueous sodium hydroxide | Reaction between dilute ethanoic acid and solid sodium hydrogencarbonate |
| Combustion of petrol | Photosynthesis |

**table A** Examples of exothermic and endothermic processes.

An example of an exothermic reaction is:

$HCl(aq) + NaOH(aq) \rightarrow NaCl(aq) + H_2O(l)$

$\Delta H = -57.1 \text{ kJ mol}^{-1}$

An example of an endothermic reaction is:

$C_6H_8O_7(aq) + 3NaHCO_3(s) \rightarrow C_6H_5O_7^{3-}(aq) + 3Na^+(aq) + 3CO_2(g) + 3H_2O(l)$
(citric acid)

$\Delta H = +70 \text{ kJ mol}^{-1}$

When considering a chemical reaction, the 'system' refers to the reaction mixture. Everything outside of the system is called the 'surroundings', which in practice is the air in the room in which the reaction is taking place.

Exothermic reactions can usually be recognised because they result in an immediate increase in temperature. For example, when hydrochloric acid is added to aqueous sodium hydroxide, the temperature of the reaction mixture increases. Similarly, when natural gas burns in oxygen, the flame produced is hot.

Conversely, endothermic reactions often produce a decrease in temperature of the reaction mixture, for example when solid sodium hydrogencarbonate is added to aqueous citric acid.

Any reaction that has to be continually heated in order for it to take place is endothermic. For instance, the thermal decomposition of calcium carbonate into calcium oxide and carbon dioxide is an endothermic reaction:

$CaCO_3(s) \rightarrow CaO(s) + CO_2(g)$

$\Delta H = +178 \text{ kJ mol}^{-1}$

## Standard conditions

In 1982, the International Union of Pure and Applied Chemistry (IUPAC) recommended that all enthalpy changes should be quoted using standard conditions of 100 kPa pressure and a stated temperature. The temperature most commonly used is 298 K.

Under these conditions, the enthalpy change measured is called the 'standard enthalpy change', and is given the symbol $\Delta H^\ominus_{298 \text{ K}}$ or simply $\Delta H^\ominus$.

### Learning tip

Some data books and textbooks quote standard enthalpy changes at a standard pressure of one atmosphere, i.e. 1 atm (1 atm = 101.325 kPa). This is not in agreement with IUPAC recommendations. For this reason, we will use 100 kPa as the standard pressure for thermochemical measurements.

## Standard enthalpy change of reaction, $\Delta_r H^\ominus$

When looking at **standard enthalpy change of reaction** ($\Delta_r H^\ominus$), it is important to recognise that the enthalpy change is for the reaction *as written*.

For the reaction

$N_2(g) + 3H_2(g) \rightarrow 2NH_3(g)$

$\Delta_r H^\ominus = -92 \text{ kJ mol}^{-1}$

But for the reaction when written as:

$\frac{1}{2}N_2(g) + 1\frac{1}{2}H_2(g) \rightarrow NH_3(g)$

$\Delta_r H^\ominus = -46 \text{ kJ mol}^{-1}$

In each case the 'per mole' refers to *one mole of equation,* and not to one mole of any reactant or product.

# Questions

1. Classify each of the following as exothermic or endothermic processes:
   (a) cooking an egg
   (b) formation of snow in clouds
   (c) burning candle wax
   (d) forming a cation from an atom in the gas phase
   (e) baking bread.

2. Sherbet is a solid mixture of sodium hydrogencarbonate and citric acid. If you put sherbet in your mouth and mix it with saliva, your mouth will feel cold. Explain why.

### Key definitions

**Exothermic** heat energy is transferred from the system to the surroundings.

**Endothermic** heat energy is transferred from the surroundings to the system.

**Standard enthalpy change of reaction** is the enthalpy change measured at 100 kPa and a stated temperature, usually 298 K, when the number of moles of substances in the equation *as written* react.

# 8.1 2 Enthalpy level diagrams

By the end of this section, you should be able to...

- construct and interpret enthalpy level diagrams showing an enthalpy change, including appropriate signs for exothermic and endothermic reactions

## How to construct and interpret enthalpy level diagrams

In an exothermic reaction, the final enthalpy of the system is less than its initial enthalpy. The reverse is true for an endothermic process. This is shown in these two enthalpy level diagrams.

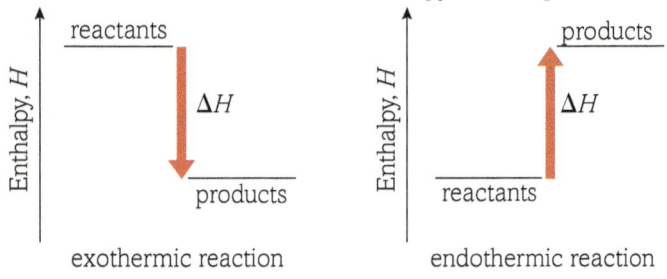

**fig A** Enthalpy level diagrams for exothermic and endothermic reactions.

The change in enthalpy, $\Delta H$, is given by:

$$\Delta H = H_{products} - H_{reactants}$$

For an exothermic reaction, $H_{reactants} > H_{products}$, so $\Delta H$ is negative.

For an endothermic reaction, $H_{reactants} < H_{products}$, so $\Delta H$ is positive.

### Learning tip

When you are labelling an enthalpy level diagram, it is important to get both the direction of the arrow for $\Delta H$ and the sign of $\Delta H$ correct.

The arrow should always point towards the products, as the change is *from* the reactants *to* the products.

### WORKED EXAMPLE 1

Draw an enthalpy level diagram for the following reaction:

$$C(s) + O_2(g) \rightarrow CO_2(g) \qquad \Delta H = -394 \text{ kJ mol}^{-1}$$

**Answer**

**fig B** Enthalpy level diagram for the reaction of carbon with oxygen.

### WORKED EXAMPLE 2

Draw an enthalpy level diagram for the following reaction:

$$C(s) + CO_2(g) \rightarrow 2CO(g) \qquad \Delta H = +172 \text{ kJ mol}^{-1}$$

**Answer**

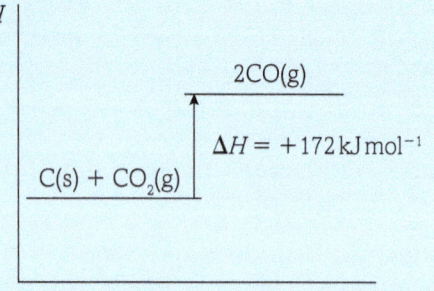

**fig C** Enthalpy level diagram for the reaction of carbon with carbon dioxide.

Here are some points to remember when constructing enthalpy level diagrams.

- You only need to label the vertical axis. It is not necessary to label the horizontal axis in an enthalpy *level* diagram (if you did, it could be labelled either 'Extent of reaction' or 'Progress of reaction'). It is, however, essential to label the axis in an enthalpy *profile* diagram (see **Section 9.1.4** on reaction kinetics).
- The formulae for both reactants and products should be given, including their state symbols.
- The values for $\Delta H$ should be given, including the correct sign.
- It is not essential to show the activation energy in an enthalpy *level* diagram, but it should be shown in an enthalpy profile diagram (see **Section 9.1.4** on reaction kinetics).

# Questions

1. Draw an enthalpy level diagram for the following reaction:

   $$CH_4(g) + 2O_2(g) \rightarrow CO_2(g) + 2H_2O(l) \qquad \Delta H = -890 \text{ kJ mol}^{-1}$$

2. (a) What information can be obtained from the following enthalpy level diagram?

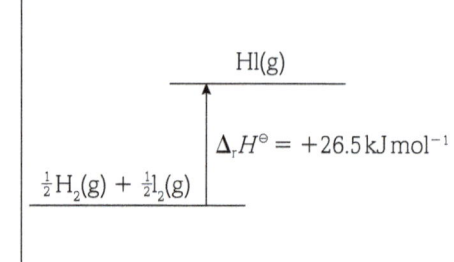

   (b) What would be the value of $\Delta_r H^\ominus$ for this reaction?

   $$H_2(g) + I_2(g) \rightarrow 2HI(g)$$

# 8.1 3 Standard enthalpy change of combustion

**By the end of this section, you should be able to...**

- define what is meant by the term standard enthalpy change of combustion
- understand simple experiments to measure enthalpy changes in terms of evaluating sources of error and assumptions made in the experiments
- process results from experiments to calculate a value for the enthalpy of combustion of a substance

## What is meant by standard enthalpy change of combustion?

The **standard enthalpy change of combustion** ($\Delta_c H^\ominus$) is the enthalpy change measured at 100 kPa and a specified temperature, usually 298 K, when one mole of a substance is completely burned in oxygen.

When writing an equation to represent the standard enthalpy change of combustion, it is important to specify one mole of the substance that is being burned. Two common ways of writing the combustion of hydrogen are:

$H_2(g) + \frac{1}{2}O_2(g) \rightarrow H_2O(l)$ and

$2H_2(g) + O_2(g) \rightarrow 2H_2O(l)$

The first equation, in which one mole of hydrogen undergoes combustion, represents $\Delta_c H^\ominus$. The enthalpy change for the second equation is $2 \times \Delta_c H^\ominus$.

## Experimental determination of enthalpy change of combustion of a liquid

To find the enthalpy change of combustion of a liquid, a known mass of the liquid is burned and the heat energy produced is used to heat a known volume of water.

The following procedure is used.

- A spirit burner containing the liquid under test is weighed.
- A known volume of water is added to a copper can.
- The temperature of the water is measured.
- The burner is lit.
- The mixture is constantly stirred with the thermometer.
- When the temperature of the water has reached approximately 20 °C above its initial temperature, the flame is extinguished and the burner is immediately reweighed.
- The final temperature is measured.

The appropriate laboratory apparatus is shown in the diagram.

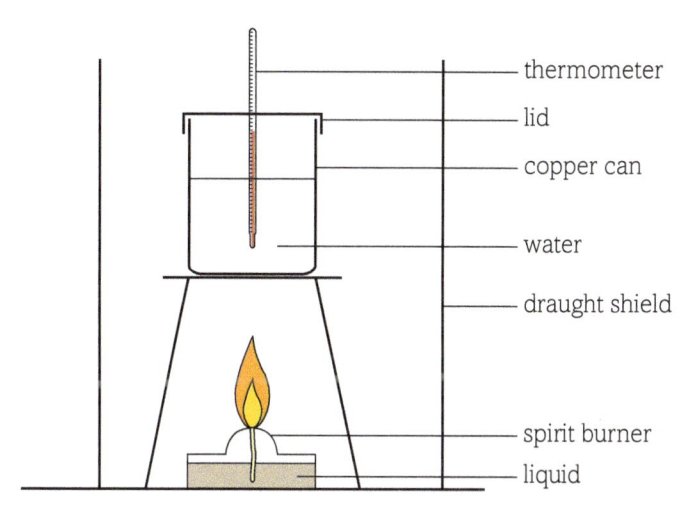

**fig A** Laboratory apparatus used to find the enthalpy change of combustion of a liquid.

A typical set of results for ethanol ($C_2H_5OH$, molar mass 46.0 g mol$^{-1}$) are:

| volume of water heated | 100.0 cm$^3$ |
|---|---|
| mass of ethanol burned | 0.42 g |
| temperature change, $\Delta T$ | +24.5 °C |

## Calculating enthalpy change of combustion

The enthalpy change of combustion is now calculated in three stages.

Stage 1: Calculate the heat energy, $Q$, transferred to the water using the equation

$Q = mc\Delta T$, where $m$ is the mass of water and $c$ is the specific heat capacity of water.

If $m$ is in grams, then $c$ is quoted as 4.18 J g$^{-1}$ K$^{-1}$. Assuming the density of water is

1 g cm$^{-3}$, then $m$ = 100.0 g.

The temperature change, $\Delta T$, has the same value in K as it does in °C

So, $Q = 100.0 \text{ g} \times 4.18 \text{ J g}^{-1} \text{K}^{-1} \times +24.5 \text{ K}$

$= +10241 \text{ J} = +10.24 \text{ kJ}$

Stage 2: Calculate the amount, $n$, of ethanol burned.

$n(C_2H_5OH) = \dfrac{0.42 \text{ g}}{46.0 \text{ g mol}^{-1}} = 9.13 \times 10^{-3} \text{ mol}$

Stage 3: Calculate the $\Delta_c H$, using the equation

$\Delta H = -\dfrac{Q}{n}$

$\Delta_c H^\ominus = -\dfrac{+10.24 \text{ kJ}}{9.13 \times 10^{-3} \text{ mol}} = -1120 \text{ kJ mol}^{-1}$ (to 3 significant figures)

It is very important that you include a sign with any value of $\Delta H$ that you quote.

## Learning tip

The two equations from stages 1 and 3, when used together, result in the correct sign for $\Delta H$. In this example, $\Delta H$ is negative, which is consistent with the exothermic reaction that is taking place.

If the temperature had decreased during the reaction, then $\Delta T$ would be negative, which in turn would make Q negative. This would lead to a positive value for $\Delta H$, consistent with an endothermic reaction.

If in doubt, always use your common sense. If the temperature increases, the reaction must be exothermic, and vice versa.

## Evaluating sources of error and assumptions made in the experiments

The value obtained from the above experiment is in reasonable agreement with the standard enthalpy of combustion of ethanol, as obtained from a data book, of $-1367 \text{ kJ mol}^{-1}$. This means that the errors in procedure were minimal.

Here are some possible sources of error.

- Some of the heat energy produced in burning is transferred to the air and not the water.
- Some of the ethanol may not burn completely to form carbon dioxide and water. (Incomplete combustion would cause soot to form on the bottom of the copper can.)
- Some of the heat energy produced in burning is transferred to the copper can and not the water.
- The conditions are not standard. For example, water vapour, not liquid water, is produced.
- As the experiment takes a long time, not all of the heat energy transferred from the water to the surroundings is compensated for.

## Questions

1. Ethanol and methoxymethane have the same molecular formula, $C_2H_6O$.

   The standard enthalpy change of combustion at 298 K of ethanol gas and methoxymethane gas are $-1367 \text{ kJ mol}^{-1}$ and $-1460 \text{ kJ mol}^{-1}$ respectively.

   (a) Write an equation to represent the standard enthalpy change of combustion of:

   (i) ethanol, and (ii) methoxymethane.

   (b) Suggest why the two compounds have different standard enthalpy changes of combustion despite having the same molecular formula.

2. The table shows the results of separately combusting 1.00 g of each of four alcohols and determining the amount of energy required to produce the same temperature rise in each reaction.

   | Alcohol | Molar mass/ g mol$^{-1}$ | Energy required/ kJ g$^{-1}$ | Energy required/ kJ mol$^{-1}$ |
   |---|---|---|---|
   | methanol | 32.0 | 22.34 | |
   | ethanol | 46.0 | 29.80 | |
   | propan-1-ol | 60.0 | 33.50 | |
   | butan-1-ol | 74.0 | 36.12 | |

   (a) Complete the table by calculating the energy required per mole for each of the four alcohols. Quote the values for $\Delta_c H$ for each of them.

   (b) Draw a graph of $\Delta_c H$ (vertical axis) against the number of carbon atoms in one molecule and use it to estimate a value for the enthalpy change of combustion of pentan-1-ol under the same conditions.

   (c) Extrapolate the graph line to 0 on the x-axis and comment on the value of $\Delta_c H$ you read off from the graph.

## Key definition

The **standard enthalpy change of combustion** ($\Delta_c H^\ominus$) is the enthalpy change measured at 100 kPa and a stated temperature, usually 298 K, when one mole of a substance is completely burned in oxygen.

# 8.1 4 Standard enthalpy change of neutralisation

By the end of this section, you should be able to...
- define what is meant by the term standard enthalpy change of neutralisation
- understand simple experiments to measure enthalpy changes in terms of evaluating sources of error and assumptions made in the experiments
- process results from experiments to calculate a value for the enthalpy change of neutralisation

## What is meant by standard enthalpy change of neutralisation?

The **standard enthalpy change of neutralisation** ($\Delta_{neut}H^\ominus$) is the enthalpy change measured at 100 kPa and a specified temperature, usually 298 K, when one mole of water is produced by the neutralisation of an acid with an alkali.

The following equation represents the standard enthalpy change of neutralisation of hydrochloric acid:

$HCl(aq) + NaOH(aq) \rightarrow NaCl(aq) + H_2O(l)$

For sulfuric acid, it is the enthalpy change for:

$\frac{1}{2}H_2SO_4(aq) + NaOH(aq) \rightarrow \frac{1}{2}Na_2SO_4(aq) + H_2O(l)$

For certain combinations of acid and alkali, the standard enthalpy change of neutralisation is remarkably constant.

| Acid | Alkali | $\Delta_{neut}H^\ominus$/kJ mol$^{-1}$ |
|---|---|---|
| HCl(aq) | NaOH(aq) | −57.9 |
| HBr(aq) | KOH(aq) | −57.6 |
| HNO$_3$(aq) | NaOH(aq) | −57.6 |

**table A** The standard enthalpy change of neutralisation is remarkably constant for certain combinations of acid and alkali.

Each of the acids and alkalis in the table above are classified as strong acids/alkalis. If we make the assumption that strong acids and alkalis are fully ionised in aqueous solution, then the reaction between them is simplified to:

$H^+(aq) + OH^-(aq) \rightarrow H_2O(l)$

Since the reaction is essentially the same in each case, it is hardly surprising that the enthalpy changes are so similar.

## Experimental determination of enthalpy change of neutralisation

Here is how to find the enthalpy change of neutralisation.
- Wear safety glasses and a lab coat.
- Using a pipette fitted with a safety filter, place 25.0 cm³ of 1.00 mol dm$^{-3}$ acid into an expanded polystyrene cup.
- Measure the temperature of the acid.
- Using a pipette fitted with a safety filter, place 25.0 cm³ of the alkali (usually dilute sodium hydroxide of a concentration slightly greater than 1.00 mol dm$^{-3}$ to make sure all the acid is neutralised) into a beaker.
- Measure the temperature of the alkali.
- Add the alkali to the acid, stir with the thermometer and measure the maximum temperature reached.

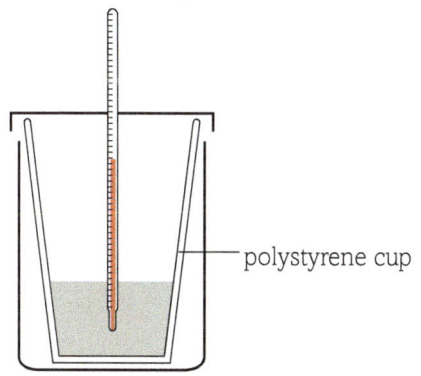

**fig A** Apparatus used to find the enthalpy change of neutralisation.

### WORKED EXAMPLE

| Volume of 1.00 mol dm$^{-3}$ HCl | 25.0 cm³ |
|---|---|
| Volume of 1.2 mol dm$^{-3}$ NaOH | 25.0 cm³ |
| Initial temperature of acid | 18.6 °C |
| Initial temperature of alkali | 18.8 °C |
| Maximum temperature reached | 25.4 °C (298.5 K) |

**Answer**

Mean starting temperature = $\frac{1}{2}$(18.6 + 18.8) = 18.7 °C (291.7 K)

Temperature change, $\Delta T$ = (298.5 − 291.7) = +6.7 K

Volume of solution heated = (25.0 + 25.0) = 50.0 cm³

Mass of solution heated = 50.0 g (assuming that the density of the solution is 1 g cm³)

Assume specific heat capacity of the solution = 4.18 J g$^{-1}$ K$^{-1}$

$Q$ = 50.0 g × 4.18 J g$^{-1}$ K$^{-1}$ × (+6.7 K) = 1400.3 J = 1.4003 kJ

Amount of acid neutralised = amount of water formed
= 1.00 mol dm$^{-3}$ × 0.025 dm³
= 0.0250 mol

$\Delta_{neut}H^\ominus$ = −1.4003 kJ/0.0250 mol = −56 kJ mol$^{-1}$

### Evaluating sources of error and assumptions made in the experiments

Answers should be given to only two significant figures, as $\Delta T$ is given to only two significant figures.

There are the usual uncertainties of measurements involved with the use of the pipette and the thermometer. Additionally, some heat energy will be transferred to the thermometer and the polystyrene cup.

## Questions

1. A thermometric titration was carried out using $1.00\,mol\,dm^{-3}$ sodium hydroxide solution and dilute hydrochloric acid of unknown concentration.

   $50.0\,cm^3$ of the sodium hydroxide solution was placed in a suitable apparatus and

   $5.00\,cm^3$ portions of the hydrochloric acid were added. The mixture was stirred after each addition of acid and the temperature of the mixture was measured. Both solutions were initially at $20\,°C$.

   The results obtained were as follows.

   | Vol. of acid/cm³ | 5.00 | 10.00 | 15.00 | 20.00 | 25.00 | 30.00 | 35.00 | 40.00 | 45.00 | 50.00 |
   |---|---|---|---|---|---|---|---|---|---|---|
   | Temperature/°C | 22.80 | 23.80 | 24.80 | 25.80 | 26.80 | 27.80 | 28.10 | 27.70 | 27.30 | 26.80 |

   (a) Plot a graph of temperature *change* against volume of acid added. Determine the concentration, in $mol\,dm^{-3}$, of the hydrochloric acid.

   (b) Calculate the enthalpy change of neutralisation, per mole of water formed, of the reaction. [Assume the specific heat capacity of the solution formed is $4.18\,J\,g^{-1}\,K^{-1}$ and that its density is $1\,g\,cm^{-3}$.]

   (c) Suggest suitable apparatus for carrying out the above experiment.

   (d) In further experiments, the following enthalpy changes of neutralisation were determined, again using $1.00\,mol\,dm^{-3}$ solutions at $20\,°C$.

   HF(aq) and NaOH(aq)         $-68.6\,kJ\,mol^{-1}$

   $CH_3COOH$(aq) and NaOH(aq)   $-55.2\,kJ\,mol^{-1}$

   Suggest why each of these values is different from that for HCl(aq) and NaOH(aq), and also why they are different from one another.

   [Note: Part (d) of this question is beyond what is expected of you at AS level. However, if you do some research into what is meant by a weak acid, and also consider the energy changes involved when acid molecules ionise in water and also when the ions are subsequently hydrated, you should be able to answer it.]

2. In the experiment described in Question 1, a $50\,cm^3$ pipette was used to measure the volume of sodium hydroxide solution, and a $50\,cm^3$ burette was used to add the hydrochloric acid. The measurement uncertainty for the pipette is $±0.10\,cm^3$. The measurement uncertainty for each reading of the burette is $±0.05\,cm^3$ and for the thermometer is $0.05\,°C$.

   (a) Calculate the total percentage measurement uncertainty in the use of:

   (i) the pipette, (ii) the burette and (iii) the thermometer.

   (b) Considering the description of how the experiment was carried out, suggest any possible procedural errors and state how you would carry out the experiment in order to minimise the effect of these errors.

### Key definition

The **standard enthalpy change of neutralisation** is the enthalpy change measured at $100\,kPa$ and a stated temperature, usually $298\,K$, when one mole of water is produced by the neutralisation of an acid with an alkali.

# 8.1  5  Standard enthalpy change of formation and Hess's Law

**By the end of this section, you should be able to...**
- define what is meant by the term standard enthalpy change of formation
- construct enthalpy cycles using Hess's Law
- calculate enthalpy changes from data using Hess's Law

## What is meant by standard enthalpy change of formation?

The **standard enthalpy change of formation** ($\Delta_f H^\ominus$) is the enthalpy change measured at 100 kPa and a specified temperature, usually 298 K, when one mole of a substance is formed from its elements in their standard states.

For the purposes of this definition, the standard state of an element is the form in which it exists at the specified temperature, usually 298 K, and a pressure of 100 kPa.

The standard enthalpy change of formation of gaseous carbon dioxide is the enthalpy change for the reaction:

$$C(s, \text{graphite}) + O_2(g) \rightarrow CO_2(g) \qquad \Delta_f H^\ominus = -394 \text{ kJ mol}^{-1}$$

The standard enthalpy change of formation of liquid ethanol is the enthalpy change for the reaction:

$$2C(s, \text{graphite}) + 3H_2(g) + \tfrac{1}{2}O_2(g) \rightarrow C_2H_5OH(l) \qquad \Delta_f H^\ominus = -278 \text{ kJ mol}^{-1}$$

### Learning tip

Enthalpy changes of formation emphasise the necessity of including state symbols in thermochemical equations. The value for the enthalpy change of formation of gaseous ethanol is different from that for liquid ethanol:

$\Delta_f H^\ominus [C_2H_5OH(l)] = -278 \text{ kJ mol}^{-1}$
$\Delta_f H^\ominus [C_2H_5OH(g)] = -235 \text{ kJ mol}^{-1}$

## Hess's Law

Most standard enthalpy changes of formation cannot be determined experimentally. For example, it is impossible to burn carbon in oxygen and form solely carbon monoxide. So, the enthalpy change for the reaction:

$$C(s) + \tfrac{1}{2}O_2(g) \rightarrow CO(g)$$

is impossible to determine directly.

Fortunately, we can make use of **Hess's Law**, which is an application of the first law of thermodynamics (the law of conservation of energy).

Hess's Law states that the enthalpy change of a reaction is independent of the path taken in converting reactants into products, provided the initial and final conditions are the same in each case.

Hess's Law allows us to calculate the standard enthalpy change of formation of carbon monoxide.

The enthalpy changes of combustion of carbon and carbon monoxide can both be determined experimentally. Their values are:

$C(s, \text{graphite}) + O_2(g) \rightarrow CO_2(g) \qquad \Delta_c H^\ominus = -394 \text{ kJ mol}^{-1}$
$CO(g) + \tfrac{1}{2}O_2(g) \rightarrow CO_2(g) \qquad \Delta_c H^\ominus = -283 \text{ kJ mol}^{-1}$

There are two ways to use these data to calculate the standard enthalpy change of formation of carbon monoxide.

### Method 1: Subtract equations

Reverse the second equation above and then add to the first:

$CO_2(g) \rightarrow CO(g) + \frac{1}{2}O_2(g)$      $\Delta H^\ominus = +283 \text{ kJ mol}^{-1}$

$C(s, \text{graphite}) + O_2(g) \rightarrow CO_2(g)$      $\Delta H^\ominus = -394 \text{ kJ mol}^{-1}$

Adding the two equations gives:

$C(s, \text{graphite}) + \frac{1}{2}O_2(g) \rightarrow CO(g)$      $\Delta H^\ominus = +283 + (-394) = -111 \text{ kJ mol}^{-1}$

So, $\Delta_f H^\ominus [CO(g)] = -111 \text{ kJ mol}^{-1}$

### Method 2: Construct an enthalpy cycle using Hess's Law

An enthalpy cycle using Hess's Law is sometimes also called a Hess's Law cycle.

$$C(s, \text{graphite}) + \tfrac{1}{2}O_2(g) \xrightarrow{\Delta_f H^\ominus} CO(g)$$

with $-394$ (with $+\tfrac{1}{2}O_2(g)$) and $-283$ (with $+\tfrac{1}{2}O_2(g)$) going down to $CO_2(g)$.

**fig A** An enthalpy cycle using Hess's Law to calculate the standard enthalpy change of formation of carbon monoxide.

By Hess's law, $\Delta_f H^\ominus + (-283) = -394$

So, $\Delta_f H^\ominus = -394 - (-283) = -111 \text{ kJ mol}^{-1}$

In general terms, this is the enthalpy cycle used to calculate enthalpy change of formation of a compound from the relevant enthalpy changes of combustion.

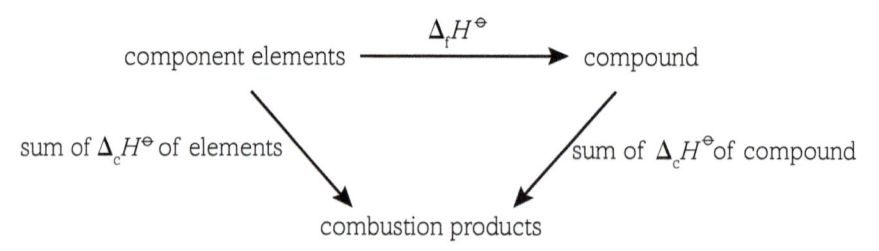

$$\text{component elements} \xrightarrow{\Delta_f H^\ominus} \text{compound}$$

sum of $\Delta_c H^\ominus$ of elements     sum of $\Delta_c H^\ominus$ of compound → combustion products

$\Delta_f H^\ominus$ = sum of $\Delta_c H^\ominus$ of elements − $\Delta_c H^\ominus$ of compound

**fig B** An enthalpy cycle to calculate enthalpy change of formation from enthalpy changes of compustion.

#### WORKED EXAMPLE

Calculate the enthalpy change of formation of methanol ($CH_3OH$), given the following enthalpy change of combustion data:

$\Delta_c H^\ominus [CH_3OH(l)] = -726 \text{ kJ mol}^{-1}$

$\Delta_c H^\ominus [C(s, \text{graphite})] = -394 \text{ kJ mol}^{-1}$

$\Delta_c H^\ominus [H_2(g)] = -286 \text{ kJ mol}^{-1}$

$$C(s, \text{graphite}) + 2H_2(g) + \tfrac{1}{2}O_2(g) \xrightarrow{\Delta_f H^\ominus} CH_3OH(l)$$

$-394 + (-286 \times 2)$ and $-715$ going down to $CO_2(g) + 2H_2O(l)$

**fig C** An enthalpy cycle using Hess's Law to calculate the enthalpy change of formation of methanol.

$\Delta_f H^\ominus = -394 + (-286 \times 2) - (-726) = -240 \text{ kJ mol}^{-1}$

## Using Hess's Law for other reactions

Hess's Law can be used to calculate the enthalpy change for many different types of reaction.

Two examples are given below.

### Example 1

Calculate $\Delta_r H^\ominus$ for the thermal decomposition of calcium carbonate into calcium oxide and carbon dioxide, given the following data:

$CaCO_3(s) + 2HCl(aq) \rightarrow CaCl_2(aq) + H_2O(l) + CO_2(g)$
$\Delta_r H^\ominus = -17 \text{ kJ mol}^{-1}$

$CaO(s) + 2HCl(aq) \rightarrow CaCl_2(aq) + H_2O(l)$
$\Delta_r H^\ominus = -195 \text{ kJ mol}^{-1}$

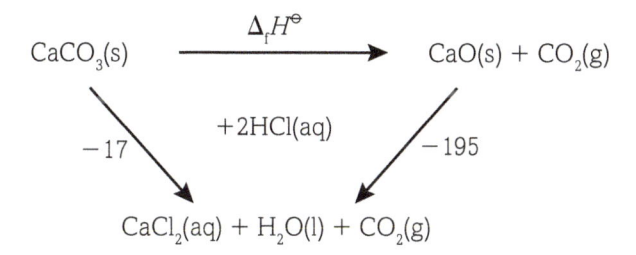

```
                    Δ_r H°
CaCO₃(s)  ─────────────────────→  CaO(s) + CO₂(g)
      \        +2HCl(aq)        /
       \                       /
    -17 \                     / -195
         \                   /
          → CaCl₂(aq) + H₂O(l) + CO₂(g)
```

**fig D** An enthalpy cycle using Hess's Law to calculate the enthalpy change of reaction for the thermal decomposition of calcium carbonate.

$\Delta_r H^\ominus = -17 - (-195) = +178 \text{ kJ mol}^{-1}$

### Example 2

Calculate $\Delta_r H^\ominus$ for the hydration of anhydrous copper(II) sulfate, given the following data:

$CuSO_4.5H_2O(s) + aq \rightarrow Cu^{2+}(aq) + SO_4^{2-}(aq) + 5H_2O(l)$
$\Delta_r H^\ominus = +11.3 \text{ kJ mol}^{-1}$

$CuSO_4(s) + aq \rightarrow Cu^{2+}(aq) + SO_4^{2-}(aq)$
$\Delta_r H^\ominus = -67.0 \text{ kJ mol}^{-1}$

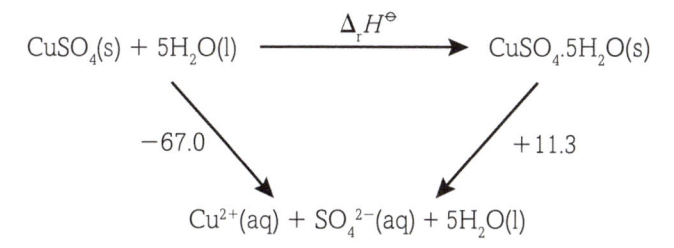

```
                        Δ_r H°
CuSO₄(s) + 5H₂O(l)  ──────────────→  CuSO₄.5H₂O(s)
         \                         /
          \                       /
     -67.0 \                     / +11.3
            \                   /
             → Cu²⁺(aq) + SO₄²⁻(aq) + 5H₂O(l)
```

**fig E** An enthalpy cycle using Hess's Law to calculate the enthalpy change of reaction for the hydration of anhydrous copper(II) sulfate.

$\Delta_r H^\ominus = -67.0 - (+11.3) = -78.3 \text{ kJ mol}^{-1}$

## Questions

1. (a) Write a chemical equation to represent the standard enthalpy change of formation of methane.

   (b) Using the following data, calculate a value for the standard enthalpy change of formation of methane.

   $\Delta_c H^\ominus [C(s, \text{graphite})] = -394 \text{ kJ mol}^{-1}$
   $\Delta_c H^\ominus [H_2(g)] = -286 \text{ kJ mol}^{-1}$
   $\Delta_c H^\ominus [CH_4(g)] = -890 \text{ kJ mol}^{-1}$

2. Using the following data, calculate the standard enthalpy change for the reaction represented by the equation:

   $AlCl_3(s) + 6H_2O(l) \rightarrow AlCl_3.6H_2O(s)$

   $\Delta_f H^\ominus [AlCl_3(s)] = -704 \text{ kJ mol}^{-1}$
   $\Delta_f H^\ominus [H_2O(l)] = -286 \text{ kJ mol}^{-1}$
   $\Delta_f H^\ominus [AlCl_3.6H_2O(s)] = -2692 \text{ kJ mol}^{-1}$

3. The data shows some values of standard enthalpy changes of formation.

   $\Delta_f H^\ominus [LiOH(s)] = -485 \text{ kJ mol}^{-1}$
   $\Delta_f H^\ominus [H_2O(l)] = -286 \text{ kJ mol}^{-1}$
   $\Delta_f H^\ominus [Li(s)] = 0 \text{ kJ mol}^{-1}$

   (a) Why is the standard enthalpy change of formation of lithium quoted as zero?

   (b) Write a chemical equation to represent the standard enthalpy change of formation of solid lithium hydroxide.

   (c) Lithium reacts with water according to the following equation:

   $2Li(s) + 2H_2O(l) \rightarrow 2Li^+(aq) + 2OH^-(aq) + H_2(g)$

   Calculate the standard enthalpy change for this reaction using the data above, together with the following information:

   $LiOH(s) + aq \rightarrow Li^+(aq) + OH^-(aq) \qquad \Delta H = -21 \text{ kJ mol}^{-1}$

### Key definitions

The **standard enthalpy change of formation** is the enthalpy change measured at 100 kPa and a specified temperature, usually 298 K, when one mole of a substance is formed from its elements in their standard states.

**Hess's Law** states that the enthalpy change of a reaction is independent of the path taken in converting reactants into products, provided the initial and final conditions are the same in each case.

## 8.2　1　Bond enthalpy and mean bond enthalpy

**By the end of this section, you should be able to...**

- know what is meant by the terms bond enthalpy and mean bond enthalpy

### What is meant by bond enthalpy?

**Bond enthalpy** ($\Delta_B H$) is the enthalpy change when one mole of a bond in the gaseous state is broken.

For a *diatomic* molecule, XY, the bond enthalpy is the enthalpy change for the following reaction:

$$XY(g) \rightarrow X(g) + Y(g)$$

Some examples are:

$Cl_2(g) \rightarrow 2Cl(g)$    $\Delta_B H = +243 \text{ kJ mol}^{-1}$

$H_2(g) \rightarrow 2H(g)$    $\Delta_B H = +436 \text{ kJ mol}^{-1}$

$HCl(g) \rightarrow H(g) + Cl(g)$    $\Delta_B H = +432 \text{ kJ mol}^{-1}$

For *polyatomic* molecules, each bond has to be considered separately. For example, with methane there are four separate bond enthalpies:

$CH_4(g) \rightarrow CH_3(g) + H(g)$    $\Delta_B H = +423 \text{ kJ mol}^{-1}$

$CH_3(g) \rightarrow CH_2(g) + H(g)$    $\Delta_B H = +480 \text{ kJ mol}^{-1}$

$CH_2(g) \rightarrow CH(g) + H(g)$    $\Delta_B H = +425 \text{ kJ mol}^{-1}$

$CH(g) \rightarrow C(g) + H(g)$    $\Delta_B H = +335 \text{ kJ mol}^{-1}$

### What is meant by mean bond enthalpy?

You will notice that the bond enthalpy of the C—H bond varies with its environment. For this reason it is often useful to quote the **mean bond enthalpy**.

The mean bond enthalpy for the C—H bond in methane is approximately $+416 \text{ kJ mol}^{-1}$.

$[\frac{1}{4}(423 + 480 + 425 + 335) = 415.75]$

The mean bond enthalpy for the C—H bond in a large number of organic compounds is $+413 \text{ kJ mol}^{-1}$. If the bond enthalpy is calculated for a particular compound, it is likely to be slightly different from the mean value. For example, the mean bond enthalpy for the C—H bond in ethane is $+420 \text{ kJ mol}^{-1}$.

Some examples of mean bond enthalpies are given in the table below.

| Bond | Mean bond enthalpy/kJ mol$^{-1}$ | Bond | Mean bond enthalpy/kJ mol$^{-1}$ |
|---|---|---|---|
| C—C | +347 | O—H | +464 |
| C=C | +612 | C—F | +467 |
| C≡C | +838 | C—Cl | +346 |
| C—O | +358 | C—Br | +290 |
| C=O | +743 | C—I | +228 |

**table A** Examples of mean bond enthalpies.

A shorthand representation for mean bond enthalpy is to use the letter $E$ followed by the bond in brackets. So the mean bond enthalpy of the C—C bond is written as $E(\text{C—C}) = +347 \text{ kJ mol}^{-1}$.

# Questions

1. (a) State what is meant by the H—I bond enthalpy.
   (b) Write an equation that represents the bond enthalpy of the H—I bond.

2. The two bonds in the water molecule can be separately broken. The enthalpy changes for the two processes are:

   H—O—H(g) → O—H(g) + H(g)    $\Delta H = +496 \text{ kJ mol}^{-1}$
   O—H(g) → O(g) + H(g)         $\Delta H = +432 \text{ kJ mol}^{-1}$

   (a) Calculate the mean bond enthalpy for the O—H bond in water.
   (b) Why is this value not the same as the value mean bond enthalpy of an O—H bond as quoted in a data book?

3. Suggest a reason why the mean bond enthalpy of the C—H bond in methane ($+416 \text{ kJ mol}^{-1}$) is different from that in ethane ($+420 \text{ kJ mol}^{-1}$).

4. (a) Write equations to show the successive breaking of the N—H bonds in ammonia, $NH_3(g)$.
   (b) Explain how the enthalpy changes for each reaction are used to calculate the mean bond enthalpy for the N—H bond in ammonia.

## Key definitions

**Bond enthalpy** is the enthalpy change when one mole of a bond in the gaseous state is broken.

**Mean bond enthalpy** is the enthalpy change when one mole of a bond, averaged out over many different molecules, is broken.

# 8.2 2 Using mean bond enthalpies

**By the end of this section, you should be able to...**

- calculate an enthalpy change of reaction using mean bond enthalpies and explain the limitations of this method of calculation
- calculate mean bond enthalpies from enthalpy changes of reaction

## Calculating an enthalpy change of reaction using mean bond enthalpies

Here is the process for calculating an enthalpy change of reaction using mean bond enthalpies.

- Step 1: Calculate the sum of the mean bond enthalpies of the bonds broken, $\Sigma$ (bonds broken).
- Step 2: Calculate the sum of the mean bond enthalpies of the bonds made, $\Sigma$ (bonds made).
- Step 3: Calculate the enthalpy change of reaction using the equation:

$$\Delta_r H = \Sigma \text{ (bonds broken)} - \Sigma \text{ (bonds made)}$$

### WORKED EXAMPLE 1

Calculate the enthalpy change of reaction for:

$$H_2(g) + Cl_2(g) \rightarrow 2HCl(g)$$

given the following data:

$E(H-H) = 436 \text{ kJ mol}^{-1}$
$E(Cl-Cl) = 244 \text{ kJ mol}^{-1}$
$E(H-Cl) = 432 \text{ kJ mol}^{-1}$

**Answer**

$\Sigma$ (bonds broken) = (436 + 244) = 680 kJ mol$^{-1}$
$\Sigma$ (bonds made) = (432 × 2) = 864 kJ mol$^{-1}$
$\Delta_r H$ = (680 - 864) = -184 kJ mol$^{-1}$

### WORKED EXAMPLE 2

Calculate the enthalpy change of reaction for:

$$H_2O_2(l) \rightarrow H_2O(l) + \tfrac{1}{2}O_2(g)$$

given the following data:

$E(O-H) = 463 \text{ kJ mol}^{-1}$
$E(O-O) = 146 \text{ kJ mol}^{-1}$
$E(O=O) = 496 \text{ kJ mol}^{-1}$

If you look at the displayed formula for each substance you will see that the two O—H bonds in hydrogen peroxide, $H_2O_2$, are also present in water:

H—O—O—H and H—O—H

So we do not have to include these in our calculations.

**Answer**

$\Sigma$ (bonds broken) = 146 kJ mol$^{-1}$
$\Sigma$ (bonds made) = $\tfrac{1}{2}$ × 496 kJ mol$^{-1}$ = 248 kJ mol$^{-1}$
$\Delta_r H$ = (146 - 248) = -102 kJ mol$^{-1}$

## Limitations of this method of calculation

The measured value for the enthalpy change of this reaction is $-98\,\text{kJ}\,\text{mol}^{-1}$. The reason for the difference is that bond enthalpies are measured in the gaseous state, and both hydrogen peroxide and water are liquids in the reaction.

> **Learning tip**
>
> If you are unsure which bonds break in the reaction, then calculate the sum of the bond enthalpies for *all* of the bonds in both the reactants and products, as in worked example 1 above.

## Calculating mean bond enthalpies from enthalpy changes of reaction

For this type of calculation you are supplied with a value for the enthalpy change of a reaction, together with all the relevant mean bond enthalpies except one; the one you are asked to calculate.

To solve the problem, simply substitute the known mean bond enthalpies and the unknown bond enthalpy into the expression

$\Delta_r H = \Sigma\,(\text{bonds broken}) - \Sigma\,(\text{bonds made})$

Rearrange the expression to make the unknown bond enthalpy the subject, and solve the problem.

### WORKED EXAMPLE 3

Calculate the bond enthalpy of the C=C bond in ethane, given the following data.

$C_2H_4(g) + H_2(g) \rightarrow C_2H_6(g) \qquad \Delta_r H = -147\,\text{kJ}\,\text{mol}^{-1}$
$E(\text{H--H}) = 436\,\text{kJ}\,\text{mol}^{-1}$
$E(\text{C--H}) = 413\,\text{kJ}\,\text{mol}^{-1}$
$E(\text{C--C}) = 347\,\text{kJ}\,\text{mol}^{-1}$

$\Sigma\,(\text{bonds broken}) = E(\text{C=C}) + E(\text{H--H}) = E(\text{C=C}) + 436\,\text{kJ}\,\text{mol}^{-1}$
$\Sigma\,(\text{bonds made}) = 2E(\text{C--H}) + E(\text{C--C}) = (2 \times 413) + 347 = 1173\,\text{kJ}\,\text{mol}^{-1}$
$\Delta_r H = \Sigma\,(\text{bonds broken}) - \Sigma\,(\text{bonds made})$
$-147 = E(\text{C=C}) + 436 - 1173$
$E(\text{C=C}) = 1173 - 147 - 436$
$E(\text{C=C}) = 590\,\text{kJ}\,\text{mol}^{-1}$

### WORKED EXAMPLE 4

In the following example, carbon is in the solid state in the reaction, so the enthalpy change for the conversion of the solid into the gas needs to be taken into account.

Calculate the mean bond enthalpy of the C–H bond in methane using the following data.

$\Delta_f H^\ominus\,[CH_4(g)] = -75\,\text{kJ}\,\text{mol}^{-1}$
$C(s,\,\text{graphite}) \rightarrow C(g)$
$\Delta H^\ominus = +715\,\text{kJ}\,\text{mol}^{-1}$
$E(\text{H--H}) = 436\,\text{kJ}\,\text{mol}^{-1}$

The equation that represents the standard enthalpy change of formation of methane is:

$C(s,\,\text{graphite}) + 2H_2(g) \rightarrow CH_4(g) \qquad \Delta_f H^\ominus = -75\,\text{kJ}\,\text{mol}^{-1}$

**Answer**

So, for the following reaction

$C(g) + 2H_2(g) \rightarrow CH_4(g)$
$\Delta_r H^\ominus = -75 + (-715) = -790\,\text{kJ}\,\text{mol}^{-1}$
$\Sigma\,(\text{bonds broken}) = 2E(\text{H--H}) = +872\,\text{kJ}\,\text{mol}^{-1}$
$\Sigma\,(\text{bonds made}) = 4E(\text{C--H})\,\text{kJ}\,\text{mol}^{-1}$
$\Delta_r H^\ominus = \Sigma\,(\text{bonds broken}) - \Sigma\,(\text{bonds made})$
$-790 = +872 - 4E(\text{C--H})$
$E(\text{C--H}) = \frac{1}{4}(790 + 872) = 415.5\,\text{kJ}\,\text{mol}^{-1}$

# Questions

1. Use the following mean bond enthalpies to calculate $\Delta_r H^\ominus$ for each of the following reactions. Assume that all species are in the gaseous state.

| Bond | H–H | C–C | C=C | C–H | C=O | O=O |
|---|---|---|---|---|---|---|
| E/kJ mol$^{-1}$ | 436 | 347 | 612 | 413 | 743 | 498 |

| Bond | H–O | F–F | C–Br | H–Br | H–F | Br–Br |
|---|---|---|---|---|---|---|
| E/kJ mol$^{-1}$ | 464 | 158 | 290 | 366 | 568 | 193 |

(a) $H_2 + F_2 \rightarrow 2HF$
(b) $CH_3CH=CH_2 + Br_2 \rightarrow CH_3CHBrCH_2Br$
(c) $2CH_3CH_3 + 7O_2 \rightarrow 4CO_2 + 6H_2O$
(d) $CH_2=CH_2 + HBr \rightarrow CH_3CH_2Br$

2. (a) Write an equation to represent the standard enthalpy change of formation of ammonia.

   (b) Calculate the mean bond enthalpy for the N–H bond in ammonia using this data:

   $\Delta_f H^\ominus\,[NH_3(g)] = -46\,\text{kJ}\,\text{mol}^{-1}$
   $E(\text{N} \equiv \text{N}) = 945\,\text{kJ}\,\text{mol}^{-1}$
   $E(\text{H--H}) = 436\,\text{kJ}\,\text{mol}^{-1}$

3. (a) The equation for the formation of sulfur hexafluoride is:

   $S(s) + 3F_2(g) \rightarrow SF_6(g)$

   Use the following data to calculate the mean bond enthalpy for the S–F bond in sulfur hexafluoride.

   $\Delta_f H^\ominus\,[SF_6(g)] = -1100\,\text{kJ}\,\text{mol}^{-1}$
   $E(\text{F--F}) = 158\,\text{kJ}\,\text{mol}^{-1}$
   $S(s) \rightarrow S(g)\quad \Delta H^\ominus = +223\,\text{kJ}\,\text{mol}^{-1}$

# THINKING BIGGER

## WHICH FUEL AND WHY?

Our dependence on organic fuels has necessitated the exploitation of renewable biofuels. One of the key aims behind the use of biofuels is to reduce carbon emissions, but with limited land space an increasing area of research is the use of waste biomass.

## ORGANIC CHEMISTS CONTRIBUTE TO RENEWABLE ENERGY

**Background – Why is this important?**

Biofuels play an essential role in reducing the carbon emissions from transportation. The development of 'drop in' fuels produced from lignocellulosic raw materials will increase both the availability of biofuels and the sustainability of the biofuel industry.

*Adrian Higson – Energy Consultant*

Biofuels can be either liquid or gaseous fuel. They can be produced from any source that can be replenished rapidly, e.g. plants, agricultural crops and municipal waste. Current biofuels are produced from sugar and starch crops such as wheat and sugar cane, which are also part of the food chain.

One of the key targets for energy researchers is a sustainable route to biofuels from non-edible lignocellulosic (plant) biomass, such as agricultural wastes, forestry residues or purpose grown energy grasses. These are examples of so-called advanced biofuels.

Current biofuels, such as ethanol, have a lower energy content (volumetric energy density) compared with conventional hydrocarbon fuels, petroleum and natural gas. The aim is to produce fuels that have a high carbon content and therefore have a higher volumetric energy density. This can be achieved by chemical reactions that remove oxygen atoms from biofuel chemical compounds. This process produces a so called 'drop-in biofuel', i.e. a fuel that can be blended directly with existing hydrocarbon fuels that have similar combustion properties.

**What did the organic chemists do?**

Levulinic acid        Furfural

**fig A** Levulinic acid and Furfural: two potential platform molecules.

Efficient synthesis of renewable fuels remains a challenging and important line of research. Levulinic acid and furfural are examples of potential 'platform molecules', i.e. molecules that can be produced from biomass and converted into biofuels. Levulinic acid can be produced in high yield (>70%) from inedible hexose bio-polymers such as cellulose, which is a polymer of glucose and the most common organic compound on Earth. Furfural has been produced industrially for many years from pentose-rich agricultural wastes and can also act as a platform molecule.

Recent reports have highlighted the use of organic chemistry to convert platform molecules like levulinic acid and furfural into potential advanced biofuels. Specifically, changing parts of the molecules that are responsible for their structure and function. This process is called 'functional group interconversion' and is part of the basic toolkit of organic chemistry. For example, researchers have described a process for converting levulinic acid into so-called 'valeric biofuels'. One of these biofuels, ethyl valerate, is claimed to be a possible advanced bio-gasoline molecule with several advantages over bio-ethanol.

A second method to create hydrocarbons involves Dumesic's approach via a decarboxylation of gamma-valerolactone, which can be produced in one step from levulinic acid by hydrogenation.

**What is the impact?**

Biodiesel is likely to be the second most important biofuel after ethanol in the short to medium term. Global production of biodiesel is expected to increase from 11 billion litres to reach 24 billion litres by 2017. For organic chemists, there are significant opportunities associated with further developing energy crops and producing advanced biofuels from new sources such as algae, industrial or post-consumer waste.

Where else will I encounter these themes?

1　　2　　3　　4　　5

Let us start by considering the nature of the writing in the article.

1. This article is written for members of an international chemistry association and thus relies on the audience having a high degree of scientific literacy. Having read the article a few times, attempt one of the following questions.

   a. Rewrite the article for a less scientifically literate reader. Can you get the main ideas across without using chemical structures and terminology to the same degree?

   b. Rewrite the article in such a way as to present a strongly positive argument in favour of biofuels. Can you present such a positive argument without altering the facts as presented?

Now we will look at the chemistry in, or connected to, this article. Don't worry if you are not ready to give answers to these questions yet. You may like to return to the questions once you have covered other topics later in the book.

2. The extract mentions the compounds levulinic acid and fufural.

   a. Calculate the molar mass of each compound.

   b. Calculate the percentage mass of oxygen in each molecule.

   c. Give a chemical test, and its result, that would enable you to distinguish between the two molecules.

3. In this question you will compare four different fuels (shown in **table A**).

| 1 | 2 | 3 | 4 | 5 | 6 |
|---|---|---|---|---|---|
| Molecule | Molecular mass in g mol$^{-1}$ | Standard molar enthalpy of combustion | Number of kJ of energy per mole of $CO_2$ produced | Number of kJ of energy per g of fuel | State at 25°C and 1 atm pressure |
| $CH_4$ | 16 | −890 | | | gas |
| $C_8H_{18}$ | 114 | −5470 | | | liquid |
| $C_2H_5OH$ | 46 | −1367 | | | liquid |
| $C_{15}H_{32}$ | 212 | −10 047 | | | liquid |

**table A** Thermochemical and physical data for four organic fuels.

   a. Write balanced equations for the complete combustion of $C_8H_{18}$ and $C_2H_5OH$.

   b. Complete columns 4 and 5 in the table.

4. Why can't the C-15 hydrocarbon synthesized by the process detailed above be described as a completely carbon neutral solution to the energy shortage problem? Refer to the text and any other web resources to help support your answer.

> **Writing scientifically**
> As you read these articles identify everything that contributes to writing a scientific article. For instance, the vocabulary, sources, the way information is presented. Make sure your own answers are written scientifically.

> This next question asks you to think more carefully about the term 'carbon neutral'. Before attempting question 4, carry out a web search for the term 'carbon neutral'. Is there a clear definition for the term?

# Activity

The metropolitan borough of East Grimshire is about to buy a fleet of 60 buses but must choose one of the four fuels in Table A on which they are to run. In groups, select one of the fuels and deliver a presentation on why your chosen fuel is the best option. Think about pros and cons of your chosen fuel. You should attempt to reference a range of source materials in support of your choice. Your presentation should consider the following:

- The pros and cons of your chosen fuel
- Energy efficiency
- Carbon footprint
- Sustainability.

Your presentation should be between 4 and 8 slides and last no more than 10 minutes but be prepared to justify your presentation in a 5 minute Q&A session!

> There will be plenty of source material on the internet but be critical about who is presenting the material. For example, an oil company will have vested interests!

- From a case study published by the Organic Division of the Royal Society of Chemistry

# 8 Exam-style questions

1. Lead forms several solid oxides, the most common of which are PbO, PbO₂ and Pb₃O₄.
   This question is about the enthalpy changes that take place during reactions involving these oxides.
   (a) (i) State what is meant by the term **enthalpy change of formation**. [2]
   (ii) State the standard conditions of temperature and pressure that are usually used in calculations involving enthalpy changes. [2]
   (iii) Write an equation representing the standard enthalpy change of formation of PbO(s). [2]
   (b) 'Red lead' (Pb₃O₄) can be made by heating PbO in oxygen:
   $$3PbO(s) + \tfrac{1}{2}O_2(g) \rightarrow Pb_3O_4(s)$$
   Calculate the standard enthalpy change for this reaction. [3]
   $\Delta_f H^\ominus$ PbO(s) = –219 kJ mol⁻¹;
   $\Delta_f H^\ominus$ Pb₃O₄(s) = –735 kJ mol⁻¹
   (c) When red lead is heated it decomposes into PbO and PbO₂.
   $$Pb_3O_4(s) \rightarrow 2PbO(s) + PbO_2(s) \quad \Delta_r H^\ominus = +20 \text{ kJ mol}^{-1}$$
   Use this information, together with the data supplied in part (b), to calculate the standard enthalpy change of formation of PbO₂(s). [3]
   **[Total: 12]**

2. Propane, C₃H₈, is a gas at room temperature. It is used as a fuel for portable gas cookers.
   (a) Give two properties of propane that make it suitable for use as a fuel. [2]
   (b) The standard enthalpy change of combustion of propane is reprented by the equation
   $$C_3H_8(g) + 5O_2(g) \rightarrow 3CO_2(g) + 4H_2O(l) \quad \Delta_c H^\ominus = -2220 \text{ kJ mol}^{-1}$$
   (i) State what is meant by the term standard enthalpy change of combustion. [3]
   (ii) Complete the enthalpy level diagram for the combustion of propane. Label $\Delta_c H^\ominus$. [3]

   Enthalpy

   C₃H₈(g) + 5O₂(g)

   (c) Hess's Law can be used to calculate enthalpy changes of formation from enthalpy changes of combustion.
   The equation for the formation of propane is
   $$3C(s) + 4H_2(g) \rightarrow C_3H_8(g)$$
   (i) Sate Hess's law. [2]
   (ii) The standard enthalpy changes of combustion of carbon and hydrogen are –394 kJ mol⁻¹ and –286 kJ mol⁻¹ respectively. Calculate the standard enthalpy change of formation of propane gas. [2]
   **[Total: 12]**

3. Bond enthalpies can be used to calculate enthalpy changes of reaction.
   Some bond enthalpies are given in the table.

   | Bond | Bond enthalpy/kJ mol⁻¹ |
   |---|---|
   | H–H | +436 |
   | F–F | +158 |
   | H–Cl | +431 |
   | H–F | +562 |

   (a) (i) State what is meant by the term **bond enthalpy**. [2]
   (ii) State why the sign for a bond enthalpy is always positive. [1]
   (b) The enthalpy change of formation of hydrogen chloride is represented by the following equation:
   $$\tfrac{1}{2}H_2(g) + \tfrac{1}{2}Cl_2(g) \rightarrow HCl(g) \quad \Delta_f H = -92 \text{ kJ mol}^{-1}$$
   Use this information, and the data in the table, to calculate the bond enthalpy of the Cl–Cl bond. [3]
   (c) Use the data in the table to calculate the enthalpy change of formation of HF(g). [2]
   **[Total: 8]**

4. The enthalpy change of neutralisation for the reaction between hydrochloric acid, HCl(aq), and sodium hydroxide, NaOH(aq), can be determined using the following method:
   - Place 50.0 cm³ of 2.00 mol dm⁻³ HCl(aq) into a polystyrene cup and measure its temperature.
   - Place 50.0 cm³ of 2.10 mol dm⁻³ NaOH(aq) into another polystyrene cup and measure its temperature.
   - Mix the two solutions, stir with the thermometer and record the highest temperature reached.
   The results of one experiment were:
   - Initial temperature of both HCl(aq) and NaOH(aq) = 20 °C
   - Highest temperature reached by the mixture = 33.6 °C
   The equation for the reaction is
   $$HCl(aq) + NaOH(aq) \rightarrow NaCl(aq) + H_2O(l)$$

(a) State what is meant by the term **enthalpy change of neutralisation**. [2]

(b) Give the name of a piece of apparatus that is suitable for measuring the volumes of acid and alkali. [1]

(c) State why the NaOH(aq) used had a slightly higher concentration than the HCl(aq). [1]

(d) (i) Calculate the heat energy transferred, using
$$Q = mc\Delta T.$$ [2]
[Assume the specific heat capacity of the final solution is $4.18\,Jg^{-1}K^{-1}$ and that its density is $1\,g\,cm^{-3}$]

(ii) Calculate the enthalpy change of neutralisation for the reaction. [3]

[Total: 9]

**5** A student used this apparatus to determine the enthalpy change of combustion of butan-1-ol, $CH_3CH_2CH_2CH_2OH$.

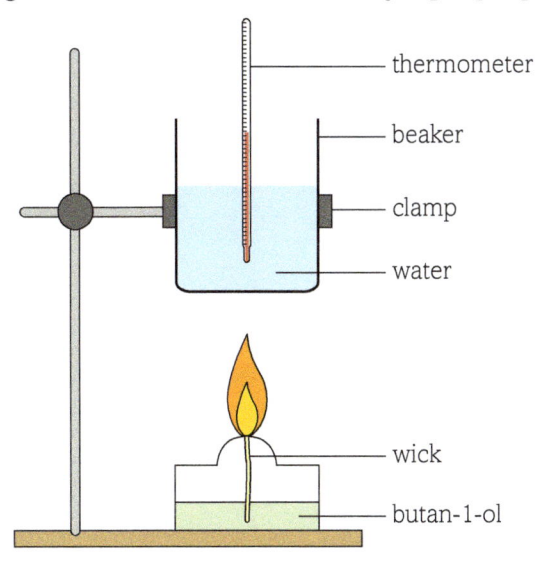

(a) Write an equation representing the standard enthalpy change of combustion of butan-1-ol. [2]

(b) The student placed $50.0\,cm^3$ of water into the beaker and lit the burner. When the temperature had risen by 10.0 °C, she found that 0.740 g of butan-1-ol has burned. Calculate the enthalpy change of combustion of butan-1-ol. [4]
[The specific heat capacity of water is $4.18\,Jg^{-1}K^{-1}$; density of water = $1\,g\,cm^{-3}$.]

(c) The student found that the value for the standard enthalpy of combustion was more exothermic than the value she obtained from her experiment.

Give two possible reasons for the difference in the two values. [2]

[Total: 8]

**6** The table gives the values of some mean bond enthalpies.

| Bond | Mean bond enthalpy/kJ mol⁻¹ |
| --- | --- |
| C–H | +412 |
| O–H | +463 |
| O=O | +496 |
| C=O | +743 |
| C–O | +360 |

The equation for the combustion of methanol ($CH_3OH$) in the gaseous state is

$$H-\underset{\underset{H}{|}}{\overset{\overset{H}{|}}{C}}-O-H + 1\tfrac{1}{2}O=O \rightarrow O=C=O + \begin{array}{c} H-O-H \\ H-O-H \end{array}$$

(a) (i) Use the data in the table to calculate the enthalpy change of combustion of gaseous methanol. [2]

(ii) Give two reasons why the standard enthalpy change of combustion of methanol is different from the value calculated in part (a)(i). [2]

(b) Which process measures the mean bond enthalpy for the C–H bond in methane? [1]

**A** $CH_4(g) \rightarrow CH_3(g) + H(g)$ $\Delta H = A$; mean bond enthalpy = A

**B** $CH_4(g) \rightarrow C(g) + 4H(g)$ $\Delta H = B$; mean bond enthalpy = B/4

**C** $CH_4(g) \rightarrow C(g) + 2H_2(g)$ $\Delta H = C$; mean bond enthalpy = C/4

**D** $CH(g) \rightarrow C(g) + H(g)$ $\Delta H = D$; mean bond enthalpy = D

(c) Calculate the mean bond enthalpy of the S–F bond in $SF_6$ given the following data. [2]

$SF_6(g) \rightarrow S(s) + 3F_2(g)$  $\Delta H = +1100\,kJ\,mol^{-1}$
$S(s) \rightarrow S(g)$  $\Delta H = +223\,kJ\,mol^{-1}$
$F_2(g) \rightarrow 2F(g)$  $\Delta H = +158\,kJ\,mol^{-1}$

[Total: 7]

# TOPIC 9

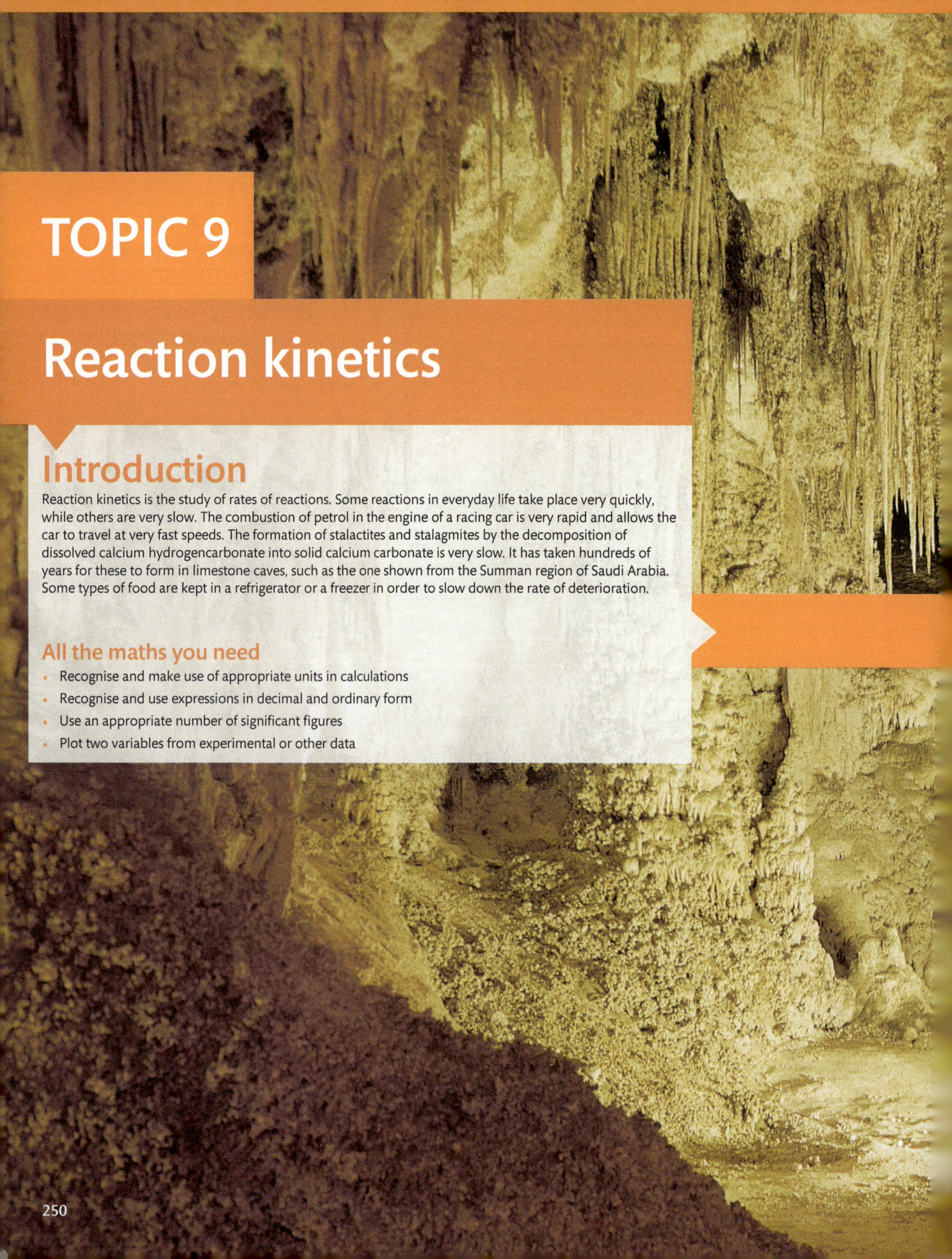

# Reaction kinetics

## Introduction

Reaction kinetics is the study of rates of reactions. Some reactions in everyday life take place very quickly, while others are very slow. The combustion of petrol in the engine of a racing car is very rapid and allows the car to travel at very fast speeds. The formation of stalactites and stalagmites by the decomposition of dissolved calcium hydrogencarbonate into solid calcium carbonate is very slow. It has taken hundreds of years for these to form in limestone caves, such as the one shown from the Summan region of Saudi Arabia. Some types of food are kept in a refrigerator or a freezer in order to slow down the rate of deterioration.

### All the maths you need
- Recognise and make use of appropriate units in calculations
- Recognise and use expressions in decimal and ordinary form
- Use an appropriate number of significant figures
- Plot two variables from experimental or other data

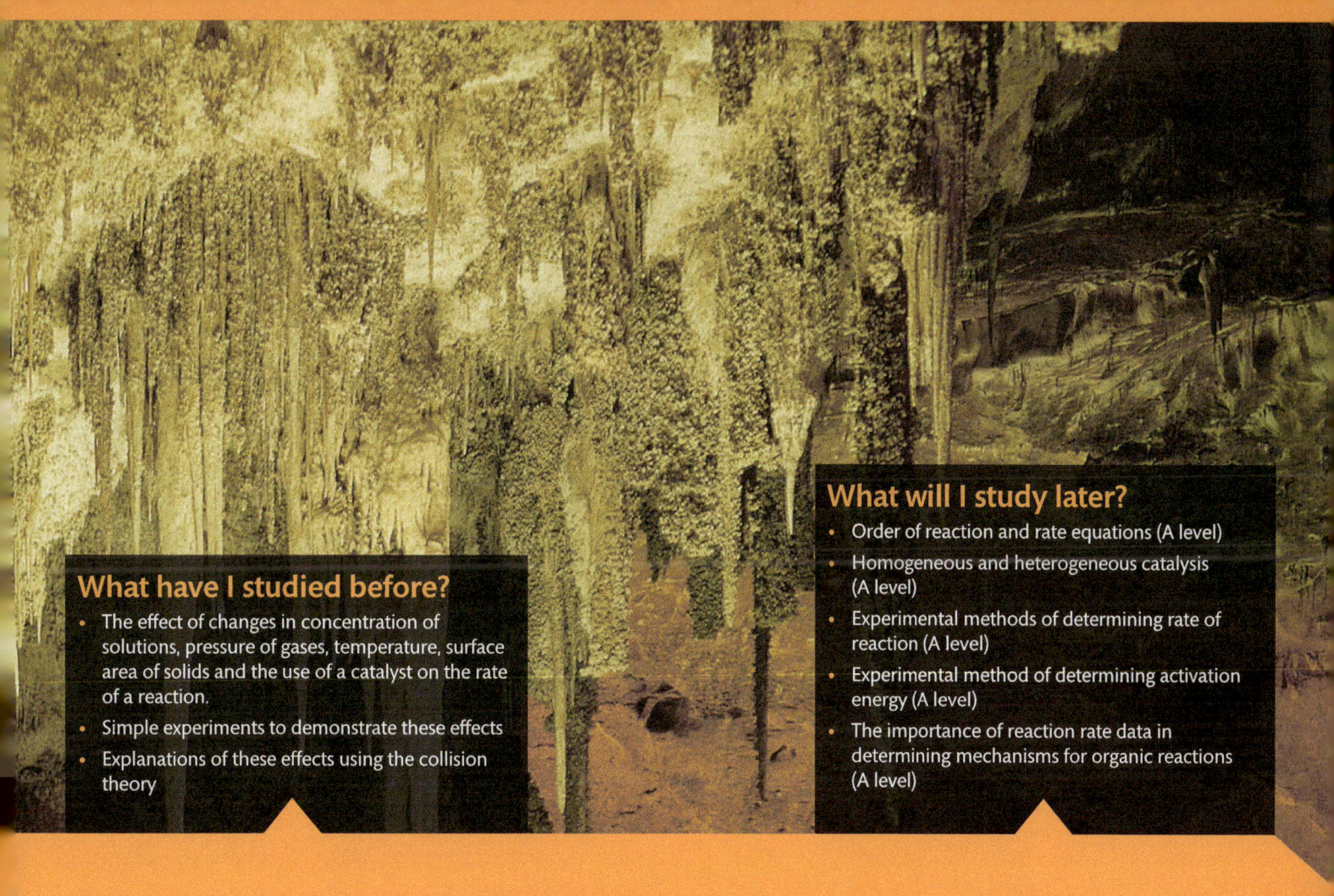

## What have I studied before?
- The effect of changes in concentration of solutions, pressure of gases, temperature, surface area of solids and the use of a catalyst on the rate of a reaction.
- Simple experiments to demonstrate these effects
- Explanations of these effects using the collision theory

## What will I study later?
- Order of reaction and rate equations (A level)
- Homogeneous and heterogeneous catalysis (A level)
- Experimental methods of determining rate of reaction (A level)
- Experimental method of determining activation energy (A level)
- The importance of reaction rate data in determining mechanisms for organic reactions (A level)

## What will I study in this topic?
- The concept of activation energy
- The Maxwell–Boltzmann model of distribution of molecular energies
- The role of catalysts in increasing the rate of chemical reactions
- Reaction profiles for both uncatalysed and catalysed reactions

# 9.1  1  Reaction rate, collision theory and activation energy

**By the end of this section, you should be able to...**
- know what is meant by the term 'rate of reaction'
- calculate the rate of a reaction from:
  - (i) the gradient of a suitable graph, by drawing a tangent, either for initial rate or a time $t$
  - (ii) data showing the time taken for a reaction

## Rate of reaction

The rate of a chemical reaction is determined by the change in concentration of a reactant or a product per unit time.

$$\text{Rate of reaction} = \frac{\text{change in concentration}}{\text{time for change to happen}}$$

So, in order to measure the rate of a reaction, we need to find out:

1. how fast one of the reactants is being used up, or
2. how fast one of the products is being formed.

In a graph of the concentration of a reactant against time, the gradient (slope) of the graph indicates the rate of the reaction.

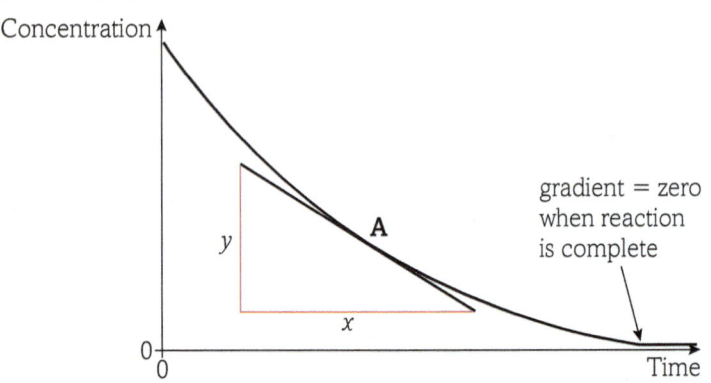

**fig A** A graph showing the concentration of a reactant against time.

The gradient decreases as the rate decreases, and becomes zero when all of the reactant is used up.

In the graph above, we determine the rate of reaction at point A by drawing a tangent to the curve at point A and measuring its gradient.

$$\text{Gradient} = \frac{y}{x}$$

where $y$ = the change in concentration and $x$ = the change in time.

If the unit of concentration is moles per cubic decimetre ($mol\,dm^{-3}$) and the unit of time is seconds (s), then the unit of rate will be moles per cubic decimetre per second ($mol\,dm^{-3}\,s^{-1}$).

Sometimes it is more convenient to measure the concentration of the product formed over a period of time.

The graph in **fig B** shows a plot of concentration against time, with gradients drawn to obtain the initial rate of reaction and the rate at time, $t$.

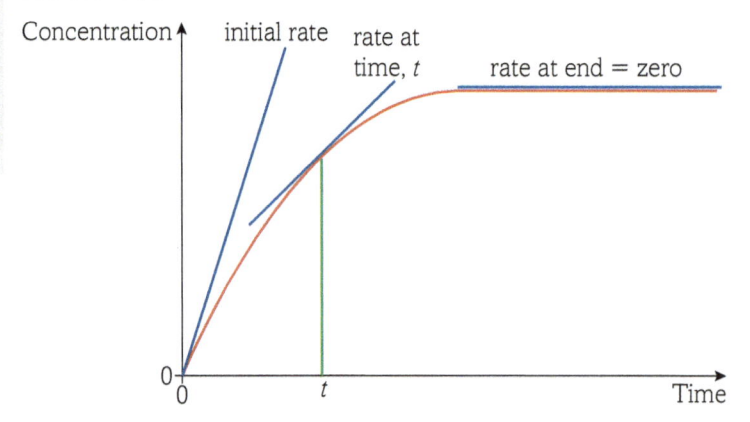

**fig B** A graph showing the concentration of a product against time.

The rate of reaction can also be calculated from the time taken for a known amount of reactant to be used up, or a known amount of product to be formed. For example if, in a series of separate experiments, the time taken to collect a given volume of gas is measured, then the rate can be calculated for each experiment using the expression:

$$\text{rate} = \text{volume of gas collected}/\text{time taken}$$

This is illustrated in the table below (**table A**).

| Time taken to collect $20.0\,cm^3$ of gas/s | Rate of reaction/$cm^3\,s^{-1}$ |
|---|---|
| 10.0 | 2.00 |
| 20.0 | 1.00 |
| 40.00 | 0.50 |
| 80.00 | 0.25 |

**table A** The rate of reaction can be calculated from the time taken for a known amount of product to be formed.

## Collision theory

Consider this reaction:

$$A + B \rightarrow C + D$$

In order for molecule **A** to react with molecule **B**, the two molecules must first of all collide with each other. If they collide they *may* react.

Why is there a possibility that the molecules may not react? This is because not all collisions between reactant molecules will result in a reaction. There are two requirements for a reaction to occur.

- The two molecules must collide with sufficient energy to cause a reaction (**activation energy**).
- The two molecules must collide in the correct orientation.

## Activation energy

If the particles collide with less energy than the activation energy, they simply bounce apart and no reaction occurs. Think of the activation energy as a barrier to the reaction. Only those collisions that have energies equal to or greater than the activation energy result in a reaction.

Any chemical reaction results in the breaking of some bonds (needing energy) and the making of new ones (releasing energy). Obviously, some bonds have to be broken before new ones can be made. Activation energy is involved in breaking some of the original bonds.

Where collisions are relatively gentle, there is not enough energy available to start the bond-breaking process, and so the particles do not react.

## Orientation

Consider the reaction between ethene and hydrogen bromide, which you met in **Section 6.2.6**:

$$H_2C=CH_2 + HBr \longrightarrow CH_3-CH_2Br$$

ethene + hydrogen bromide $\longrightarrow$ bromoethane

The reaction can only happen if the hydrogen end of the H—Br molecule approaches the C=C of the ethene molecule. Any other collision between the two molecules will result in the molecules simply bouncing off each other.

Of the collisions shown in the figure below, only collision 1 may possibly lead to a reaction.

collision 1 — possible reaction
collision 2 — no reaction
collision 3 — no reaction

## Steric hindrance

When the shapes of molecules influence reactions, we say that there is a 'steric factor' involved in the reaction. In some cases, the atoms (or groups of atoms) in a molecule can hinder the course of a reaction. If an atom or group of atoms is particularly large, then it can get in the way of an attacking species. If this happens, we say that the reaction suffers from 'steric hindrance'.

For example, the tertiary halogenoalkane 2-bromomethylpropane, $(CH_3)_3CBr$, is hydrolysed rapidly when added to water, but the mechanism of the reaction does not involve attack by the water molecule on the $\delta+$ carbon atom attached to the bromine, as it does with bromomethane, $CH_3Br$.

This is because the three methyl groups are so large that they prevent the water molecule approaching the carbon atom sufficiently close to interact with it. So, the mechanism for the reaction is totally different.

If you would like to find out more, look up $S_N1$ and $S_N2$ mechanisms for the hydrolysis of halogenoalkanes.

**fig C** The three methyl groups do not allow the water molecule to approach the carbon atom of the C—Br bond.

# Questions

1. At room temperature and pressure, in each cubic decimetre of gas there are about $1 \times 10^{32}$ collisions every second. Why, therefore, are reactions between gases not completed in a fraction of a second?

2. Why do many reactions involving organic compounds have to be heated or refluxed for long periods of time?

3. Chloroalkanes, such as chloromethane ($CH_3Cl$), can be hydrolysed to alcohols by heating with aqueous sodium hydroxide. The hydroxide ion acts as a nucleophile attacking the $\delta+$ carbon atom and replacing the chlorine:

    $$CH_3Cl + OH^- \rightarrow CH_3OH + Cl^-$$

    When tetrachloromethane ($CCl_4$) is heated under reflux with aqueous sodium hydroxide, no reaction takes place. Suggest a reason why.

4. What will happen if a hydrogen atom and a chlorine atom collide with a total energy of $1.0 \times 10^{-18}$ J? [$E$(H—Cl) = 431 kJ mol$^{-1}$].

### Key definition

**Activation energy**, $E_a$, is the minimum energy that colliding particles must possess for a reaction to occur.

# 9.1 2 Making a reaction go faster – Part 1

By the end of this section, you should be able to...
- explain qualitatively how changes in concentration of solution, pressure of a gas and surface area of a solid can affect the rate of a chemical reaction

According to the collision theory, reactant particles have to collide with sufficient energy before they can react. It is sensible, therefore, to suggest that the rate of a reaction can be increased by increasing the frequency of collisions with sufficient energy between reactant particles. We will call collisions that result in a reaction *successful* collisions.

This is often found to be the case, but as in so many situations in chemistry, there are exceptions. These will be dealt with in **Book 2**.

## The effect of concentration

For reactions in solution, an increase in concentration often causes an increase in reaction rate. For many reactions, if the concentration of a solution is increased, then the frequency of collision between reacting solute particles also increases. This is because they are closer together as there are more of them in a given volume of solution. So, the frequency of successful collisions increases (i.e. there are more successful collisions per second), which in turn produces an increase in the rate of reaction.

The graph below shows the effect of the change in volume of carbon dioxide given off with time for the reaction between calcium carbonate and excess dilute hydrochloric acid.

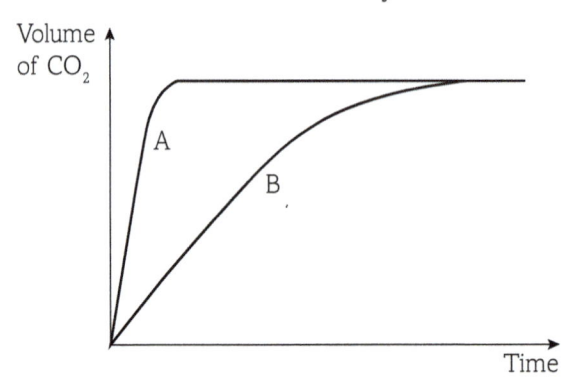

**fig A** A concentration–time graph showing the effect of the change in volume of carbon dioxide given off with time for the reaction between calcium carbonate and excess dilute hydrochloric acid.

Curve A represents the higher concentration of acid. You will notice that the gradient of curve A is always greater than that of curve B, and that curve A levels off before curve B. Both of these factors indicate that the rate of reaction is greater for the higher concentration of acid.

## The effect of pressure

For a reaction in which molecules collide and react in the gas phase, an increase in pressure will bring about an increase in the rate of reaction.

The explanation is similar to that for concentration of solution. If the pressure of the gaseous mixture is increased, there will be more reactant molecules in a given volume of mixture. So, the frequency of collisions will increase, with a resulting increase in the rate of reaction.

Changing the pressure has virtually no effect on reactions in the solid or liquid phase. This is because the volume of solids and liquids changes very little when they are put under pressure, so their particles do not move closer together.

### Additional reading

#### The Haber process
It is interesting to look at the situation that exists when a reaction between two gases takes place solely on the surface of a solid that is in contact with the reaction mixture. This is the case with the catalytic conversion of hydrogen gas and nitrogen gas to make ammonia in the Haber process.

$$3H_2(g) + N_2(g) \xrightarrow{Fe(s)} 2NH_3(g)$$

The reaction between the hydrogen and nitrogen molecules takes place at the surface of the iron catalyst, but only at selected sites called 'active' sites. At the high pressures used in the Haber process, there are always more reactant molecules than there are active sites. So, for this particular reaction, an increase in pressure will not increase the rate of reaction, since all of the active sites are already occupied.

## The effect of surface area

For heterogeneous reactions involving a solid, a large surface area of the solid will result in a faster reaction.

The reaction between magnesium and dilute hydrochloric acid is represented by this ionic equation:

$$Mg(s) + 2H^+(aq) \rightarrow Mg^{2+}(aq) + H_2(g)$$

Only collisions between the hydrogen ions and magnesium atoms on the *surface* of the magnesium can result in reaction. If the magnesium is powdered, the surface area is increased and hydrogen is given off more quickly.

The effectiveness of solid catalysts is also improved if they are finely divided. For example, the rate of the catalysed decomposition of hydrogen peroxide by manganese(IV) oxide is increased significantly if the catalyst is a powder rather than lumps. Here is the equation for the reaction:

$$H_2O_2(aq) \xrightarrow{MnO_2(s) \text{ catalyst}} H_2O(l) + \tfrac{1}{2}O_2(g)$$

> **Did you know?**
> A heterogeneous reaction is one in which the reactants are in more than one phase; for example, a solid and a gas, or a solid and a solution.

# Questions

1. Ammonia and hydrogen chloride react in the gas phase to form ammonium chloride.

   $$NH_3(g) + HCl(g) \rightarrow NH_4Cl(s)$$

   State and explain the effect, if any, on the rate of reaction of:
   (a) halving the volume of the container at constant temperature
   (b) increasing the pressure by adding more ammonia at constant volume and temperature.

2. (a) Draw a concentration–time graph for the reaction between marble chips (calcium carbonate) and excess dilute hydrochloric acid. Place the concentration of the hydrochloric acid on the vertical axis.
   (b) On the same axes, draw the curve obtained if the reaction was repeated at the same temperature, with the same mass of powdered marble.
   (c) Explain the reason for the different-shaped curves obtained.

# 9.1 3 Making a reaction go faster – Part 2

By the end of this section, you should be able to...

- explain how changes in temperature can affect the rate of a reaction, in terms of a qualitative understanding of the Maxwell–Boltzmann model of the distribution of molecular energies

## Maxwell–Boltzmann distribution curves

The molecules in a sample of gas have a wide range of energies. In order to estimate what fraction of collisions will have the required activation energy, we need to know the energy distribution of the molecules.

This was first calculated in 1860 by James Clerk Maxwell and verified in 1872 by Ludwig Boltzmann. The graph below shows the distribution of molecular energies at two temperatures: $T_1$ (in blue) and $T_2$ (in red). $T_2$ is a higher temperature than $T_1$.

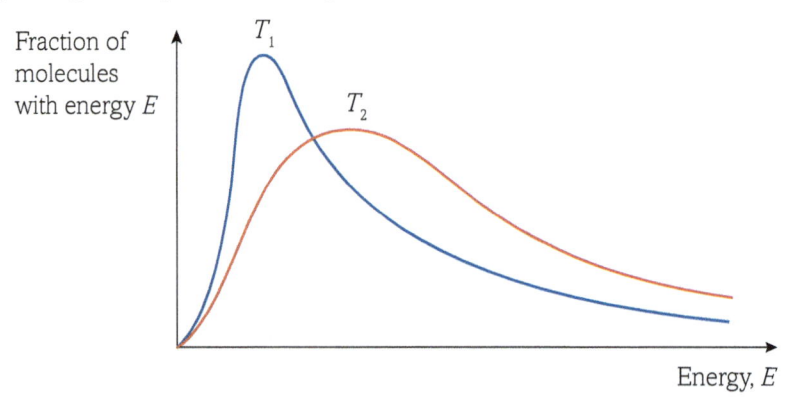

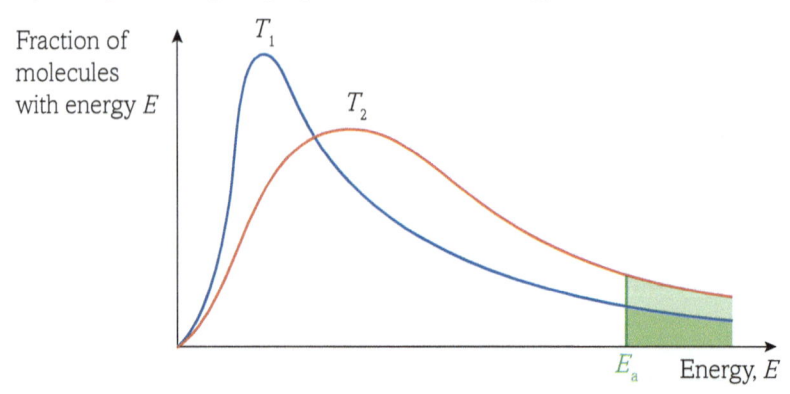

**fig A** A graph showing the distribution of molecular energies at two temperatures. $T_2$ is a higher temperature than $T_1$.

There are four important points to note about the curves;

- Neither curve is symmetrical.
- Both curves start at the origin and finish by approaching the x-axis asymptotically.
- The area under each curve is the same, since the number of molecules has not changed.
- The peak of $T_2$ is displaced to the right and is lower than that for the peak of $T_1$.

The curves in the graph below show, once again, the molecular distribution of energies at two temperatures, $T_1$ and $T_2$, where $T_2 > T_1$. $E_a$ is the activation energy for the reaction.

**fig B** A graph showing the distribution of molecular energies at two temperatures. $T_2 > T_1$.

The area shaded dark green represents the fraction of molecules that have the required energy to react at $T_1$.

The combined area shaded in dark green and light green represents the fraction of molecules that has the required energy to react at $T_2$.

It is easy to see that the fraction of molecules that can react at the higher temperature is greater.

## How changes in temperature can affect the rate of a reaction

We can now explain the effect of an increase in temperature on the rate of a reaction.

An increase in temperature increases the fraction of molecules that possesses the required activation energy. The rate of the reaction increases because the number of successful collisions per second increases.

The argument used above assumes that the fraction of collisions with energy greater than or equal to $E_a$ is the same as the fraction of molecules with this energy. This is not strictly true. However, the difference is very small at high energies when the fraction is very small. Under these circumstances, it is not unreasonable to draw and use the molecular energy distribution curve instead of the collision distribution curve.

## Collisions in solution

In the gas phase, molecules are moving around at high speeds and so frequently collide with one another. At a pressure of 100 kPa and a temperature of 298 K, a single molecule might have somewhere between $10^9$ and $10^{10}$ collisions per second, depending on its size.

In solution, the picture is rather different. The molecules are much more closely packed together so there is not a lot of 'space' between them. In low-to-medium concentrations, most of the solution is solvent, so solute molecules tend to be entirely surrounded by solvent molecules. These solute molecules are said to be trapped in a 'solvent cage'.

You may think that collisions between solute molecules would be far less frequent than similar collisions in the gas phase because the solvent molecules will get in the way. However, there are situations in which a number of solute molecules become trapped in the same solvent cage. This increases the collision rate between the solute molecules and, if the collisions are sufficiently energetic, they may react, just as in the gas phase.

There are, therefore, two distinct stages to a reaction in solution:

- First of all, the molecules have to come together, by a process of diffusion, into the same solvent cage.
- Secondly, they have to react.

There may, of course, be some reactions between solute molecules that have 'jumped out' of their cage and happen to encounter one another.

Although the picture of what is happening in a solution phase reaction is rather different to what is happening in the gas phase, the resulting kinetics are the same. For simple reactions involving two species, the chance of them encountering each other in solution is proportional to their concentrations, just as in the gas phase.

### Learning tip

The Maxwell–Boltzmann diagram showing the effect of temperature consists of *two* curves and *one* activation energy line.

If you are asked to use the Maxwell–Boltzmann distribution curves as part of your explanation, you must make it clear that the area under the curve to the right of the activation energy line represents the fraction of molecules that has enough energy for the collisions to be successful.

### Learning tip

When explaining the increase in reaction rate owing to an increase in temperature, it is important to mention that the number of *successful* collisions per second increases, and not just that the *overall* number of collisions per second increases.

Although there is an increase in the overall collision frequency, the effect of this is negligible compared to the increase in frequency of collisions that have energy equal to or greater than the activation energy line.

# Questions

1. State the effect of increasing the temperature on the rate of:
   (a) an exothermic reaction
   (b) an endothermic reaction.

2. (a) Draw Maxwell–Boltzmann distribution curves for a gas at two temperatures, $T_c$ and $T_h$, where $T_c < T_h$.
   (b) Indicate a suitable value for the activation energy of the reaction and use the curves to explain the effect of lowering the temperature on the rate of reaction.

# 9.1 4 Making a reaction go faster – Part 3

By the end of this section, you should be able to...

- explain how the addition of a catalyst can affect the rate of a reaction, in terms of a qualitative understanding of the Maxwell–Boltzmann model of the distribution of molecular energies
- understand the role of catalysts in providing alternative reaction routes of lower activation energy
- draw the reaction profiles of both an uncatalysed and a catalysed reaction
- understand the economic benefits of the use of catalysts in industrial reactions
- understand the use of a solid (heterogeneous) catalyst for industrial reactions involving gases, in terms of providing a surface for the reaction

## The effect of catalysts

A **catalyst** works by providing an alternative route for the reaction. This alternative route has a lower activation energy than the original route.

The diagram below shows the effect on the fraction of molecules that have the required energy to react when a route of lower activation energy is available.

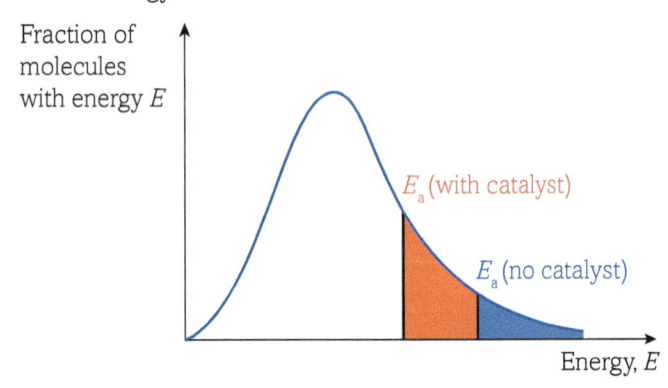

**fig A** A graph showing the effect on the fraction of molecules that have the required energy to react when a catalyst is present.

The blue shaded area represents the fraction of molecules that have $E \geqslant E_a$ when no catalyst is present.

The combined blue and red shaded area represents the fraction of molecules that have $E \geqslant E_a$ when a catalyst is present.

### Learning tip

The Maxwell–Boltzmann diagram showing the effect of a catalyst consists of *one* curve and *two* activation energy lines. Do not confuse this with the diagram used to show the effect of temperature.

## Reaction profile diagrams

A reaction profile diagram is an extension of an enthalpy level diagram (see **Section 8.1.2**). In addition to showing the relative enthalpy levels of reactants and products, it includes the activation energy for the reaction.

A typical reaction profile, not to scale, for the combustion of methane is shown below.

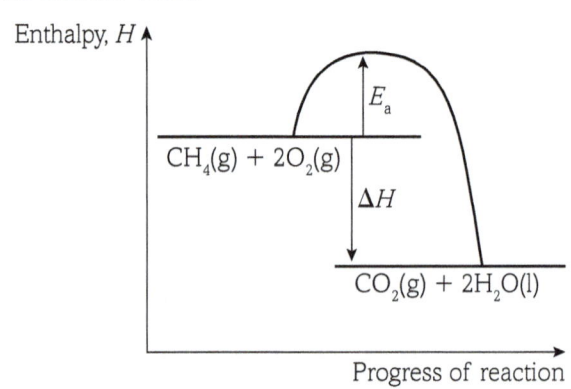

**fig B** A reaction profile for the combustion of methane.

For an endothermic reaction, the enthalpy level of the products is above that of the reactants, but otherwise the profile is the same.

The diagram below shows the simplified reaction profiles for an uncatalysed and a catalysed reaction.

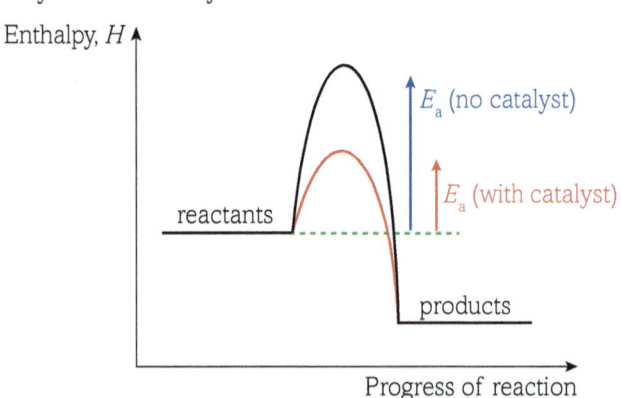

**fig C** Simplified reaction profiles for an uncatalysed and a catalysed reaction.

# Catalysts in industry

The first recorded use of a catalyst in industry was in 1746, when John Roebuck developed the lead chamber process for manufacturing sulfuric acid. Since then catalysts have been increasingly used in a large section of the chemical industry. The development of catalysts has been mostly for economic reasons, but increasingly of late for political and environmental reasons as well.

The two major economic advantages of the use of catalysts are:

- they increase the rate of a chemical reaction meaning that more of the desired product can be made in given time period
- reactions can take place at lower temperatures resulting in a decrease in the energy costs to the manufacturer.

Most catalysts used in industry are **heterogenous catalysts**. A heterogeneous catalyst is one that is in a different phase to that of the reactants. Solids are commonly used as heterogeneous catalysts for reactions involving gases. The most well known examples are the use of iron in the Haber process and vanadium(V) oxide in the contact process (see **Section 10.2.2**).

A solid catalyst provides a surface on which the gas molecules can adsorb and then react. The product molecules then desorb from the surface and more reactant molecules take their place. This process is described in more detail in **Book 2**.

### Did you know?

Environmental catalysts have also been developed to remove toxic products from industrial waste materials.

## Question

1. An aqueous solution of hydrogen peroxide decomposes very slowly at room temperature into water and oxygen. The decomposition is catalysed by solid manganese(IV) oxide.

    $$H_2O_2(aq) \rightarrow H_2O(l) + \tfrac{1}{2}O_2(g) \quad \Delta_r H^\ominus = -196 \text{ kJ mol}^{-1}$$

    (a) Using the same axes, draw labelled reaction profiles for the reaction both with and without manganese(IV) oxide.

    (b) Explain the change in rate of reaction that occurs when manganese(IV) oxide is added to the hydrogen peroxide solution.

### Key definitions

A **catalyst** is a substance that increases the rate of a chemical reaction but is chemically unchanged at the end of the reaction.

A **heterogeneous catalyst** is one that is in a different phase to that of the reactants.

# THINKING BIGGER

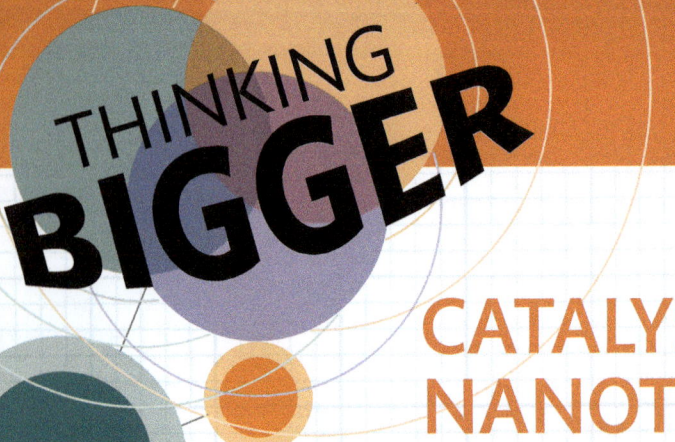

## CATALYSTS AND NANOTECHNOLOGY

Two aspects of intense research in today's chemical industry are the ongoing search for more effective and specific catalysts and the use of nanotechnology. The following extract considers an application of both of these features.

### TINY TUBES SET CHEMICAL REACTIONS RACING

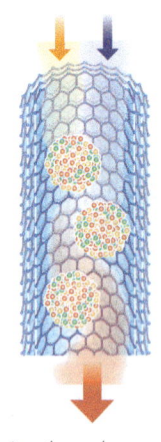

**fig A** Reactant molecules passing through a cardboard nanotube make contact with the catalyst inside the tube.

The efficiency of a fuel-generating chemical reaction can be boosted dramatically by mixing chemicals and catalysts inside nanoscopic test tubes, Chinese researchers have shown.

A team from the Dalian Institute of Chemical Physics in China found that a mixture of carbon monoxide and hydrogen (known as syngas) makes ethanol more quickly, with the help of tiny particles of a metal catalyst, when poured into carbon nanotubes.

The catalytic reaction was found to be more than 10 times faster when performed inside nanotubes, despite the lack of space for fresh reactants to circulate. The team showed multi-wall carbon nanotubes – 4 and 8 nanometres wide on the inside and 400 nm in length – boost the metal catalyst's efficiency.

The researchers think the finding could be important for two reasons: Firstly, it could prove useful for generating ethanol fuel using syngas extracted from natural gas, coal or even biomass. Secondly, they suspect that other catalysts should get a speed boost from being inside nanotubes.

**Inside and out**

Ethanol yield was measured when metal catalyst particles were attached to both the outside and the inside of the nanotubes. The process occurred more than 10 times more quickly when they were on the inside.

Reactions were also performed inside different types of tube – such as glass – the results of which suggest that it is not just the shape of the environment that makes the difference, but the material too.

The team believes the unique way electrons are arranged inside carbon nanotubes may be responsible for the effect. They are more sparse on the inside of a tube and this may make it easier for the catalyst to loosen the bond that holds carbon and oxygen molecules together.

### Did you know?

Everyone knows that chemical reactions speed up when the reaction temperature is increased, but is that always true? Well there are some rare exceptions where reaction rate slows down as the temperature increases. One such example is the reaction between nitrogen monoxide and oxygen to form nitrogen dioxide. The overall reaction can be summarised as follows:

$$2NO(g) + O_2(g) \rightarrow 2NO_2(g)$$

The fact that the rate of reaction decreases as temperature increases is thought to be due to the fact that the rate determining step of the reaction involves the formation of a $N_2O_2$ dimer which is more likely to break apart at higher temperatures. If this dimer does not last for a sufficiently long period then there is less time for the second oxidation part of the reaction to take place. You will learn more about rate determining steps in the second part of the course.

Where else will I encounter these themes?

1  2  3  4  5

Let us start by considering the nature of the writing in the article.

1. This article is taken from the *New Scientist,* which is aimed at those with a high degree of general scientific literacy. Suppose that you now had to re-write this article for the BBC 'Bitesize' website which is aimed at 11–16 year olds preparing for their GCSEs.

   Prepare an article of between 250–300 words to convey the main themes of this article to this audience. You might like to consider the following:
   - How can you put across the idea of what a catalyst is?
   - How can you convey ideas of the scale involved (i.e. the nanometer scale)?
   - How might the researchers build on their initial experiments?

> It would be worth looking at the Bitesize website before you attempt this question to see how the science is tackled and how language is used.

Now we will look at the chemistry in, or connected to, this article. Don't worry if you are not ready to give answers to these questions yet. You may like to return to the questions once you have covered other sections later in the book.

2. Taking the length of each carbon nanotube to be 400 nm, calculate how many of these tubes, laid end to end, would stretch to 1 cm.
3. One possible reaction for the conversion of syngas into ethanol is given by the equation:

   $6CO + 6H_2 \rightarrow 2C_2H_5OH + 2CO_2$

   a. Calculate the atom economy by mass of this reaction, assuming that the only desired product is ethanol.
   b. Calculate the volume of carbon monoxide (measured at RTP, 1 mole = 24 $dm^3$) required to make 5000 kg of ethanol.
4. What two factors are being altered in the process detailed above to increase the rate of reaction?
5. What evidence is there in the text that suggests that the carbon nanotubes are providing something other than just a matrix on which to put the catalyst?

# Activity

Many transition metals and their compounds are used as catalysts in the chemical industry. These catalysts can be either heterogeneous or homogeneous. Choose a transition metal or a transition metal compound that is used to catalyse a particular reaction in the chemical industry and deliver a presentation. You should discuss:

- Which catalyst have you chosen and which reaction is it used to catalyse?
- Is the catalyst heterogeneous or homogeneous?
- Does the catalyst need to be recovered or treated in any way?
- Is there any indication of how the catalyst is involved in the reaction mechanism?
- Have other catalysts been tried and rejected? If so why?

The presentation should be between 5–8 slides and last for no more than 10 minutes.

> You will have covered the Haber process at GCSE so it might be worth looking back at the role of the catalyst in this reaction.

● From an article in *New Scientist* magazine

# 9 Exam-style questions

1. A student investigated the effect of concentration on the reaction between solid magnesium carbonate and dilute hydrochloric acid. The equation for the reaction is

    $MgCO_3 + 2HCl \rightarrow MgCl_2 + CO_2 + H_2O$

    (a) Write an ionic equation for this reaction. Include state symbols. [2]

    (b) The student added an excess of dilute hydrochloric acid to some magnesium carbonate. She collected the gas, measuring its volume at regular intervals. She then plotted a graph of her results.

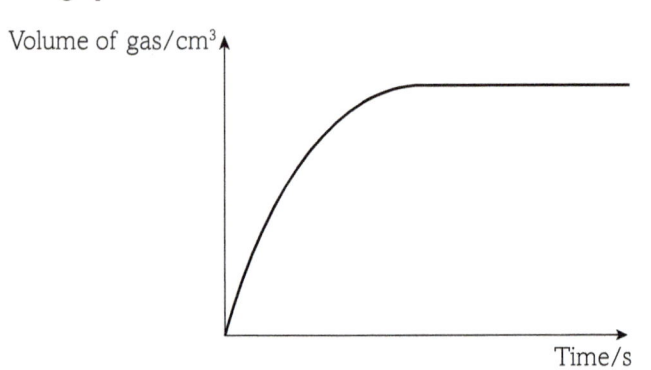

    **graph A**

    Using the collision theory, explain changes in the rate of reaction as it takes place. [4]

    (c) The student repeated the experiment using the same amount of magnesium carbonate but with hydrochloric acid of a lower concentration, still in excess. Sketch on the graph the curve you would expect to obtain. [2]

    [Total: 8]

2. This question is about some graphs that you may have seen during your course.

    (a) **Graph B** is a Maxwell-Boltzmann distribution curve.

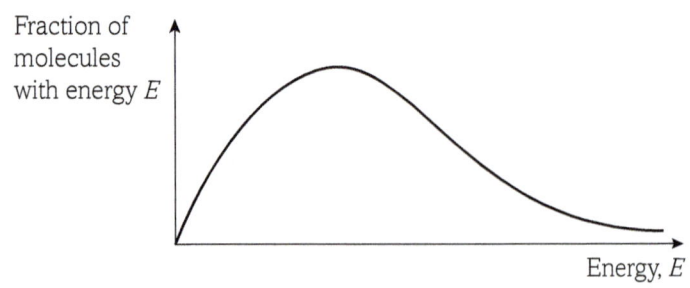

    **graph B**

    Explain, using **graph B**, the effect of adding an effective catalyst on the rate of reaction. [4]

    (b) **Graph C** shows the change in mass observed in an experiment to investigate the rate of reaction between marble chips and dilute hydrochloric acid.

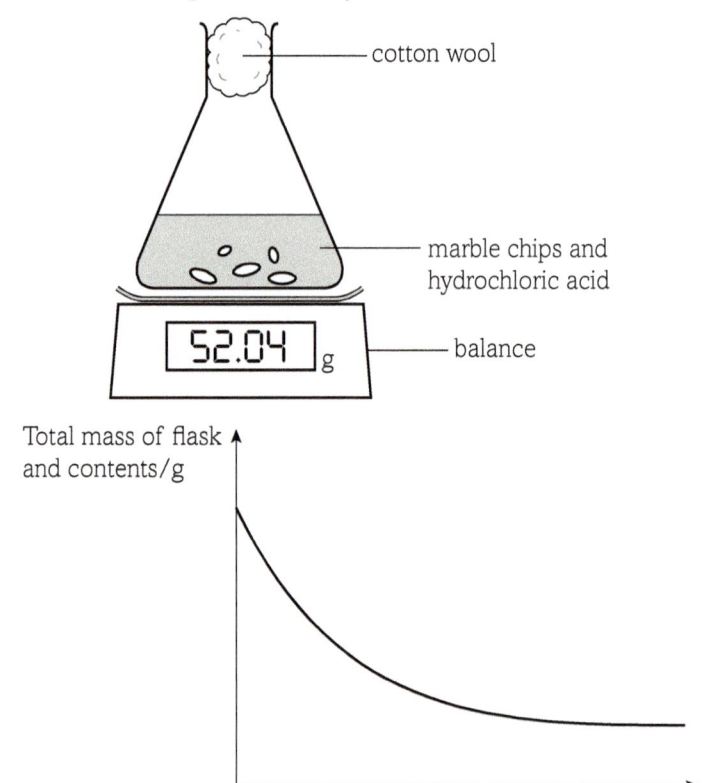

    **graph C**

    (i) State the purpose of the cotton wool in the neck of the conical flask. [1]

    (ii) Explain why the mass of the flask and contents decreases during the course of the experiment. [2]

    (iii) The experiment was repeated at the same temperature, with the same volume and concentration of acid and with the same mass of powdered marble. Sketch, on **graph B**, the curve you would expect to obtain. [2]

    [Total: 9]

3. This question is about the effect that an increase in temperature has on the rate of a chemical reaction.

    (a) A student found this statement in a text book:
    'The rate of a chemical reaction increases as the temperature increases because the particles collide more often'.
    Discuss the extent to which this statement is true. [5]

(b) The diagram shows a Maxwell-Boltzmann distribution curve for a gas at temperature $T_1$.

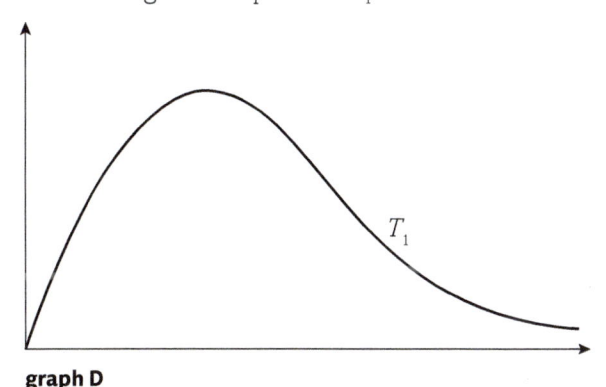

**graph D**

(i) Label the axes on the diagram. [2]
(ii) Sketch, on the diagram, a curve to represent the distribution at higher temperature. Label this curve $T_2$. [2]

[Total: 9]

4 The diagram is the reaction profile for the uncatalysed decomposition of hydrogen peroxide ($H_2O_2$) into water and oxygen.

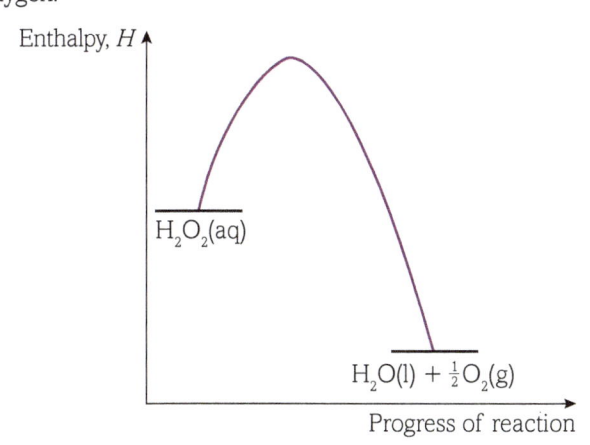

**graph E**

(a) Label the diagram with the activation energy ($E_a$) and the enthalpy change of reaction ($\Delta H$) [2]
(b) The decomposition is catalysed by the addition of solid manganese(IV) oxide ($MnO_2$).
  (i) Draw the curve to represent the catalysed decomposition. [2]
  (ii) State, in principle, how the catalyst works in this reaction. [2]
(c) State whether the reaction is exothermic or endothermic. Justify your answer. [1]

[Total: 7]

5 A student performed an investigation into the rate of reaction between zinc and dilute hydrochloric acid.

$$Zn(s) + 2HCl(aq) \rightarrow ZnCl_2(aq) + H_2(g)$$

He initially performed two experiments.

Experiment 1

He added 0.002 mol of zinc powder to an excess of dilute hydrochloric acid and measured the time taken for the total volume of hydrogen to be given off.

Experiment 2

He repeated the experiment with exactly the same quantities of the same batch of zinc and hydrochloric acid, but this time he dissolved 0.001 mol of copper(II) sulfate in the acid before adding the zinc.

His results are shown in the table.

| Experiment | Total volume of hydrogen evolved/m³ | Time taken to produce total volume of hydrogen/s |
|---|---|---|
| 1 | 48 | 300 |
| 2 | 44 | 80 |

(a) When the reaction in experiment 2 had finished, a red powder remained undissolved. The student identified this as copper.
  (i) Write an ionic equation for formation of copper in experiment 2. Include state symbols. [2]
  (ii) Explain another observation made in experiment 2 that would suggest that copper had been formed in the reaction. [2]
(b) The student concluded that the copper(II) sulfate had is some way increased the rate of the reaction between zinc and dilute hydrochloric acid. He found an article on the internet that suggested that copper acts as a catalyst in this reaction.

In order to test this, he repeated the experiment under identical conditions, this time mixing some copper powder with the zinc powder before adding the mixture to the acid.

It took 140 s for 44 cm³ and 180 s for 48 cm³ of hydrogen to be produced.

Comment on this result in the light of the results produced in experiments 1 and 2. [3]

[Total: 7]

# TOPIC 10

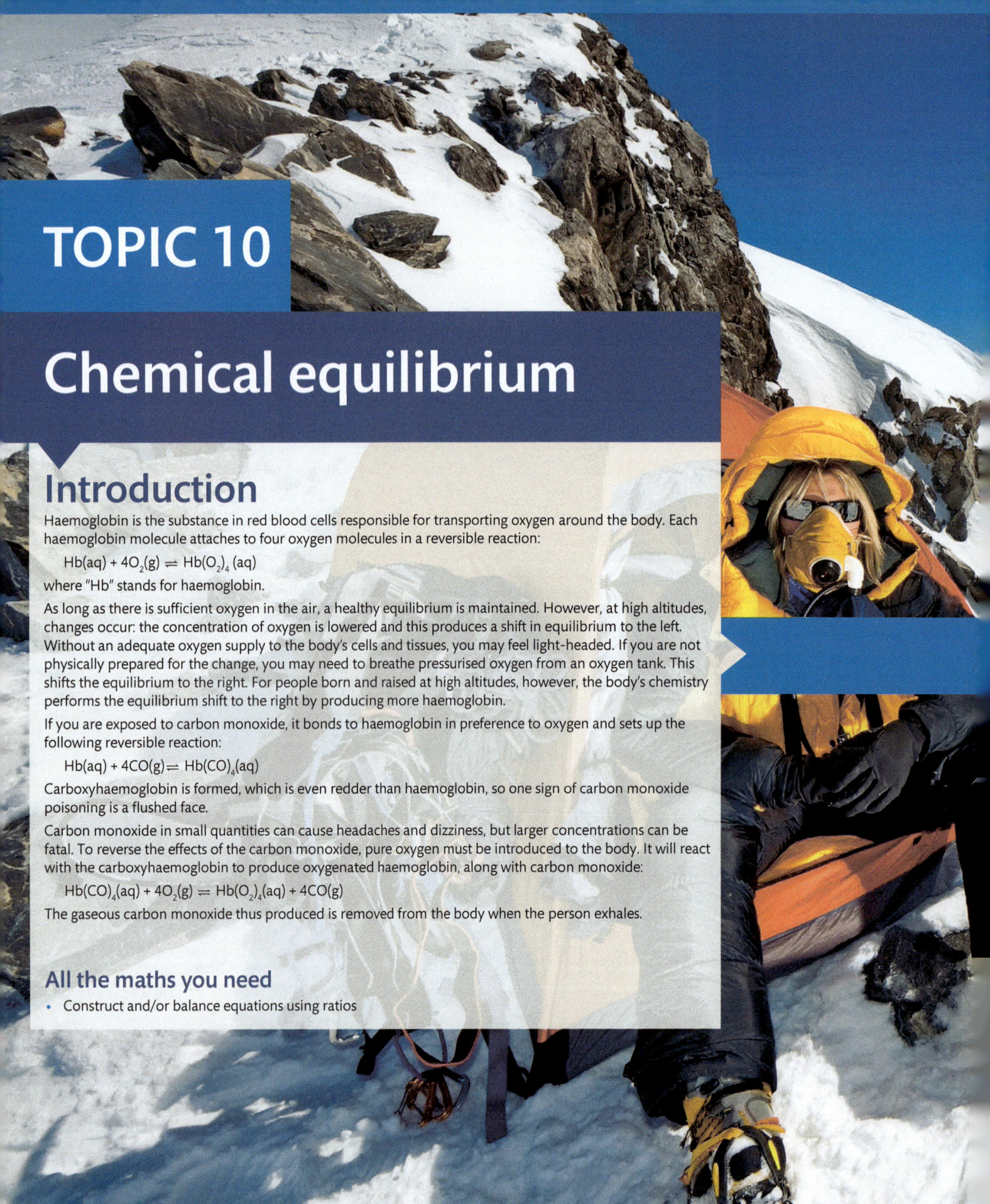

# Chemical equilibrium

## Introduction

Haemoglobin is the substance in red blood cells responsible for transporting oxygen around the body. Each haemoglobin molecule attaches to four oxygen molecules in a reversible reaction:

$$Hb(aq) + 4O_2(g) \rightleftharpoons Hb(O_2)_4(aq)$$

where "Hb" stands for haemoglobin.

As long as there is sufficient oxygen in the air, a healthy equilibrium is maintained. However, at high altitudes, changes occur: the concentration of oxygen is lowered and this produces a shift in equilibrium to the left. Without an adequate oxygen supply to the body's cells and tissues, you may feel light-headed. If you are not physically prepared for the change, you may need to breathe pressurised oxygen from an oxygen tank. This shifts the equilibrium to the right. For people born and raised at high altitudes, however, the body's chemistry performs the equilibrium shift to the right by producing more haemoglobin.

If you are exposed to carbon monoxide, it bonds to haemoglobin in preference to oxygen and sets up the following reversible reaction:

$$Hb(aq) + 4CO(g) \rightleftharpoons Hb(CO)_4(aq)$$

Carboxyhaemoglobin is formed, which is even redder than haemoglobin, so one sign of carbon monoxide poisoning is a flushed face.

Carbon monoxide in small quantities can cause headaches and dizziness, but larger concentrations can be fatal. To reverse the effects of the carbon monoxide, pure oxygen must be introduced to the body. It will react with the carboxyhaemoglobin to produce oxygenated haemoglobin, along with carbon monoxide:

$$Hb(CO)_4(aq) + 4O_2(g) \rightleftharpoons Hb(O_2)_4(aq) + 4CO(g)$$

The gaseous carbon monoxide thus produced is removed from the body when the person exhales.

### All the maths you need
- Construct and/or balance equations using ratios

### What have I studied before?
- Examples of reversible reactions such as the action of heat on ammonium chloride
- The concept of dynamic equilibrium
- Predicting the effect of changing the temperature and pressure of the equilibrium position

### What will I study later?
- The equilibrium constant in terms of partial pressures, $K_p$ (A level)
- Calculating values for $K_c$ and $K_p$ (A level)
- The effect of changing the temperature on the value of $K_c$ and $K_p$ (A level)
- Acid-base equilibria (A level)
- The relationship between Gibbs energy and equilibrium constant (A level)

### What will I study in this topic?
- The effect of concentration of a reactant on the position of equilibrium
- Reversible reactions in industry
- The equilibrium constant in terms of concentrations, $K_c$

# 10.1  1  Reversible reactions and dynamic equilibrium

**By the end of this section, you should be able to...**

- know that many reactions are readily reversible
- know that reversible reactions can reach a state of dynamic equilibrium in which:
  (i) both forward and backward reactions are still occurring
  (ii) the rate of the forward reaction is equal to the rate of the backward reaction
  (iii) the concentrations of reactants and products remain constant

## Irreversible and reversible reactions

When a mixture of hydrogen and oxygen in a 2 : 1 molar ratio is ignited, water is produced. There is very little, if any, uncombined hydrogen or oxygen remaining at the end of the reaction. We often describe such reactions as 'irreversible'.

Most combustion reactions fall into this category since they are highly exothermic. That is, $\Delta H$ is large and negative. However, there are many reactions, particularly in organic chemistry, for which $\Delta H$ is small. These reactions may not go to completion. At the end of the reaction, detectable amounts of the reactants remain, mixed with the product. Such reactions are called 'reversible' reactions.

## How to decide whether a reaction is reversible

Deciding whether a reaction is reversible or not depends on how carefully we measure the concentrations of reactants and products. For example, the reaction between dilute hydrochloric acid and aqueous sodium hydroxide appears to go to completion. Both acid and alkali are almost completely ionised in water, so the equation for the reaction is:

$$H^+(aq) + OH^-(aq) \rightarrow H_2O(l)$$

Pure water has a slight electrical conductivity, which results from the ionisation of water molecules:

$$H_2O(l) \rightarrow H^+(aq) + OH^-(aq)$$

This indicates that the reverse reaction is taking place to a small extent. Since only one molecule in approximately 550 million is ionised, we usually ignore this small extent of ionisation, but it becomes important when we study the pH scale of acidity.

In practical terms, if a reaction is more than 99% complete, we usually consider it to have gone to completion.

## The reaction between hydrogen and iodine

If a mixture of hydrogen and iodine vapour in a 1 : 1 molar ratio is heated to 573 K in a closed container, about 90% of the hydrogen and iodine reacts to form hydrogen iodide. Provided the reaction mixture remains in the closed container at 573 K, 10% of the hydrogen and iodine will remain unreacted no matter how long we leave the reaction mixture.

If a sample of hydrogen iodide is heated to 573 K in a closed container, it partially decomposes and the mixture formed is identical to that produced when starting with an equimolar mixture of hydrogen and iodine. The reaction is clearly reversible, and when there is no further change in the concentrations of the reactants and products, the system is said to be in 'equilibrium'.

The symbol $\rightleftharpoons$ is used in an equation to represent a reversible reaction. The equation for the reaction between hydrogen and iodine is therefore written as follows:

$$H_2(g) + I_2(g) \rightleftharpoons 2HI(g)$$

When the equation is written in this way, the reaction between hydrogen and iodine is referred to as the *forward* reaction. The decomposition of hydrogen iodide into hydrogen and iodine is called the *backward* reaction.

## How is equilibrium established?

When the mixture of hydrogen and iodine is heated, the two gases start to react and form hydrogen iodide. With increasing time, the concentrations of hydrogen and iodine decrease, so the rate of the forward reaction decreases.

As soon as some hydrogen iodide is formed, it starts to slowly decompose. With increasing time, however, the concentration of hydrogen iodide increases, so the rate of the backward reaction increases.

Eventually, the rates of the forward and the backward reactions become equal and after this point there is no further change in concentrations of reactants and products. The system is now in equilibrium. It is referred to as a 'dynamic' equilibrium since both forward and backward reactions are taking place at the same time, and also at the same rate.

## Dynamic equilibrium

The figure below shows dynamic equilibrium.

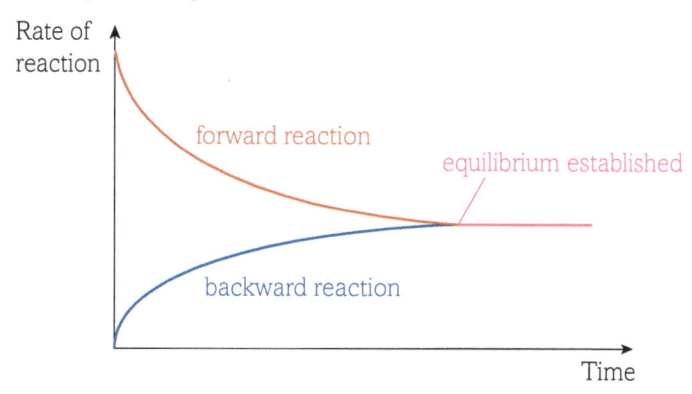

**fig A** Dynamic equilibrium.

Two conditions must be met for dynamic equilibrium to be established:

1 the reaction must be reversible
2 the reaction mixture must be in a closed container.

Three important features define a system that is in dynamic equilibrium:

- Both forward and backward reactions are continuously occurring.
- The rate of the forward reaction is equal to the rate of the backward reaction.
- The concentrations of reactants and products remain constant.

# Questions

1 Explain what is meant by the term 'dynamic equilibrium' with reference to the following reaction:
   $H_2(g) + I_2(g) \rightleftharpoons 2HI(g)$

2 Nitrogen and hydrogen react reversibly to form ammonia:
   $N_2(g) + 3H_2(g) \rightleftharpoons 2NH_3(g)$
   1 mol of nitrogen and 3 mol of hydrogen were mixed in a closed container and allowed to reach equilibrium. Twenty per cent of the nitrogen and hydrogen were converted into ammonia.
   Draw three graphs, using a single set of axes, to show how the number of moles of nitrogen, hydrogen and ammonia vary with time.

# 10.1 2 The effect of changes in conditions on equilibrium composition

By the end of this section, you should be able to...

- predict and justify the qualitative effect of a change in concentration, pressure or temperature, or the addition of a catalyst, on the composition of an equilibrium mixture

## Changing the composition of an equilibrium mixture

When a reaction mixture reaches a position of equilibrium, the composition of the equilibrium mixture (i.e. the concentration of each component) will not alter as long as the conditions remain the same.

However, if we make a change in condition (i.e. add some more of or remove one of the components, or change the temperature of the system) then the composition may change. This is often referred to as 'changing the position of equilibrium', and we refer to the position being moved to the right, to the left or not changed.

For example, if acid is added to a yellow solution containing chromate(VI) ions, $CrO_4^{2-}$, the solution turns orange, owing to an increase in the amount of dichromate(VI) ions, $Cr_2O_7^{2-}$.

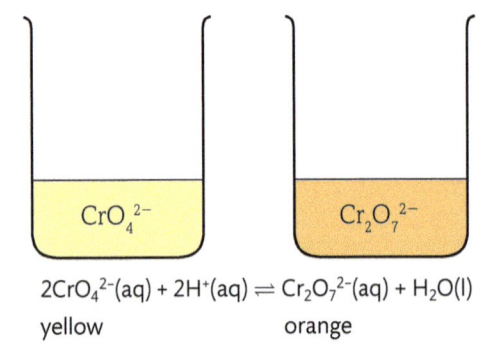

$$2CrO_4^{2-}(aq) + 2H^+(aq) \rightleftharpoons Cr_2O_7^{2-}(aq) + H_2O(l)$$
yellow                                      orange

**fig A** Beakers containing solutions of chromate(VI) ions and dichromate(VI) ions.

The equilibrium position moves to the right when the acid ($H^+$) is added. If sufficient alkali is added to the orange solution, it will turn yellow as the amount of $CrO_4^{2-}$ ions increases and exceeds the amount of $Cr_2O_7^{2-}$ ions. The equilibrium position moves to the left when alkali ($OH^-$) is added.

We will consider four factors that may affect the position of equilibrium of a reaction mixture. These are:

1. concentration of a component
2. pressure of the system
3. temperature of the system
4. addition of a catalyst.

## Effect of a change in concentration

If we increase the concentration of one of the reactants in a system in equilibrium, the rate of the forward reaction will increase and more products will form. As the concentration of the products increases, the rate of the backward reaction increases and eventually a new equilibrium is established. The equilibrium position has moved to the right, with slightly more product being present than at the original position of equilibrium.

If the concentration of one of the reactants is *decreased*, the position of equilibrium moves to the left. Similar changes occur if the concentration of the product is increased or decreased.

The changes that occur are summarised in the following table.

| Concentration of reactants | Concentration of products | Change in position of equilibrium |
|---|---|---|
| increased | | to the right |
| decreased | | to the left |
| | increased | to the left |
| | decreased | to the right |

**table A** The change in the position of equilibrium when the concentration of reactants or products is increased or decreased.

## Effect of a change in pressure

The effect of pressure only applies to reversible reactions involving gases. At a given temperature, the pressure of a gaseous mixture depends solely on the number of gas molecules in a given volume. So, the pressure of a gaseous mixture may be increased by reducing the volume and reduced by increasing the volume.

Alternatively, the pressure at which the reaction is carried out can be:

- increased by initially using more moles of the reactants in the same volume
- decreased by using less moles of the reactants in the same volume.

The effect of a change in pressure (at constant temperature) caused by changing the volume of the reaction mixture can be studied using a gas syringe and pushing in, or pulling out, the plunger. The effect depends on the total number of moles of gas on each side of the balanced equation, and is summarised in the following table.

| Number of moles of reactants | Number of moles of products | Change in position of equilibrium when the pressure is increased |
|---|---|---|
| more | fewer | to the right |
| fewer | more | to the left |
| same | same | no change |

**table B** The change in the position of equilibrium when the pressure is increased and the number of moles of reactants or products is increased, decreased or not changed.

The reverse changes are true for a decrease in pressure.

The effect of changes in pressure as a result of using different amounts of gaseous reactants in a fixed volume container is shown in the graph below. The reaction is between nitrogen and hydrogen, forming ammonia.

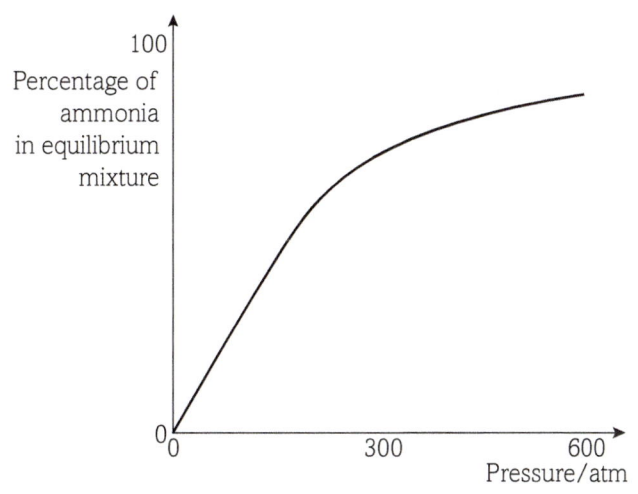

**fig B** The effect of changes in pressure from using different amounts of nitrogen and hydrogen to form ammonia in a fixed volume container.

The equation for the reaction is:

$$N_2(g) + 3H_2(g) \rightleftharpoons 2NH_3(g)$$

In the balanced equation, there are 4 moles of gas on the left-hand side and 2 moles of gas on the right-hand side.

The graph shows that the higher the pressure, the more ammonia there is in the equilibrium mixture formed. This agrees with the prediction in table B which states that an increase in pressure shifts the position of equilibrium to the side that has fewer moles of gas.

### Effect of a change in temperature

If the temperature of an equilibrium mixture is raised, the rates of both the forward and the backward reactions will increase. However, the increase in the rate of the endothermic reaction will be greater than the increase in the rate of the exothermic reaction. Therefore, an increase in temperature will shift the position of equilibrium in the direction of the endothermic reaction.

So, the change in position of equilibrium will depend on whether the forward reaction is exothermic or endothermic.

The effects of temperature change are summarised in the following table.

| Temperature change | Thermicity of forward reaction | Change in position of equilibrium |
|---|---|---|
| increased | exothermic ($\Delta H$ –ve) | to the left |
| decreased | exothermic ($\Delta H$ –ve) | to the right |
| increased | endothermic ($\Delta H$ +ve) | to the right |
| decreased | endothermic ($\Delta H$ +ve) | to the left |

**table C** The change in the position of equilibrium and the thermicity of the forward reaction when the temperature changes.

> **Learning tip**
>
> When answering questions dealing with the effect of temperature change, avoid phrases such as 'an increase in temperature favours the endothermic direction'. An increase in temperature 'favours' (i.e. increases the rate of) both the endothermic and the exothermic reaction.
>
> The important point to remember is that it increases the rate of the endothermic reaction *more* than it increases the rate of the exothermic reaction.

### Effect of the addition of a catalyst

If a catalyst is added to a reaction mixture that is in equilibrium, the rate of both the forward and the backward reactions will increase. However, unlike the effect of increasing the temperature, the increase in rate will be the same for both reactions. So, the position of equilibrium is not altered.

The advantage of adding a catalyst at the beginning of the reaction is that it will reduce the time required to establish equilibrium.

### Limitations of making qualitative predictions

The first thing to recognise is that the qualitative predictions we have made about the effect of concentration, pressure and temperature on the position of equilibrium are just that – predictions.

The arguments we have used are *not* explanations of why changes sometimes occur. In fact, there are occasions when it is impossible to predict the direction of change, or indeed when the prediction turns out to be incorrect.

For example, if an equilibrium mixture of $NO_2(g)$ (brown) and $N_2O_4(g)$ (colourless) is placed into a beaker of hot water in a closed container at room temperature, both the temperature and the pressure of the gaseous mixture will rise.

The equation for the reaction is:

$$2NO_2(g) \rightleftharpoons N_2O_4(g) \qquad \Delta H = -57.2 \text{ kJ mol}^{-1}$$

Since the forward reaction is exothermic, we would predict that an increase in temperature would shift the equilibrium to the left.

However, since there are fewer moles of gas on the right-hand side of the equation, we would predict that an increase in pressure would shift the equilibrium to the right.

We do not know which effect is greater, so we cannot make a prediction of which way the equilibrium will shift. In practice, the mixture becomes darker in colour, so the temperature effect must be greater than the pressure effect because the equilibrium must have shifted to the left to produce more brown $NO_2(g)$.

It might be interesting for you to do further research into the limitations of qualitative predictions. For example, what would you expect to happen if you add more nitrogen at constant pressure and temperature to an equilibrium mixture of nitrogen, hydrogen and ammonia? What would you expect to happen if you added an inert gas at constant volume and temperature to an equilibrium mixture of sulfur dioxide, oxygen and sulfur trioxide? The results may surprise you!

## Questions

1. Ethanoic acid and ethanol react reversibly to produce ethyl ethanoate and water:

    $CH_3COOH(l) + CH_3CH_2OH(l) \rightleftharpoons CH_3COOCH_2CH_3(l) + H_2O(l)$

    Assume $\Delta H = 0\,kJ\,mol^{-1}$

    (a) Predict the effect on the position of equilibrium of increasing the temperature. Justify your answer.

    (b) Suggest why the reaction mixture is heated when used to prepare ethyl ethanoate.

2. When carbon dioxide dissolves in water, it forms a solution containing some carbonic acid:

    $CO_2(g) + H_2O(l) \rightleftharpoons H_2CO_3(aq)$

    Carbon dioxide is less soluble in hot water than in cold water. Is $\Delta H$ for the forward reaction negative or positive? Justify your answer.

3. A sample of the insoluble solid lead(II) chloride, $PbCl_2$, is shaken with some dilute hydrochloric acid and left until an equilibrium mixture containing some undissolved lead(II) chloride is established:

    $PbCl_2(s) + 2Cl^-(aq) \rightleftharpoons PbCl_4^{2-}(aq)$

    State what would be observed if concentrated hydrochloric acid was added to the mixture. Justify your answer.

4. In each case, predict whether the equilibrium position shifts to the right or the left, or is unaltered when the pressure is increased at constant temperature.

    (a) $H_2(g) + I_2(g) \rightleftharpoons 2HI(g)$

    (b) $2SO_2(g) + O_2(g) \rightleftharpoons 2SO_3(g)$

    (c) $2O_3(g) \rightleftharpoons 3O_2(g)$

    (d) $4NH_3(g) + 5O_2(g) \rightleftharpoons 4NO(g) + 6H_2O(g)$

### Learning tip

A principle known as 'Le Chatelier's Principle' is sometimes recommended as a useful way of working out the possible change in position of equilibrium when a condition is altered. There is no requirement to learn or use this principle for the Edexcel specification.

Le Chatelier's Principle does not offer an explanation as to why a position of equilibrium alters. Explanations are only possible through the use of the equilibrium constant, $K$. This concept will be introduced in **Section 10.2.1** and developed in **Book 2**.

# 10.2 1 The equilibrium constant

By the end of this section, you should be able to...
- deduce an expression for the equilibrium constant, $K_c$, for both homogeneous and heterogeneous systems, in terms of equilibrium concentrations

## What are homogeneous and heterogeneous systems?

A **homogeneous system** is one in which all components are in the same phase.

Here are some examples.

| | |
|---|---|
| Gas phase | $H_2(g) + I_2(g) \rightleftharpoons 2HI(g)$ |
| Liquid phase | $CH_3COOH(l) + CH_3CH_2OH(l) \rightleftharpoons CH_3COOCH_2CH_3(l) + H_2O(l)$ |
| Aqueous phase | $CH_3COOH(aq) \rightleftharpoons CH_3COO^-(aq) + H^+(aq)$ |

In contrast, a **heterogeneous system** is where at least two different phases are present.

Here are a few examples.

| | |
|---|---|
| Gas and solid | $CaCO_3(s) \rightleftharpoons CaO(s) + CO_2(g)$ |
| Solid and aqueous | $PbCl_2(s) + 2Cl^-(aq) \rightleftharpoons PbCl_4^{2-}(aq)$ |
| Liquid and aqueous | $H_2O(l) \rightleftharpoons H^+(aq) + OH^-(aq)$ |

## The equilibrium constant

Look at this reaction:

$$H_2(g) + I_2(g) \rightleftharpoons 2HI(g)$$

**Table A** shows the equilibrium concentrations of each component at 730 K for four different experiments for the reaction.

| Equilibrium concentrations/mol dm$^{-3}$ | | |
|---|---|---|
| $H_2(g)$ | $I_2(g)$ | $HI(g)$ |
| $1.14 \times 10^{-2}$ | $0.12 \times 10^{-2}$ | $2.52 \times 10^{-2}$ |
| $0.92 \times 10^{-2}$ | $0.22 \times 10^{-2}$ | $3.08 \times 10^{-2}$ |
| $0.86 \times 10^{-2}$ | $0.86 \times 10^{-2}$ | $5.86 \times 10^{-2}$ |
| $0.77 \times 10^{-2}$ | $0.31 \times 10^{-2}$ | $3.34 \times 10^{-2}$ |

**table A** The equilibrium concentrations of $H_2(g)$, $I_2(g)$ and $HI(g)$ at 730 K, for four different experiments for the reversible reaction.

Each experiment was carried out in a sealed tube with different starting amounts of hydrogen and iodine.

Where $[H_2(g)]_{eq}$, $[I_2(g)]_{eq}$ and $[HI(g)]_{eq}$ represent the equilibrium concentrations of hydrogen, iodine and hydrogen iodide, respectively, we will now calculate values for the relationship:

$$\frac{[HI(g)]^2_{eq}}{[H_2(g)]_{eq}[I_2(g)]_{eq}}$$

The values for the four experiments are 46.4, 46.9, 46.4 and 46.7, respectively.

Within the limits of experimental error, the four values are constant.

The value of this ratio is known as the equilibrium constant. The equilibrium constant is represented by the symbol $K_c$.

For the reaction $H_2(g) + I_2(g) \rightleftharpoons 2HI(g)$,

$$K_c = \frac{[HI(g)]^2_{eq}}{[H_2(g)]_{eq}[I_2(g)]_{eq}}$$

The subscript 'eqm' is usually omitted from the concentration terms, so the expression for $K_c$ is shortened to:

$$K_c = \frac{[HI(g)]^2}{[H_2(g)][I_2(g)]}$$

## The Equilibrium Law

The equilibria in many other chemical reactions have been studied. In each case, the equilibrium constant relates to the stoichiometric equation in a similar manner to that for the hydrogen, iodine and hydrogen iodide system.

If an equilibrium mixture contains substances A, B, C and D related by the equation:

$$aA + bB \rightleftharpoons cC + dD$$

it is found experimentally that:

$$K_c = \frac{[C]^c[D]^d}{[A]^a[B]^b}$$

This general statement is known as the Equilibrium Law.

It is important when writing the equilibrium constant for a reaction that the concentrations of the right-hand side of the equation are written in the numerator, and that the concentrations of the left-hand side are written in the denominator.

It is therefore essential to relate any numerical value for an equilibrium constant to the particular equation concerned.

## Homogeneous systems

Suppose for the reaction $H_2(g) + I_2(g) \rightleftharpoons 2HI(g)$ that:

$$K_c = \frac{[HI(g)]^2}{[H_2(g)][I_2(g)]} = 50 \text{ (at a given temperature)}$$

Then, for the reaction $2HI(g) \rightleftharpoons H_2(g) + I_2(g)$:

$$K'_c = \frac{[H_2(g)][I_2(g)]}{[HI(g)]^2} = 0.02 \text{ (at the same temperature)}$$

For the reaction $\frac{1}{2}H_2(g) + \frac{1}{2}I_2(g) \rightleftharpoons HI(g)$:

$$K''_c = \frac{[HI(g)]}{[H_2(g)]^{\frac{1}{2}}[I_2(g)]^{\frac{1}{2}}} = \sqrt{50} = 7.07$$

## Heterogeneous systems

For heterogeneous systems, the situation is slightly different.

Consider the interconversion of liquid water and water vapour, which can be represented as follows:

$$H_2O(l) \rightleftharpoons H_2O(g)$$

For a liquid in a closed container existing in equilibrium, with its vapour at a given temperature, we can write:

$$\frac{[H_2O(g)]}{[H_2O(l)]} = \text{a constant}$$

At a given temperature, $[H_2O(l)]$ is constant, irrespective of the amount of water present. This is because:

$$[H_2O(l)] = \frac{m}{MV} = \frac{\text{density}}{M}$$

where $m$ = mass of water, $V$ = volume of water and $M$ = molar mass of water.

Since both the density of water and its molar mass are constant at a given temperature, $[H_2O(l)]$ must also be constant.

For water at 4°C, the density is $1000\,g\,dm^{-3}$ and $M = 18\,g\,mol^{-1}$.

So, $[H_2O(l)] = \dfrac{100\,g\,dm^{-3}}{18\,g\,mol^{-1}} = 55.56\,mol\,dm^{-3}$

We can absorb this constant into the equilibrium constant, giving us a value of $K_c$ as follows:

$$K_c = [H_2O(g)]$$

This argument can be applied to any heterogeneous system involving a solid.

For example, consider the equilibrium $CaCO_3(s) \rightleftharpoons CaO(s) + CO_2(g)$.

Since $[CaCO_3(s)]$ and $[CaO(s)]$ are both constant,

$$K_c = [CO_2(g)]$$

## Using $K_c$ to predict the effect of concentration on position of equilibrium

Consider the equilibrium $2NO_2(g) \rightleftharpoons N_2O_4(g)$, for which

$$K_c = \frac{[N_2O_4(g)]}{[NO_2(g)]^2}$$

If the concentration of $NO_2(g)$ is suddenly increased, then the value of the denominator in the above expression increases.

Suddenly, the numerical value of $\dfrac{[N_2O_4(g)]}{[NO_2(g)]^2}$ is less than $K_c$.

In order to restore its value to that of $K_c$, the value of $[N_2O_4(g)]$ must increase. The only way this can be achieved in a closed system is for the equilibrium position to shift to the right. A new equilibrium will be established in which $[NO_2(g)]$ and $[N_2O_4(g)]$ are both greater than in the previous equilibrium, but the ratio of $\dfrac{[N_2O_4(g)]}{[NO_2(g)]^2}$ is still the same as before.

## Questions

1. Write the expression for the equilibrium constant for each of the following reactions:
   (a) $CH_3COOH(l) + CH_3CH_2OH(l) \rightleftharpoons CH_3COOCH_2CH_3(l) + H_2O(l)$
   (b) $N_2(g) + O_2(g) \rightleftharpoons 2NO(g)$
   (c) $2N_2O(g) \rightleftharpoons 2N_2(g) + O_2(g)$
   (d) $NH_4HS(s) \rightleftharpoons NH_3(g) + H_2S(g)$
   (e) $3Fe(s) + 4H_2O(g) \rightleftharpoons Fe_3O_4(s) + 4H_2(g)$

2. The equilibrium constant, $K_c$, for the reaction
   $$2NO_2(g) \rightleftharpoons N_2O_4(g)$$
   has a numerical value of 200 at 298 K.
   (a) Write an expression for the equilibrium constant for the reaction.
   (b) If the concentration of $N_2O_4(g)$ in the equilibrium mixture at 298 K is $0.002\,mol\,dm^{-3}$, what is the concentration of $NO_2(g)$?
   (c) Calculate the numerical value of $K_c$ at 298 K for the reactions:
      (i) $N_2O_4(g) \rightleftharpoons 2NO_2(g)$
      (ii) $NO_2(g) \rightleftharpoons \frac{1}{2}N_2O_4(g)$

3. Use the value of the equilibrium constant for the reaction
   $$H_2(g) + I_2(g) \rightleftharpoons 2HI(g)$$
   to explain the effect of the addition of more $HI(g)$ to a reaction mixture in equilibrium in a closed container of fixed volume at constant temperature.

### Key definitions

A **homogeneous system** is a system in which all components are in the same phase.

A **heterogeneous system** is where at least two different phases are present.

# 10.2 2 Reversible reactions in industry

### By the end of this section, you should be able to...

- evaluate data to explain the necessity, for many industrial processes, of reaching a compromise between the yield and the rate of reaction

## Applying the principles of reaction rates and reversibility to industrial processes

The principles of reaction rates and reversibility play an important role in the design and conditions for many industrial processes. In order to maximise profits, the major problems confronting chemists are to convert the reactants into the products:

- as quickly as possible
- as completely as possible.

The first problem relates to kinetics (rate of reaction) and the second problem relates to reversibility.

The solution to each of these problems requires a careful choice of reaction conditions. This is easily demonstrated by considering the Haber process for the manufacture of ammonia.

## The Haber process

Ammonia is manufactured in industry by direct synthesis of nitrogen and hydrogen:

$$N_2(g) + 3H_2(g) \rightleftharpoons 2NH_3(g) \qquad \Delta H = -92 \text{ kJ mol}^{-1}$$

If the reaction mixture were to reach equilibrium, the maximum yield of ammonia would be obtained by using a low temperature (forward reaction is exothermic) and a high pressure (4 mol of gas on the left; 2 mol of gas on the right).

However, the reaction mixture does not reach a position of equilibrium in the reaction chamber. You can see this by analysing the following graphs.

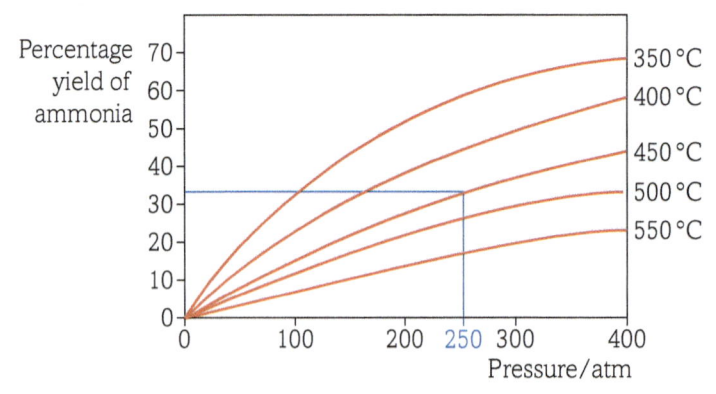

**fig A** Graphs showing the equilibrium yield of ammonia at different temperatures and pressures.

The conditions employed in the Haber process are typically 450 °C and 250 atmospheres (atm) pressure. If the reaction mixture were to reach equilibrium, then the yield of ammonia would be just over 30%. In practice, the yield is approximately 15%. This is because the reaction mixture does not remain in the reaction chamber long enough for equilibrium to be established.

## Why are conditions of 450 °C and 250 atm used?

The reaction between nitrogen and hydrogen is extremely slow at room temperature, largely because of the very strong nitrogen to nitrogen triple bond [$E(N\equiv N) = 945\,kJ\,mol^{-1}$] producing a high activation energy for the reaction. Even at high temperatures, the rate of reaction is low in the absence of a suitable catalyst. Many metals will catalyse this reaction, including tungsten and platinum. However, these metals are very expensive, so iron is used in the industrial process.

The catalyst does not function very efficiently at low temperatures, so a relatively high temperature is necessary. However, a very high temperature would be uneconomical, owing to the extra energy costs involved. It might also result in a decreased yield, since the increase in rate of the backward reaction will be greater than the increase in rate of the forward reaction. This is debatable, since the reaction mixture may not be in the reaction chamber for long enough for this to make a significant difference. For these reasons, a compromise temperature of 450 °C is used.

Under these conditions, the actual yield is about 50% of the equilibrium yield. If we assume this to be true for other pressures at a temperature of 450 °C, then a pressure of 100 atm would give a yield of around 12–13%, while a pressure of 400 atm would give a yield of around 27–28%. The higher the pressure, the larger the energy costs of compressing the gases: the lower the pressure, the lower the yield. Once again, a compromise is reached between yield and cost and a pressure of 250 atm is used.

In order to increase the efficiency of the process, the unreacted nitrogen and hydrogen, after separation from the ammonia, is mixed with fresh nitrogen and hydrogen and fed into the reaction chamber.

## The Contact process

Sulfuric acid is manufactured by the Contact process. This process gets its name from the stage of the process that involves the reaction between sulfur dioxide and oxygen at the surface of a solid vanadium(V) oxide, $V_2O_5$, catalyst, to form sulfur trioxide.

$$SO_2(g) + \tfrac{1}{2}O_2(g) \rightleftharpoons SO_3(g) \qquad \Delta H = -96\,kJ\,mol^{-1}$$

The forward reaction is exothermic, so a low temperature would favour a high yield of $SO_3(g)$. This is shown by the graph below, which shows the percentage yield of $SO_3(g)$ against temperature at a pressure of 1 atm.

Once again, the catalyst would not be very effective at low temperatures, so a moderately high temperature of 450 °C is used.

At 1 atm pressure, the yield of $SO_3(g)$ is already very high at around 97%. Higher pressures would increase the yield since there are fewer moles of gas on the right-hand side of the equation.

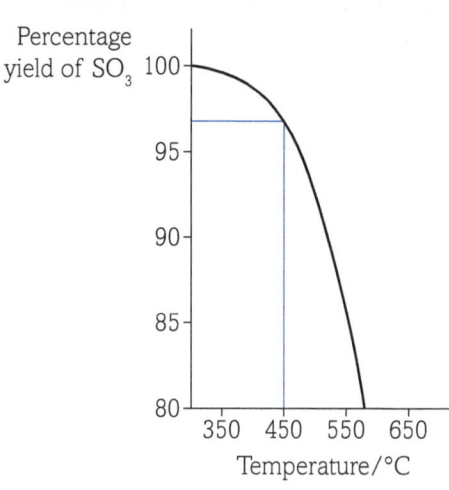

**fig B** A graph showing the percentage yield of sulfur trioxide against temperature at a pressure of 1 atm.

The pressure employed in the Contact process is about 2 atm. This is high enough to maintain a constant flow of gases through the reaction chamber. Pressures higher than this are unnecessary since the yield is already very high.

It is interesting to look at the mechanism of action of the vanadium(V) oxide. Unlike the iron in the Haber process, the vanadium changes its oxidation state during the reaction, but it then converts back to its original oxidation state at the end.

The mechanism for the reaction is:

$$SO_2(g) + V_2O_5(s) \rightleftharpoons SO_3(g) + V_2O_4(s)$$
$$V_2O_4(s) + \tfrac{1}{2}O_2(g) \rightleftharpoons V_2O_5(s)$$

# Questions

1. The first step in the manufacture of nitric acid, $HNO_3(l)$, from ammonia involves the exothermic reaction of ammonia with oxygen gas to form nitrogen monoxide $NO(g)$ and steam. This is a reversible reaction that can reach a position of equilibrium.
   (a) Write an equation for the reaction between ammonia and oxygen to form nitrogen monoxide and steam.
   (b) Assuming the reaction mixture reaches a position of equilibrium, make a qualitative prediction of the conditions of temperature and pressure that would produce the maximum yield of $NO(g)$.
   (c) In industry, the conditions used are high temperature and a pressure of around 7 atm. Suggest why these conditions are different to those you predicted in (b) to obtain a high yield of $NO(g)$ in the equilibrium mixture. Again, assume that the reaction reaches a position of equilibrium.

2. The densities of diamond and graphite are 3.5 and 2.3 g cm$^{-3}$, respectively.

   The change from graphite to diamond can be represented by the equation:

   $$C(graphite) \rightleftharpoons C(diamond) \qquad \Delta H = +2\,kJ\,mol^{-1}$$

   Suggest the conditions of temperature and pressure that would favour the formation of diamond from graphite. Justify your answers.

# THINKING BIGGER

## IN AT THE DEEP END

In this spread we will consider the role of chlorine in the treatment of water; specifically in swimming pools. Water-borne pathogens are still responsible for millions of deaths across the world and so effective treatment is important both for drinking water and for recreation, but the article below raises some additional issues about its use.

## SOMETHING IN THE WATER...

**fig A** 'A better way to cut the chloroform levels would be to reduce the amount of organic matter in the water.'

Public swimming pools contain high levels of a chemical linked to miscarriage.

A team led by Mark Nieuwenhuijsen at Imperial College, London, found that the chloroform* content of eight pools in the city was on average 20 times as high as in drinking water. The chloroform is formed when chlorine disinfectants react with organic compounds in the water.

Some studies in the US have suggested there's a correlation between the amount of chlorinated tap water drunk daily by pregnant women and their risk of miscarriage. Chloroform and related chemicals in the water have been blamed. Nieuwenhuijsen accepts that the studies linking chlorination and miscarriage are inconsistent. 'But pregnant women are advised to go swimming', he says. 'And there is a much higher level of chloroform in pools than in drinking water, so it could be a bigger pathway for exposure.'

Chloroform is produced when chlorine in the water reacts with flecks of skin, body-care products and other organic materials. Swimmers absorb the chemical through their skin, by swallowing water or by inhaling the gas.

Nieuwenhuijsen found an average of 113.3 micrograms of chloroform per litre in 40 samples from eight different pools. He also found that the more people using the pool and the warmer the water, the higher the concentration of chloroform.

But stopping chlorination is not the answer, says Nieuwenhuijsen, as alternative disinfectants such as ozone or ultraviolet light don't work so well. 'Chlorine is very effective', he says. 'A better way to cut the chloroform levels would be to reduce the amount of organic matter in the water by making sure people shower before a swim, and by improving filtration.'

*Chloroform = trichloromethane

### Did you know?

Chlorine gas is an irritant that causes serious damage to the eyes and respiratory system by reacting with water to form hydrochloric acid. For this reason, it was used as a chemical weapon in the First World War. The strong odour and the green colour of the gas made it easy to detect so more effective alternatives were soon developed. One of these was phosgene – a gas with the chemical formula $COCl_2$. The gas was formed by the reaction of chlorine and carbon monoxide in the presence of light so it was named by combining the Greek 'phos' (meaning light) and 'genesis' (meaning birth). Since then, over 190 states have signed up to the Geneva Protocol, which prohibits the use of chemical and biological weapons.

Where else will I encounter these themes?

1　2　3　4　5

# Thinking Bigger | 10

Let us start by considering the nature of the writing in the article.

**1.** You can have some fun with this first question! Imagine you are the editor of one of the more (in)famous 'red top' British newspapers. It is a slow news day and you have to lead with this article and give the story maximum impact. Using some illustrations and the key parts of this article write the most sensational article you can, using only the information in the article above.

Now we will look at the chemistry in, or connected to, this article. Don't worry if you are not ready to give answers to these questions immediately. Read through the article again, trying to identify the most important pieces of information as you go. This will help you to answer the questions.

**2. a.** The IUPAC name for chloroform is trichloromethane. Draw out the displayed formula for trichloromethane

**b.** Describe the shape of the trichloromethane molecule and, using diagrams, explain why the molecule has a permanent dipole.

**3.** Convert 113.3 µg/l into a micromolar concentration. (1 µg = $1 \times 10^{-6}$ g)

**4.** When chlorine is added to water the following equilibrium is set up:

$$Cl_2(aq) + H_2O(l) \rightleftharpoons HOCl(aq) + HCl$$

**a.** The reaction shown is an example of a disproportionation reaction. State what this means.

**b.** There are two acids formed in the reaction on the left: HCl is a strong acid, and HOCl is a weak acid. Explain the difference.

**5.** When HOCl is dissolved in water, the following equilibria is set up.

$$HOCl(aq) + H_2O(l) \rightleftharpoons H_3O^+(aq) + OCl^-(aq)$$

**a.** How would an increase in the pH value of this solution affect the above equilibrium? Explain your answer.

**b.** How would a greater dilution of HOCl(aq) affect the above equilibrium? Explain.

**6.** A second chloro-organic molecule detected in the analysis had the molecular formula $C_2H_3Cl_3$. Draw both structural isomers of $C_2H_3Cl_3$ and name them.

> **Command word**
> When you are asked to state the meaning of a word or phrase, you'll need to describe the term clearly and concisely in words. There might be more than one way it can be described.

## Activity

Design an experiment to investigate how a change of pH in a sample of swimming pool water might influence bacterial growth. Your plan should include:
- All the important variables that need to be considered
- How bacterial growth might be quantified experimentally
- How you would ensure experimental reliability.

It might be worth looking at other experimental protocols that you can find either in science books or on the internet. This may help you appreciate the variables that are important and how they can be controlled.

> **Thinking scientifically**
> Think of everything you have learned so far in the course and how it fits together to inform your understanding as a scientist. By now, you should be able to use correct terminology, analyse sources, write scientifically, and think like a scientist.

- From an article in *New Scientist* magazine

# 10 Exam-style questions

1. When potassium dichromate(VI), $K_2Cr_2O_7$, dissolves in water the following dynamic equilibrium is established in solution:

    $Cr_2O_7^{2-}(aq) + H_2O(l) \rightleftharpoons 2CrO_4^{2-}(aq) + 2H^+(aq)$
    (orange)                              (yellow)

    (a) One feature of a system that is in dynamic equilibrium is that both the forward and the backward reactions are simultaneously occurring. State two other features of a system in dynamic equilibrium. [2]
    (b) Explain why a solution of potassium dichromate(VI) turns from orange to yellow when aqueous sodium hydroxide is added. [3]
    [Total: 5]

2. Many chemical reactions are easily reversible and can form equilibrium mixtures. Hydrogen and iodine react together in the gaseous state to form hydrogen iodide:

    $H_2(g) + I_2(g) \rightleftharpoons 2HI(g)$   $\Delta H = +52\,kJ\,mol^{-1}$
    (colourless) (purple) (colourless)

    Hydrogen and iodine vapour are placed into a sealed container at 400 °C and allowed to reach equilibrium with hydrogen iodide. The resulting mixture has a very pale purple colour.

    (a) When the reaction mixture is cooled to 200 °C, the reaction mixture becomes notably darker in colour. Give a reason for this observation in terms of the change to the equilibrium composition of the mixture. [2]
    (b) In another experiment, all three gases were mixed together in a sealed tube and once again allowed to establish an equilibrium. The concentrations of the gases were monitored and the results were plotted on a graph. At time t, a change was made to the composition of the mixture.

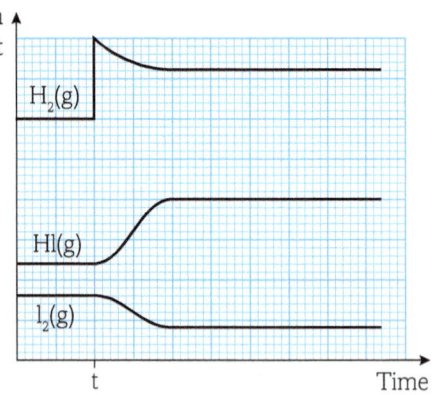

    (i) State the change that was made to the reaction mixture at time t. [1]
    (ii) Describe the changes that occur to the reaction mixture after time t. [3]
    [Total: 6]

3. Many industrial processes involve reversible reactions. Chemists were investigating the production of a chemical $XY_2$ that can be formed from $X_2$ and $Y_2$ as shown in the equation

    $X_2(g) + 2Y_2(g) \rightleftharpoons 2XY_2(g)$

    The chemists carried out a series of experiments at different temperatures, each time allowing the reaction mixture to establish equilibrium.
    The chemists measured the percentage conversion of $Y_2$ at each temperature.
    The results are shown in the graph.

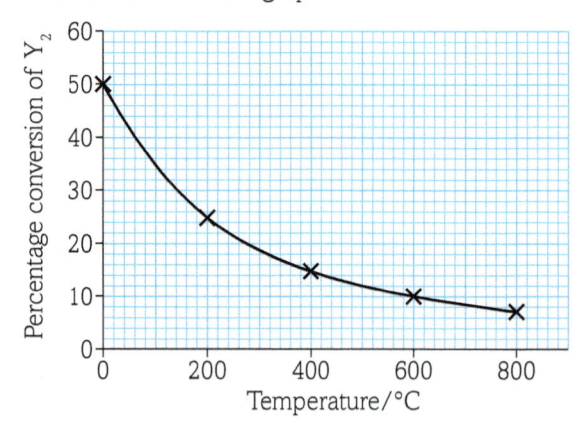

    (a) Explain how the graph shows that the forward reaction is exothermic. [2]
    (b) The chemists decide to use a catalyst in the process. Explain the effect of a catalyst on:
       (i) the rate at which equilibrium is established
       (ii) the percentage conversion of $Y_2$ at equilibrium. [4]
    [Total: 6]

4. Methanol ($CH_3OH$) can be manufactured from carbon monoxide and hydrogen.

    $CO(g) + 2H_2(g) \rightleftharpoons CH_3OH(g)$   $\Delta H = -128\,kJ\,mol^{-1}$

    (a) Predict how the composition of the equilibrium mixture is affected by:
       (i) carrying out the reaction at a higher temperature but keeping the pressure the same
       (ii) carrying out the reaction at a higher pressure but keeping the temperature the same.
       In each case justify your answer. [4]
    (b) Assuming the reaction takes place by collision of molecules in the gas phase, explain the effect on the rate of reaction of increasing the pressure. [2]
    [Total: 6]

5 Graphite and diamond are two allotropes of carbon. Artificial diamonds can be made from graphite.

$$C(graphite) \rightleftharpoons C(diamond) \quad \Delta H = +1.8 \text{ kJ mol}^{-1}$$

| Substance | Density / g cm$^{-3}$ |
|---|---|
| graphite | 2.25 |
| diamond | 3.51 |

(a) Complete the following enthalpy profile diagram for the conversion of graphite into diamond. Label the enthalpy change, $\Delta H$. [2]

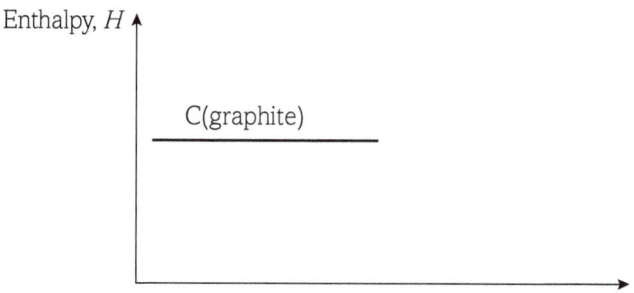

Enthalpy, $H$

C(graphite)

Progress of reaction

(b) State and justify which allotrope is the more thermodynamically stable. [1]
(c) State and justify in which allotrope the atoms take up less space per mole of substance. [1]
(d) Predict the conditions of temperature and pressure that are required to convert graphite into diamond. Justify your predictions. [4]

[Total: 8]

6 Ammonia is manufactured by the Haber process in which nitrogen and hydrogen are directly combined.

$$N_2(g) + 3H_2(g) \rightleftharpoons 2NH_3(g) \quad \Delta H = -92 \text{ kJ mol}^{-1}$$

(a) State the pressure used in the Haber process. [1]
(b) The pressure used is often described as a 'compromise' pressure.
  (i) State the main disadvantage of using higher pressures. [1]
  (ii) State the main disadvantage of using lower pressures. [1]
(c) The temperature used is 450 °C. State why a lower temperature is not used, even though it may result is a higher yield of ammonia. [1]
(d) Under the conditions employed the yield of ammonia is about 15%. State what happens to the unreacted nitrogen and hydrogen. [1]

[Total: 5]

7 An acid-base indicator, HIn, dissolves in water to form a green solution. The following equilibrium is established:

$$HIn(aq) \rightleftharpoons H^+(aq) + In^-(aq)$$
yellow          blue

The indicator appears green because it contains sufficient amounts of both the yellow and blue species.

(a) Hydrochloric acid is added to the indicator solution. State what you would observe. Give a reason for your answer. [2]
(b) Aqueous sodium hydroxide is added dropwise to the resulting solution from (a) until no further colour change takes place. State all of the colours that would be observed. Give reasons for your answers. [4]

[Total: 6]

8 The hydrogen required for the manufacture of ammonia and margarine is made by reacting methane with steam. The two gases react in a reversible reaction as shown by the equation

$$CH_4(g) + H_2O(g) \rightleftharpoons CO(g) + 3H_2(g) \quad \Delta H = +210 \text{ kJ mol}^{-1}$$

(a) State the conditions of temperature and pressure that would provide a high yield of hydrogen. Justify your answers. [4]
(b) State the conditions of temperature and pressure that would produce a fast rate of reaction. Justify your answers. Assume that the reaction takes place by molecules colliding in the gas phase. [4]
(c) Use your answers to (a) and (b) to explain why a temperature of 800 °C and a pressure of 30 atm are used for the manufacture of hydrogen from methane. [3]

[Total: 11]

# Maths skills

In order to be able to develop your skills, knowledge and understanding in Chemistry, you will need to have developed your mathematical skills in a number of key areas. This section gives more explanation and examples of some key mathematical concepts you need to understand. Further examples relevant to your AS/A level Chemistry studies are given throughout the book.

## Arithmetic and numerical computation

### Using standard form

Dealing with very large or small numbers can be difficult. For example, Avogadro's constant is an important value in Chemistry which is approximately equal to 602 000 000 000 000 000 000 000 mol$^{-1}$. To make such numbers easier to handle, we can write them in the format $a \times 10^b$. This is called standard form. Using standard form, Avogadro's constant can be written as $6.02 \times 10^{23}$ mol$^{-1}$.

To change a number from decimal form to standard form:

Count the number of positions you need to move the decimal point by until it is directly to the right of the first number which is not zero.

This number is the index number that tells you how many multiples of 10 you need. If the original number was a decimal, your index number must be negative.

Here are some examples:

| Decimal notation | Standard form notation |
| --- | --- |
| 0.000 000 012 | $1.2 \times 10^{-8}$ |
| 15 | $1.5 \times 10^{1}$ |
| 1000 | $1 \times 10^{3}$ |
| 3 700 000 | $3.7 \times 10^{6}$ |

## Using ratios, fractions and percentages

Ratios, fractions and percentages help you to express one quantity in relation to another with precision. Ratios compare like quantities using the same units. Fractions and percentages are important mathematical tools for calculating proportions.

### Ratios

A ratio is used to compare quantities. You can simplify ratios by dividing each side by a common factor. For example 12 : 4 can be simplified to 3 : 1 by dividing each side by 4.

> **WORKED EXAMPLE**
>
> *Divide 180 into the ratio 3 : 2*
> **Answer**
> Our strategy is to work out the total number of parts then divide 180 by the number of parts to find the value of one part.
>
> Total number of parts = 3 + 2 = 5
> Value of one part = 180 ÷ 5 = 36
> Answer = 3 × 36 : 2 × 36 = 108 : 72
>
> Check your answer by making sure the parts add up to 180
> 108 : 72 = 180

## WORKED EXAMPLE

*An excess of magnesium is added to $0.2\ dm^3$ of $1\ mol\ dm^{-3}$ dilute hydrochloric acid. The equation for the reaction is:*

$Mg + 2HCl \rightarrow MgCl_2 + H_2$

*How many moles of hydrogen are formed?*

Since there is an excess of magnesium, we know that all of the hydrochloric acid will react. We can use the following equation to calculate the number of moles of hydrochloric acid:

$$\text{concentration in mol dm}^{-3} = \frac{\text{amount in moles}}{\text{volume in dm}^3}$$

$$1\ \text{mol dm}^{-3} = \frac{\text{amount in moles of HCl}}{0.2\ \text{dm}^3}$$

amount of HCl reacted = 0.2 mol

The ratio for $HCl:H_2$ in this reaction is $2:1$, so when two moles of HCl react, one mole of $H_2$ is formed.

number of moles of $H_2$ formed = 0.5 × number of moles of HCl reacted

= 0.5 × 0.2 mol

= 0.1 mol

## Fractions

When using fractions, make sure you know the key strategies for the four operators:

To add or subtract fractions, find the lowest common multiple (LCM) and then use the golden rule of fractions. The golden rule states that a fraction remains unchanged if the numerator and denominator are multiplied or divided by the same number.

### WORKED EXAMPLE

$\frac{1}{2} + \frac{1}{5} = \frac{5}{10} + \frac{2}{10} = \frac{7}{10}$

To multiply fractions together, simply multiply the numerators together and multiply the denominators together.

### WORKED EXAMPLE

$\frac{2}{7} \times \frac{4}{9} = \frac{8}{63}$

To divide fractions, simply invert or flip the second fraction and multiply.

### WORKED EXAMPLE

$\frac{2}{3} \div \frac{7}{9} = \frac{2}{3} \times \frac{9}{7} = \frac{18}{21} = \frac{6}{7}$

## Percentages

When using percentages, it is useful to recall the different types of percentage questions.

To increase a value by a given percentage, use a percentage multiplier.

### WORKED EXAMPLE

*Increase 30 mg by 23%.*

If we increase by 23%, our new value will be 123% of the original value. We therefore multiply by 1.23.

**Answer** = 30 × 1.23 = 36.9 mg

To decrease a value by a given percentage, you need to focus on the part that is left over after the decrease.

### WORKED EXAMPLE

*Decrease 30 mg by 23%.*

If we decrease by 23%, our new value will be 100 − 23 = 77% of the original value. We therefore multiply by 0.77.

Answer = 30 × 0.77 = 23.1 mg

To calculate a percentage increase, use the following equation:

$$\text{Percentage change} = \frac{\text{difference between values}}{\text{original value}} \times 100$$

To calculate percentage decrease, use the same equation but remember that your answer should be negative.

**WORKED EXAMPLE**

*The volume of a solution increased from 40 ml to 50 ml. Calculate the percentage increase.*

Change in volume = 10 ml

Percentage increase = $\frac{10}{40} \times 100 = 25\%$.

## Algebra

### Changing the subject of an equation

It can be very helpful to rearrange an equation to express the variable that you are interested in in terms of other variables. Always remember that any operation that you apply to one side of the equation must also be applied to the other side.

**WORKED EXAMPLE**

*A sample of 2.5 mol of a substance has a mass of 12.5 g.*
*What is the molar mass of the substance?*

The equation for calculating moles is $n = \frac{m}{M}$ where $m$ = mass in grams, $M$ = molar mass and $n$ = amount of substance in moles.

If we wished to rearrange this equation to make $M$ the subject, we would first multiply each side by $M$ to obtain:

$nM = m$

Now to obtain the formula in terms of $n$, we divide each side by $M$ to obtain:

$M = \frac{m}{n}$

We can now simply substitute in the values for the question:

$m = 12.5\,g$
$n = 2.5\,mol$

$M = \frac{m}{n} = \frac{12.5\,g}{2.5\,mol} = 5\,g\,mol^{-1}$

## Handling data

### Using significant figures

Often when you do a calculation, your answer will have many more figures than you need. Using an appropriate number of significant figures will help you to interpret results in a meaningful way.

Remember the 'rules' for significant figures:

1. The first significant figure is the first figure which is not zero.
2. Digits 1–9 are always significant.
3. Zeros which come after the first significant figure are significant unless the number has already been rounded.

Here are some examples:

| Exact number | To one s.f. | To two s.f.s. | To three s.f.s. |
|---|---|---|---|
| 45 678 | 50 000 | 46 000 | 45 700 |
| 45 000 | 50 000 | 45 000 | 45 000 |
| 0.002 755 | 0.003 | 0.002 8 | 0.002 76 |

# Maths skills

# Applying your skills

You will often find that you need to use more than one maths technique to answer a question. In this section, we will look at three example questions and consider which maths skills are required and how to apply them.

## WORKED EXAMPLE

*10 g of potassium was added to excess water, producing potassium hydroxide and hydrogen gas, the latter was set alight because of the exothermic nature of the reaction.*

(i) *Calculate the number of moles of potassium used in the reaction.*
(ii) *Calculate the number of moles of $H_2(g)$.*
(iii) *Calculate the volume of $H_2(g)$ produced in $dm^3$.*

This type of question is very common. The key is to be familiar with the three different types of moles equations and when to use them. You will also usually have to work with ratios in order to find the correct number of moles of a reactant or product, and this tends to be where lots of mistakes are made.

The relevance of this question to the wider world is that, as chemists, we must know how much product we are going to make. This makes sense both for economic reasons, and in the case above where we are producing a highly flammable gas, for our safety as well!

Moles equations

(a) $\text{moles} = \dfrac{\text{mass}}{\text{molar mass}}$
   This equation tends to be used when dealing with solids or masses.

(b) moles = volume × concentration.
   This equation can be used when dealing with solutions or titrations.

(c) $\text{moles} = \text{volume in } \dfrac{dm^3}{24}$ or $\text{volume in } \dfrac{cm^3}{24\,000}$.
   This equation is used for gases.

If you are confident with mathematics then you can also combine the above equations and rearrange them to solve the missing value.

$$\dfrac{\text{mass}}{\text{molar mass}} = \text{volume} \times \text{concentration}$$

$$\dfrac{\text{mass}}{\text{molar mass}} = \text{volume in } \dfrac{dm^3}{24}$$

Using the information above:

Step 1: Calculate the number of moles of potassium used in the reaction.

In order to answer this question we first need to know which equation to use. We are given a mass, and know we are dealing with solids, so we should use equation (a). Knowing this, it is now a simple case of substituting in the numerical values and solving the equation.

$$\text{moles} = \dfrac{\text{mass}}{\text{molar mass}}$$

$$\text{moles} = \dfrac{10\,g}{39.1\,g\,mol^{-1}} = 0.258\,mol$$

Step 2: Calculate the number of moles of $H_2(g)$.

To find the number of moles of $H_2(g)$ requires us to work with ratios. First we write the chemical equation and then make sure it is balanced; sometimes this will already be done for you or it could be part of a previous question. Either way, balancing the chemical equation is key to determining the ratio of reactants to products etc. From this, we can now determine the ratio of K to $H_2$, which tells us how many moles of K there are compared to $H_2$.

$$2K(s) + 2H_2O(l) \rightarrow 2KOH(aq) + H_2(g)$$

K : $H_2$ = 2 : 1 therefore 0.2558 mol : 0.1279 mol

Step 3: Calculate the volume of $H_2(g)$ produced in $dm^3$.

The final part of the question requires us to calculate the volume of $H_2$ produced. To do this we must decide which moles equation to use. As we are dealing with gases it leaves us with one obvious choice.

$$\text{moles} = \text{volume in } \dfrac{dm^3}{24} \text{ or moles} = \text{volume in } \dfrac{cm^3}{24\,000}$$

Always check what units are required as you could lose valuable marks otherwise (this is usually stated either in the question or next to the space provided to write the answer). This question has asked for the answer to be shown in dm³. Rearranging this equation, substituting in the numerical values before solving provides us with the answer.

$$\text{volume in dm}^3 = 0.1279 \text{ mol} \times 24 \text{ mol dm}^{-3} = 3.069 \text{ dm}^3$$

It is always good to check your answer following a long question like this. The best way is to work backwards, and make sure we get 10 g of K as the end point if you have time!

### WORKED EXAMPLE

*In the United Kingdom there are currently 16 operational nuclear reactors located at 9 power stations. However, as the country's energy demands increase this number could rise. Recent surveys have found public opinion is not in favour of this increase, due to risk of leaks and radiation exposure following an explosion. One possible radioactive contaminant following an explosion is radioactive iodine, I\*, which is taken up by a person's thyroid gland. To counter this, potassium iodide, KI, is given to patients. This blocks the uptake of radioactive iodine by the thyroid gland by swamping the body with non-radioactive iodine instead.*

The current recommended dose of iodine, $I_2$, is 130 mg.
1 mg = 1 × $10^{-3}$ g

Calculate the mass of potassium iodide required to obtain the recommended dose of iodine. Show your answer in mg.

This style of question requires you to think carefully about what the question is asking. Whilst there are only two steps to finding the correct answer, this can still prove troublesome. This question is made harder by the fact that we are using iodine, which is $I_2$, rather than just $I^-$. This means the molar mass is 2 × 126.9 g mol$^{-1}$.

We must first determine the number of moles of iodine required, using the basic moles equation to do so. We know it is this equation because we are dealing with solids and masses. Therefore the equation to solve is

$$\text{moles} = \frac{\text{mass}}{\text{molar mass}}$$

$$\frac{0.130 \text{ g}}{253.8 \text{ g mol}^{-1}} = 5.122 \times 10^{-4} \text{ mol}$$

However, it takes two moles of potassium iodide, KI, to release one mole of $I_2$. Therefore, the required moles of KI is 2 × 5.122 × $10^{-4}$ = 1.024 × $10^{-3}$ mol. We can now work out the mass of potassium iodide by simply changing the subject of the equation and solving it using the molar mass of KI as 166.0 g mol$^{-1}$.

**mass = moles × molar mass**

However, the question has asked for the mass to be shown in units of mg. This means we must multiply the answer by 1000. When answering any question is it better to leave the rounding to the end, in order to avoid rounding errors.

$$(1.024 \times 10^{-3} \text{ mol} \times 166.0) \times 1000 = 170 \text{ mg}$$

## WORKED EXAMPLE

*Copper(II) chloride solution, $CuCl_2(aq)$, can be prepared by reacting excess powdered copper(II) oxide, $CuO(s)$, with hot hydrochloric acid:*

$$CuO(s) + 2HCl(aq) \rightarrow CuCl_2(aq) + H_2O(l)$$

*Unreacted copper(II) is then removed by filtration, and water is evaporated from the filtrate to leave crystals of copper(II) chloride-2-water, $CuCl_2.2H_2O$.*

*In one such preparation, $50.0\ cm^3$ of $0.500\ mol\ dm^{-3}$ hydrochloric acid is used.*

(a) *Calculate the amount (in moles) of hydrochloric acid used.*

(b) *Calculate the minimum amount (in moles) of copper(II) oxide needed.*

(c) *Calculate the mass of copper(II) oxide needed, if an excess of 20% is necessary. Give you answer to two significant figures.*

(d) *Calculate the molar mass of copper(II) chloride-2-water, $CuCl_2.2H_2O$.*

(e) *$1.81\ g$ of copper(II) chloride-2-water is obtained. Calculate the percentage yield.*

In this type of question, you will need to perform a series of calculations, often using answers from earlier in the question. Pay particular attention to units and follow the 'NAUTE' rule – no approximation until the end! Give your final answer to an appropriate number of significant figures.

(a) $50.0\ cm^3 = 0.50\ dm^3$

   $0.50\ dm^3 \times 0.5\ mol\ dm^{-3} = 0.025\ mol$

(b) From the equation we can see that one mole of CuO reacts with two moles of HCl. We therefore need half the number of moles of CuO than of HCl.

   $0.025\ mol \div 2 = 0.0125\ mol$

(c) To obtain an excess of 20%, we need 120% of the minimum number of moles we found in part (b). To calculate 120%, we can multiply by 1.2.

   excess = $0.0125 \times 1.2 = 0.015\ mol$

   We can now find the mass required by multiplying the number of moles by the molar mass of CuO.

   molar mass of CuO = $63.5 + 16.0 = 79.5$
   mass of CuO required = $0.015 \times 79.5 = 1.1925\ g$

   The data in the question was given to three significant figures, so we give our final answer to two significant figures:

   mass of CuO required = $1.2\ g$

(d) We can calculate the molar mass using the relative atomic masses of the elements:

   molar mass of $CuCl_2.2H_2O$ = $63.5 + (2 \times 35.5) + (2 \times 18.0) = 170.5$

(e) From part (b) we know that the expected number of moles is 0.0125. We can use our answer from (d) to calculate the mass of the expected yield.

   mass of expected yield = $0.0125 \times 170.5 = 2.13\ g$
   percentage yield = $1.82 \div 2.13 \times 100 = 85\%$

# Preparing for your exams

## Introduction

The way that you are assessed will depend on whether you are studying for the AS or the A level qualification. Here are some key differences:

- AS students will sit two exam papers, each covering 50% of the content of the AS specification.
- A level students will sit three exam papers, each covering content from both years of A level learning. The third paper will include synoptic questions that may draw on two or more different topics.
- A level students will also have their competency in key practical skills assessed by their teacher in order to gain the Science Practical Endorsement. The endorsement will not contribute to the overall grade but the result (pass or fail) will be recorded on the certificate.

The tables below give details of the exam papers for each qualification.

### AS exam papers

| Paper | Paper 1: Core Inorganic and Physical Chemistry | Paper 2: Core Organic and Physical Chemistry |
|---|---|---|
| Topics covered | Topics 1–5 | Topic 2<br>Topics 5–10 |
| % of the AS qualification | 50% | 50% |
| Length of exam | 1 hour 30 minutes | 1 hour 30 minutes |
| Marks available | 80 marks | 80 marks |
| Question types | multiple-choice<br>short open<br>open-response<br>calculation<br>extended writing | multiple-choice<br>short open<br>open-response<br>calculation<br>extended writing |
| Experimental methods? | Yes | Yes |
| Mathematics | A minimum of 20% of the marks across both papers will be awarded for mathematics at Level 2 or above. | |

### A level exam papers

| Paper | Paper 1: Advanced Inorganic and Physical Chemistry | Paper 2: Advanced Organic and Physical Chemistry | Paper 3: General and Practical Principles in Chemistry |
|---|---|---|---|
| Topics covered | Topics 1–5<br>Topic 8<br>Topics 10–15 | Topics 2–3<br>Topics 5–7<br>Topic 9<br>Topic 16–19 | Topics 1–19 |
| % of the A level qualification | 30% | 30% | 40% |
| Length of exam | 1 hour 45 minutes | 1 hour 45 minutes | 2 hours 30 minutes |
| Marks available | 90 marks | 90 marks | 120 marks |
| Question types | multiple-choice<br>short open<br>open-response<br>calculation<br>extended writing | multiple-choice<br>short open<br>open-response<br>calculation<br>extended writing | short open<br>open-response<br>calculation<br>extended writing<br>synoptic |
| Experimental methods? | No | No | Yes |
| Mathematics | A minimum of 20% of the marks across all three papers will be awarded for mathematics at Level 2 or above. | | |
| Science Practical Endorsement | Assessed by teacher throughout course.<br>Does not count towards A level grade but result (pass or fail) will be reported on A level certificate. | | |

# Exam strategy

## Arrive equipped

Make sure you have all of the correct equipment needed for your exam. As a minimum you should take:

- pen (black ink or ball-point pen)
- pencil (HB)
- ruler (ideally 30 cm)
- rubber (make sure it's clean and doesn't smudge the pencil marks or rip the paper)
- calculator (scientific).

## Ensure your answers can be read

Your handwriting does not have to be perfect but the examiner must be able to read it! When you're in a hurry it's easy to write key words that are difficult to decipher.

## Plan your time

Note how many marks are available on the paper and how many minutes you have to complete it. This will give you an idea of how long to spend on each question. Be sure to leave some time at the end of the exam for checking answers. A rough guide of a minute a mark is a good start, but short answers and multiple choice questions may be quicker. Longer answers might require more time.

## Understand the question

Always read the question carefully and spend a few moments working out what you are being asked to do. The command word used will give you an indication of what is required in your answer.

Be scientific and accurate, even when writing longer answers. Use the technical terms you've been taught.

Always show your working for any calculations. Marks may be available for individual steps, not just for the final answer. Also, even if you make a calculation error, you may be awarded marks for applying the correct technique.

## Plan your answer

In questions marked with an *, marks will be awarded for your ability to structure your answer logically showing how the points that you make are related or follow on from each other where appropriate. Read the question fully and carefully (at least twice!) before beginning your answer.

## Make the most of graphs and diagrams

Diagrams and sketch graphs can earn marks – often more easily and quickly than written explanations – but they will only earn marks if they are carefully drawn.

- If you are asked to read a graph, pay attention to the labels and numbers on the $x$ and $y$ axes. Remember that each axis is a number line.
- If asked to draw or sketch a graph, always ensure you use a sensible scale and label both axes with quantities and units. If plotting a graph, use a pencil and draw small crosses or dots for the points.
- Diagrams must always be neat, clear and fully labelled.

## Check your answers

For open-response and extended writing questions, check the number of marks that are available. If three marks are available, have you made three distinct points?

For calculations, read through each stage of your working. Substituting your final answer into the original question can be a simple way of checking that the final answer is correct. Another simple strategy is to consider whether the answer seems sensible. Pay particular attention to using the correct units.

# Sample answers with comments

## Question type: multiple choice

Solutions containing chlorate (I) ions are used as household bleaches and disinfectants.
These solutions decompose on heating as shown in the equation below.

$$3ClO^- \rightarrow ClO_3^- + 2Cl^-$$

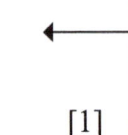

Which oxidation state is shown by chlorine in each of these three ions?

|   | $ClO^-$ | $ClO_3^-$ | $Cl^-$ |
|---|---|---|---|
| ☐ A | +1 | +3 | −1 |
| ☐ B | −1 | +3 | +1 |
| ☐ C | −1 | +5 | +1 |
| ☐ D | +1 | +5 | −1 |

[1]

> Three oxidation states have to be correctly calculated before an answer can be selected. However, before any answers are given, the question is very clearly stated: 'Which oxidation state is shown by chlorine in each of these three ions?' Calculate the three oxidation states before looking at any of the answers and then select the correct answer from the list based on the three oxidation states that you have already determined.

> Multiple choice questions always have one mark and the answer is given! For this reason students often make the mistake of thinking that they are the easiest questions on the paper. Unfortunately, this is not the case. These questions often require several answers to be worked out and error in one of them will lead to the wrong answer being selected. The three incorrect answers supplied (distractors) will feature the answers that students arrive at if they make typical or common errors. The trick is to answer the question before you look at any of the answers.

### Question analysis

- Multiple choice questions look easy until you try to answer them. Very often they require some working out and thinking.
- In multiple choice questions you are given the correct answer along with three incorrect answers (called distractors). You need to select the correct answer and put a cross in the box of the letter next to it.
- If you change your mind, put a line through the box (⊠) and then mark your new answer with a cross (☒).

### Average student answer

|   | $ClO^-$ | $ClO_3^-$ | $Cl^-$ |
|---|---|---|---|
| ☒ A | +1 | +3 | −1 |

> In $ClO^-$ chlorine has an oxidation state of +1, in $Cl^-$ −1 and in $ClO_3^-$ is +5. The correct answer is D. However, the student has selected answer A, and so does not gain any marks.

### Verdict

- This is an incorrect answer because:
- The student has calculated the oxidation state of chlorine in $ClO^-$ and $2Cl^-$ correctly, however, the oxidation state of chlorine in $ClO_3^-$ is +5. The student should have selected answer D.

> If you have any time left at the end of the paper go back and check your answer to each part of a multiple choice question so that a slip like this does not cost you a mark.

## Question type: short open

Compounds in Group 2 show trends in their properties.
State the block in the Periodic Table in which the Group 2 elements are found.

[1]

> State, give or name are used interchangeably when you are required to simply recall and write down a piece of information. These are usually simple short answers, often one word, requiring you to recollect the chemistry you have been taught.

### Question analysis

- The command word in this question is state. It requires you to recall and write down one or more pieces of information.
- Generally one piece of information is required for each mark given in the question. There is one mark available for this question and so one piece of information is required.

# Preparing for your exams

- Clarity and brevity are the keys to success on short open questions. For one mark, it is not always necessary to write complete sentences.

**Average student answer**

S-block

> The Periodic Table is split into blocks; s, p, d and f according to the orbitals that their highest energy electrons occupy. For Group 1 and 2 elements in the s-block, the highest energy electrons are in the s-block as the student has correctly stated.

## Verdict

This is a strong answer because:

- The student clearly understands that there are different blocks on the Periodic Table and has correctly identified Group 2 elements as s-block elements.

---

### Question type: open response

*This question is about Group 7 and redox chemistry.*

*Explain the trend in the boiling temperatures of the elements down Group 7, from fluorine to iodine.* [4]

> The command word in this question is explain. This requires a justification of a point, which in this case is the trend in boiling temperatures down Group 7. The question doesn't actually give the trend in boiling temperatures so it is a reasonable assumption that one mark will be given for simply stating how they change from fluorine to iodine, so start by doing that. There are four marks available for this question and so you should then aim to make three valid points as to why the boiling temperatures increase down Group 7.

### Question analysis

- With any question worth three or more marks, think about your answer and the points that you need to make before you write anything down. Keep your answer concise, and the information you write down relevant to the question. You will not gain marks for writing down chemistry that is not relevant to the question (even if correct) but it will cost you time.
- Remember that you can use bullet points or diagrams in your answer.

**Average student answer**

The boiling temperature of the elements increases as you go down the group. This is because the intermolecular forces called London forces get bigger from fluorine to iodine and so the boiling temperature increases. In some groups such as Group 1 the boiling temperature decreases because the intermolecular bonding is different and so that has a different effect on the boiling temperature. In other groups the boiling temperature increases and then decreases.

> At this level, your answers need technical terms and clarity in expression otherwise you will find yourself losing marks.

## Verdict

This is a weak answer because:

- Although the student has correctly stated that the boiling temperature increases down Group 7 and that this is due to increasing London forces, they have not explained why London forces increase down Group 7.
- The student has written about the trend in boiling temperatures in Group 1 and other groups. The question specifically asks about Group 7 so this information cannot gain any marks.

## Question type: extended writing

*An experiment was carried out involving the addition of aqueous silver nitrate, followed by aqueous ammonia, to distinguish between aqueous solutions containing chloride, bromide or iodide ions.*

*State the observations that would be made in this experiment and show how they can be used to deduce which ion is in each solution.* [6]

> It is reasonable to assume that the marks will be divided equally between the two parts of the question; three marks for describing the result you would expect to see for each ion (one for each) when silver nitrate is added with an explanation of how this is used to distinguish each ion, and three marks for describing what you would observe when aqueous ammonia is added to each of these solutions.

### Question analysis

- There will be questions in your exams which assess your understanding of practical skills and draw on your experience of the core practicals. For these questions, think about:
  - how apparatus is set up
  - the method of how the apparatus is to be used
  - how readings are to be taken
  - how to make the readings reliable
  - how to control any variables.

- It helps with extended writing questions to think about the number of marks available and how they might be distributed. For example, if the question asked you to give the arguments for and against a particular case, then assume that there would be equal numbers of marks available for each side of the argument and balance the viewpoints you give accordingly. However, you should also remember that marks will also be available for giving an overall conclusion so you should be careful not to omit that.

- It is vital to plan out your answer before you write it down. There is always space given on an exam paper to do this so just jot down the points that you want to make before you answer the question in the space provided. This will help to ensure that your answer is coherent and logical and that you don't end up contradicting yourself. However, once you have written your answer go back and cross these notes out so that it is clear they do not form part of the answer.

### Average student answer

When the aqueous silver nitrate is added chloride ions a white precipitate of silver chloride forms. This dissolves when ammonia is added.

When it is added to bromide ions a yellow precipitate of silver bromide forms. This dissolves when concentrated ammonia is added.

When it is added to iodide ions a yellow precipitate forms. This doesn't dissolve when ammonia is added however concentrated it is.

> Be very careful about making statements with the word 'it'. In this answer, it is clear that the student means 'silver nitrate' but if there is ambiguity you may not gain the mark.

## Verdict

This is an average answer because:
- The student has given results for all of the tests. However, in several cases the results are not given precisely enough for the marks to be scored. For example, for chloride ions the student does not specify whether the silver chloride precipitate will dissolve in dilute or concentrated ammonia and so does not score the second mark.

## Question type: calculation

*But-1-ene reacts with hydrogen bromide to form two different products. Analysis of one of the products showed that it contains 35.0% carbon, 6.6% hydrogen and 58.4% bromine by mass.*
*Calculate the empirical formula of this product.* [2]

> The command word here is calculate. This means that you need to obtain a numerical answer to the question, showing relevant working. If the answer has a unit, this must be included.

### Question analysis

- The important thing with calculations is that you must show your working clearly and fully. The correct answer on the line will gain all the available marks. However, an incorrect answer can gain all but one of the available marks if your working is shown and is correct.
- Show the calculation that you are performing at each stage and not just the result. When you have finished, look at your result and see if it is sensible. In an empirical formula calculation such as this one, if you do not end up with a whole number ratio for the atoms then you probably have an error in your calculation (have you divided the percentage of each element by the atomic number rather than the relative mass?). Go back and check.

### Average student answer

| carbon | hydrogen | bromine |
|---|---|---|
| 35 | 6.6 | 58.4 |
| $\frac{35}{12} = 2.9$ | $\frac{6.6}{1} = 6.6$ | $\frac{58.4}{80} = 0.73$ |
| $\frac{2.9}{0.73} = 4$ | $\frac{6.6}{0.73} = 9$ | $\frac{0.73}{0.73} = 1$ |

Ratio = 4 : 9 : 1

Empirical formula $C_4H_9Br$

> The first mark is awarded for dividing each of the percentage masses given in the question by the relative mass ($M_r$) of the element to obtain the number of moles. You can see that the student has clearly done this in the third line of the answer above. The second mark is awarded for calculating the ratio of each of the elements present by dividing by 0.73 and then using the result to work out the empirical formula. Again this is clearly shown in the student's answer.

### Verdict

This is a strong answer because:
- The student has correctly divided each of the percentage masses by the $M_r$ of the element to obtain the number of moles and so gained the first mark.
- The student has used the moles to correctly work out the ratio of the number of atoms of each element and so has obtained the second of the available marks.
- The answer has been laid out very clearly so that even if an error had been made then a mark could have been awarded for part of the calculation.

# Glossary

**Accuracy** is a measure of how close values are to the accepted or correct value.

**Activation energy**, $E_a$, is the minimum energy that colliding particles must possess for a reaction to occur.

The **actual yield** in a reaction is the actual mass obtained.

An **addition reaction** is a reaction in which two molecules combine to form one molecule.

**Aldehydes** are a homologous series of organic compounds formed by the partial oxidation of primary alcohols.

**Atom economy** is the molar mass of the desired product divided by the sum of the molar masses of all the products, expressed as a percentage.

The **atomic number** ($Z$) of an element is the number of protons in the nucleus of an atom of that element.

The **Avogadro constant** is the number of atoms of $^{12}C$ in exactly $12\,g$ of $^{12}C$.

**Avogadro's law** states that equal volumes of gases under the same conditions of temperature and pressure contain the same numbers of molecules.

The **base peak** indicates the peak with the greatest abundance.

**Basic oxides** are oxides of metals that react with water to form metal hydroxides, and with acids to form salts and water.

**Bioalcohols** are fuels made from plant matter, often using enzymes or bacteria.

A **biodegradable** polymer is one that can be broken down by microbes.

**Biodiesel** is a fuel made from vegetable oils obtained from plants.

**Biofuels** are fuels obtained from living matter that has died recently.

The Periodic Table is divided into **blocks**.

**Bond enthalpy** is the enthalpy change when one mole of a bond in the gaseous state is broken.

**Bond length** is the distance between nuclei of the two atoms that are covalently bonded together.

A **carbocation** is a positive ion in which the charge is shown on a carbon atom.

**Carboxylic acids** are a homologous series of organic compounds formed by the complete oxidation of primary alcohols.

A **catalyst** is a substance that increases the rate of a chemical reaction but is chemically unchanged at the end of the reaction.

**Coefficients** are the numbers written in front of species when balancing an equation.

**Complete combustion** means that all of the atoms in the fuel are fully oxidised.

**Concordant titres** are those that are close together (usually within $0.20\,cm^3$ of each other).

**Cracking** is the breakdown of molecules into shorter ones by heating with a catalyst.

**Curly arrows** represent the movement of electron pairs.

A **dehydration** reaction results in the removal of the hydroxyl group in an alcohol molecule, together with a hydrogen atom from an adjacent carbon atom, forming a C=C double bond.

**Delocalised electrons** are electrons that are not associated with any single atom or any single covalent bond.

A **diol** is a compound containing two OH (alcohol) groups.

A **dipole** is said to exist when two charges of equal magnitude but opposite signs are separated by a small distance.

A **discrete (simple) molecule** is an electrically neutral group of two or more atoms held together by chemical bonds.

A **displacement reaction** is a reaction in which one element replaces another element in a compound.

A **displayed (full structural) formula** shows each bonding pair as a line drawn between the two atoms involved.

A **displayed formula** shows every atom and every bond.

**Disproportionation** is the simultaneous oxidation and reduction of an element in a single reaction.

**Distillation with addition** involves heating a reaction mixture, but adding another liquid and distilling off the product as it forms.

**Electronegativity** is the ability of an atom to attract a bonding pair of electrons in a covalent bond.

The **electronic configuration** of an atom shows the number of electrons in each sublevel in each energy level of the atom.

An **electron-releasing** group is one that pushes electrons towards the atom it is joined to.

An **electrophile** is a species that is attracted to a region of high electron density.

**Electrophilic addition** is a reaction in which two molecules form one molecule and the attacking molecule is an electrophile.

An **elimination** reaction is one in which a molecule loses atoms attached to adjacent carbon atoms, forming a C=C double bond.

An **empirical formula** shows the numbers of each atom in the simplest whole-number ratio.

The **end point** is the point at which the indicator just changes colour. Ideally, the end point should coincide with the equivalence point.

**Endothermic** heat energy is transferred from the surroundings to the system.

# Glossary

The **equivalence point** is the point at which there are exactly the right amounts of substances to complete the reaction.

An **error** is the difference between an experimental value and the accepted or correct value.

An **ethanolic** solution is one in which ethanol is the solvent.

**Exothermic** heat energy is transferred from the system to the surroundings.

Use as a **feedstock** involves converting polymer waste into chemicals that can be used to make new polymers.

The **first ionisation energy** of an element is the energy required to remove an electron from each atom in one mole of atoms in the gaseous state.

**Fractional distillation** is the process used to separate a liquid mixture into fractions by boiling and condensing.

**Fragmentation** occurs when the molecular ion breaks into smaller pieces.

A **functional group** is an atom or group of atoms in a molecule that is responsible for its chemical reactions.

**Geometric isomers** are compounds containing a C=C bond with atoms or groups attached at different positions.

The vertical columns in the Periodic Table are called **groups**.

**Halogenation** involves the addition of a halogen.

**Heating under reflux** involves heating a reaction mixture with a condenser fitted vertically.

**Hess's Law** states that the enthalpy change of a reaction is independent of the path taken in converting reactants into products, provided the initial and final conditions are the same in each case.

A **heterogeneous catalyst** is one that is in a different phase to that of the reactants.

A **heterogeneous system** is where at least two different phases are present.

**Heterolytic fission** is the breaking of a covalent bond so that both bonding electrons are taken by one atom.

A **homogeneous system** is a system in which all components are in the same phase.

A **homologous series** is a family of compounds with the same functional group, which differ in formula by $CH_2$ from the next member.

**Homolytic fission** is the breaking of a covalent bond where each of the bonding electrons leaves with one species, forming a radical.

**Hund's rule** states that electrons will occupy the orbitals singly before pairing takes place.

**Hydrates** are compounds containing water of crystallisation, represented by formulae such as $CuSO_4.5H_2O$.

**Hydration** involves the addition of water (or steam).

A **hydrocarbon** is a compound that contains only carbon and hydrogen atoms.

The **hydrogen bond** is an intermolecular interaction (in which there is some evidence of bond formation) between a hydrogen atom of a molecule (or molecular fragment) bonded to an atom which is more electronegative than hydrogen and another atom in the same or a different molecule.

**Hydrogenation** involves the addition of hydrogen.

A **hydrolysis** reaction is one in which water or hydroxide ions replace an atom in a molecule with an —OH group.

An **incinerator** converts polymer waste into energy.

**Incomplete combustion** means that some of the atoms in the fuel are not fully oxidised.

**Infrared radiation** is the part of the electromagnetic spectrum with frequencies below that of red light.

An **initiation** step involves the formation of radicals, usually as a result of bond breaking caused by ultraviolet radiation.

The **intensity** of an infrared absorption describes the amount of infrared radiation absorbed.

**Isotopes** are atoms of the same element with different masses.

**Ketones** are a homologous series of organic compounds formed by oxidation of secondary alcohols.

A **locant** is a number used to indicate which carbon atom in the chain an atom or group is attached to.

The **mass concentration** of a solution is the mass (in g) of the solute divided by the volume of the solution.

The **mass number** of an atom is the sum of the number of protons and the number of neutrons in the nucleus of that atom.

**Mean bond enthalpy** is the enthalpy change when one mole of a bond, averaged out over many different molecules, is broken.

A **measurement uncertainty** is the potential error involved when using a piece of apparatus to make a measurement.

A **mechanism** is the sequence of steps in an overall reaction. Each step shows what happens to the electrons involved in bond breaking or bond formation.

The **meniscus** is the curving of the upper surface in a liquid in a container. The lowest (horizontal) part of the meniscus should be read.

**Metallic bonding** is the electrostatic force of attraction between the nuclei of metal cations and delocalised electrons.

The **molar concentration** of a solution is the amount (in mol) of the solute divided by the volume of the solution.

The **molar mass** is the mass per mole of a substance. It has the symbol $M$ and the units $g\,mol^{-1}$.

**Molar volume** is the volume occupied by 1 mol of any gas.

A **mole** is the amount of substance that contains the same number of particles as the number of carbon atoms in exactly 12 g of the carbon-12 isotope.

A **molecular formula** shows the actual number of atoms of each element in a molecule.

The **molecular ion peak** is the peak with the highest $m/z$ ratio in the mass spectrum, the $M$ peak.

**Monomers** are the small molecules that combine together to form a polymer.

A **multiple bond** is two or more covalent bonds between two atoms.

**Nitriles** are organic compounds containing the C—CN group.

**Non-renewable** energy sources are not being replenished, except over geological timescales.

A **nucleophile** is a species that donates a lone pair of electrons to form a covalent bond with an electron-deficient atom.

A **nucleophilic substitution** reaction is one in which an attacking nucleophile replaces an existing atom or group in a molecule.

An **orbital** is a region within an atom that can hold up to two electrons with opposite spins.

**Oxidation** is the loss of electrons.

**Oxidation number** is the charge that an ion has or the charge that it would have if the species were fully ionic.

An element is **oxidised** when its oxidation number *increases*.

An **oxidising agent** is a species (atom, molecule or ion) that oxidises another species by removing one or more electrons. When an oxidising agent reacts it gains electrons and is, therefore, reduced.

The **Pauli Exclusion Principle** states that two electrons cannot occupy the same orbital unless they have opposite spins. Electron spin is usually shown by the use of upward and downward arrows: ↑ and ↓.

The **percentage uncertainty** in an experiment is the actual measurement uncertainty multiplied by 100 and divided by the value recorded.

The **percentage yield** is 100 × the actual yield divided by the theoretical yield.

**Periodicity** is a regularly repeating pattern of atomic, physical and chemical properties with increasing atomic number.

The horizontal rows in the Periodic Table are called **periods**.

**Pi bonds** are covalent bonds formed when electron orbitals overlap sideways.

A **polar covalent bond** is a type of covalent bond between two atoms where the bonding electrons are unequally distributed. Because of this, one atom carries a slight negative charge and the other a slight positive charge.

**Precipitation reactions** are reactions in which an insoluble solid is one of the products.

**Precision** is a measure of how close values are to each other.

A **prefix** is a set of letters written at the beginning of a name.

**Primary amines** are compounds containing the C—$NH_2$ group.

**Primary standards** are substances used to make a standard solution by weighing.

The **propagation** steps are the two steps that, when repeated many times, convert the starting materials into the products of a reaction.

A **quantum shell** defines the energy level of an electron.

A **radical** is a species that contains an unpaired electron.

**Random errors** are errors caused by unpredictable variations in conditions.

**Recycling** involves converting polymer waste into other materials.

A **redox reaction** is a reaction that involves both reduction and oxidation.

An element is **reduced** when its oxidation number *decreases*.

A **reducing agent** is a species that reduces another species by adding one or more electrons. When a reducing agent reacts it loses electrons and is, therefore, oxidised.

**Reduction** is the gain of electrons.

**Reforming** is the conversion of straight-chain hydrocarbons into branched-chain and cyclic hydrocarbons.

The weighted mean (average) mass of an atom of an element compared to $\frac{1}{12}$ of the mass of an atom of carbon-12 is called the **relative atomic mass** ($A_r$) of the element.

The mass of an individual atom of a particular isotope relative to $\frac{1}{12}$ of the mass of an atom of carbon-12 is called the **relative isotopic mass**.

**Renewable** energy sources use sources that can be continuously replaced.

The **repeat unit** of a polymer is the set of atoms that are joined together in large numbers to produce the polymer structure.

**Restricted rotation** around a C=C bond fixes the position of the atoms or groups attached to the C=C atoms.

**Saturated** refers to a compound containing only single bonds.

The **second ionisation energy** of an element is the energy required to remove an electron from each singly charged positive ion in one mole of positive ions in the gaseous state.

**Sigma bonds** are covalent bonds formed when electron orbitals overlap axially (end-on).

**Simple distillation** is used to separate liquids with very different boiling temperatures.

A **skeletal formula** shows all the bonds between carbon atoms.

A **solute** is a substance that is dissolved.

A **solution** is a solute dissolved in a solution.

A **solvent** is a substance that dissolves a solute.

**Solvent extraction** is used to separate a liquid from a mixture by causing it to move from the mixture to the solvent.

**Spectator ions** are the ions in an ionic compound that do not take part in a reaction.

The **standard enthalpy change of combustion** ($\Delta_c H^\ominus$) is the enthalpy change measured at 100 kPa and a stated temperature, usually 298 K, when one mole of a substance is completely burned in oxygen.

The **standard enthalpy change of formation** is the enthalpy change measured at 100 kPa and a specified temperature, usually 298 K, when one mole of a substance is formed from its elements in their standard states.

The **standard enthalpy change of neutralisation** is the enthalpy change measured at 100 kPa and a stated temperature, usually 298 K, when one mole of water is produced by the neutralisation of an acid with an alkali.

# Glossary

**Standard enthalpy change of reaction** is the enthalpy change measured at 100 kPa and a stated temperature, usually 298 K, when the number of moles of substances in the equation *as written* react.

A **standard solution** is a solution whose concentration is accurately known.

**Stereoisomers** are compounds with the same structural formula (and the same molecular formula), but with the atoms or groups arranged differently in three dimensions.

**Stretching** occurs when a bond absorbs infrared radiation and uses it to alter the length of the bond.

A **structural formula** shows (unambiguously) how the atoms are joined together.

**Structural isomers** are compounds with the same molecular formula but with different structural formulae.

A **substitution reaction** is one in which an atom or group is replaced by another atom or group.

A **suffix** is a set of letters written at the end of a name.

**Systematic errors** are errors that are constant or predictable, usually because of the apparatus used.

A **termination** step involves the formation of a molecule from two radicals.

The **theoretical yield** in a reaction is the maximum possible mass of a product, assuming complete reaction and no losses.

**Thermal stability** is a measure of the extent to which a compound decomposes when heated.

The **titre** is the volume added from the burette during a titration.

The **transmittance** value in an infrared spectrum represents the amount of radiation absorbed at a particular wavenumber.

**Unsaturated** refers to a compound containing one or more multiple bonds.

The **wavenumber** of an infrared absorption represents the frequency of infrared radiation absorbed by a particular bond in a molecule.

# Periodic Table

**Key**
Atomic (proton number)
**Atomic symbol**
Name
Relative atomic mass

| Period | Group 1 (1) | 2 (2) | (3) | (4) | (5) | (6) | (7) | (8) | (9) | (10) | (11) | (12) | 3 (13) | 4 (14) | 5 (15) | 6 (16) | 7 (17) | 8 (18) |
|---|---|---|---|---|---|---|---|---|---|---|---|---|---|---|---|---|---|---|
| 1 | 1 **H** Hydrogen 1.0 | | | | | | | | | | | | | | | | | 2 **He** Helium 4.0 |
| 2 | 3 **Li** Lithium 6.9 | 4 **Be** Beryllium 9.0 | | | | | | | | | | | 5 **B** Boron 10.8 | 6 **C** Carbon 12.0 | 7 **N** Nitrogen 14.0 | 8 **O** Oxygen 16.0 | 9 **F** Fluorine 19.0 | 10 **Ne** Neon 20.2 |
| 3 | 11 **Na** Sodium 23.0 | 12 **Mg** Magnesium 24.3 | | | | | | | | | | | 13 **Al** Aluminium 27.0 | 14 **Si** Silicon 28.1 | 15 **P** Phosphorus 31.0 | 16 **S** Sulfur 32.1 | 17 **Cl** Chlorine 35.5 | 18 **Ar** Argon 39.9 |
| 4 | 19 **K** Potassium 39.1 | 20 **Ca** Calcium 40.1 | 21 **Sc** Scandium 45.0 | 22 **Ti** Titanium 47.9 | 23 **V** Vanadium 50.9 | 24 **Cr** Chromium 52.0 | 25 **Mn** Manganese 54.9 | 26 **Fe** Iron 55.8 | 27 **Co** Cobalt 58.9 | 28 **Ni** Nickel 58.7 | 29 **Cu** Copper 63.5 | 30 **Zn** Zinc 65.4 | 31 **Ga** Gallium 69.7 | 32 **Ge** Germanium 72.6 | 33 **As** Arsenic 74.9 | 34 **Se** Selenium 79.0 | 35 **Br** Bromine 79.9 | 36 **Kr** Krypton 83.8 |
| 5 | 37 **Rb** Rubidium 85.5 | 38 **Sr** Strontium 87.6 | 39 **Y** Yttrium 88.9 | 40 **Zr** Zirconium 91.2 | 41 **Nb** Niobium 92.9 | 42 **Mo** Molybdenum 95.9 | 43 **Tc** Technetium (98) | 44 **Ru** Ruthenium 101.1 | 45 **Rh** Rhodium 102.9 | 46 **Pd** Palladium 106.4 | 47 **Ag** Silver 107.9 | 48 **Cd** Cadmium 112.4 | 49 **In** Indium 114.8 | 50 **Sn** Tin 118.7 | 51 **Sb** Antimony 121.8 | 52 **Te** Tellurium 127.6 | 53 **I** Iodine 126.9 | 54 **Xe** Xenon 131.3 |
| 6 | 55 **Cs** Caesium 132.9 | 56 **Ba** Barium 137.3 | 57 **La*** Lanthanum 138.9 | 72 **Hf** Hafnium 178.5 | 73 **Ta** Tantalum 180.9 | 74 **W** Tungsten 183.8 | 75 **Re** Rhenium 186.2 | 76 **Os** Osmium 190.2 | 77 **Ir** Iridium 192.2 | 78 **Pt** Platinum 195.1 | 79 **Au** Gold 197.0 | 80 **Hg** Mercury 200.6 | 81 **Tl** Thallium 204.4 | 82 **Pb** Lead 207.2 | 83 **Bi** Bismuth 209.0 | 84 **Po** Polonium (209) | 85 **At** Astatine (210) | 86 **Rn** Radon (222) |
| 7 | 87 **Fr** Francium (223) | 88 **Ra** Radium (226) | 89 **Ac*** Actinium (227) | 104 **Rf** Rutherfordium (261) | 105 **Db** Dubnium (262) | 106 **Sg** Seaborgium (266) | 107 **Bh** Bohrium (264) | 108 **Hs** Hassium (277) | 109 **Mt** Meitnerium (268) | 110 **Ds** Darmstadtium (271) | 111 **Rg** Roentgenium (272) | 112 **Cn** Copernicium 112 | | 114 **Fl** flerovium | | 116 **Lv** livermorium | | |

Lanthanides:
| 58 **Ce** Cerium 140.1 | 59 **Pr** Praseodymium 140.9 | 60 **Nd** Neodymium 144.2 | 61 **Pm** Promethium 144.9 | 62 **Sm** Samarium 150.4 | 63 **Eu** Europium 152.0 | 64 **Gd** Gadolinium 157.2 | 65 **Tb** Terbium 158.9 | 66 **Dy** Dysprosium 162.5 | 67 **Ho** Holmium 164.9 | 68 **Er** Erbium 167.3 | 69 **Tm** Thulium 168.9 | 70 **Yb** Ytterbium 173.0 | 71 **Lu** Lutetium 175.0 |
|---|---|---|---|---|---|---|---|---|---|---|---|---|---|

Actinides:
| 90 **Th** Thorium 232.0 | 91 **Pa** Protactinium (231) | 92 **U** Uranium 238.1 | 93 **Np** Neptunium (237) | 94 **Pu** Plutonium (242) | 95 **Am** Americium (243) | 96 **Cm** Curium (247) | 97 **Bk** Berkelium (245) | 98 **Cf** Californium (251) | 99 **Es** Einsteinium (254) | 100 **Fm** Fermium (253) | 101 **Md** Mendelevium (256) | 102 **No** Nobelium (254) | 103 **Lr** Lawrencium (257) |
|---|---|---|---|---|---|---|---|---|---|---|---|---|---|

# Index

accuracy 144–5, 147
acids, reactions 97, 158–9
activation energy 252–3
addition reactions 188–93
    polymerisation 194–5
alcohols 171
    boiling temperatures 60–1
    naming 173
    oxidation 206–7
    reactions 204–5
aldehydes 206
alkaline earth elements 92–5
    compounds 96–103
alkalis 106–8, 158–9
alkanes 170, 171
    boiling temperatures 59–60, 61
    combustion 180–1
    from crude oil 178–9
    halogenoalkanes 198–203, 205
    naming 173
    properties 64
    substitution reactions 184–5
alkenes
    addition reactions 188–93
    bonding 186–7
    from dehydration of alcohols 205
    polymerisation 194–5
ammonia 113, 255, 274–5
ammonium ions 103, 114
atom economy 152–3
atomic number 10–11
atoms
    counting 125
    orbitals 16–19
    radius 26
    structure 10–11
Avogadro constant 125–6
Avogadro's law 132

balanced equations 84–5, 130–1
barium meal 97
biodegradable polymers 197
biofuels 182–3, 246–7
blocks in the Periodic Table 24
boiling temperatures
    across the Periodic Table 27
    alcohols 60–1
    alkanes 59–60, 61
    halogens 104–5
    and hydrogen bonding 62
    organic compounds 211
bond angles 51–2

bond enthalpy 242–5
bonds/bonding
    alkenes 186–7
    covalent 42–4, 46, 50, 68
    hydrogen 42–3, 56–8, 62, 63–4
    ionic 38–41, 68
    length and strength 44
    metals 36–7, 68
    polar 46

carbocation 191–3
carbon dioxide, test for 96, 156
carbon monoxide 180, 264
carbon neutrality 182–3
carbonates 99–100, 159
carboxylic acids 206
cars and pollution 181
catalysts 181, 212, 258–61, 270, 275
chlorine 276–7
    reactions 94, 106–8, 184–5, 205
*cis–trans* isomers 176–7
collision theory 252–3, 257
combustion 74
    alcohols 205
    alkanes 180–1
    analysis 121
    enthalpy change of 235–6
concentration
    calculating 142–3
    and equilibrium position 268–9
    mass 135, 136
    and reaction rate 254
    solutions 136–7
Contact process 275
covalent bonding 42–4, 46, 50, 68
covalent compounds 65–6
cracking 179
crude oil 178–9

d orbitals 17
dative covalent bonds 50
dehydration 205
diamond 65
diols 190
dipoles 54, 55–6, 104–5
displacement reactions 106, 154–5
displayed formulae 49, 50, 168
disproportionation 77, 107
distillation 207, 209–10
    fractional 178–9, 210
dot-and-cross diagrams 48–9, 50
double bonds 186–7

dynamic equilibrium 267

*E/Z* isomers 177
electron density map 45–6
electron pair repulsion 51
electronegativity 45–7, 105
electrons 10–11
    configurations 18–19
    energy levels 20–3
    loss and gain 76–7
electrophiles 191
electrophilic addition 191–3
elimination reactions 203
empirical formulae 120–1, 169
endothermic reactions 232–3
enthalpy 232
    cycle 240
enthalpy change 232–3
    bond enthalpies 242–5
    of combustion 235–6
    of formation 239–41
    of neutralisation 237–8
enthalpy level diagrams 234
equations
    balanced 84–5, 130–1
    changes of state 105
    half-equations 84–5, 109–10, 129
    ionic 84–5, 128–9
    polymerisation reactions 194–5
    writing 127–9
equilibrium constant 272–3
equilibrium position 267–75
errors 146–9, 236, 237–8
ethanol 182–3
exothermic reactions 232–3

flame tests 101–3
fluorine 43, 108
formulae
    empirical 120–1, 169
    molecular 122–4, 168–9, 225
    for names 127
    organic compounds 167, 168–9
    from oxidation numbers 82–3
    skeletal 60, 168
    structural 49, 50, 168, 174
fossil fuels 183
fractional distillation 178–9, 210
fragmentation 218–19, 220–1
free radicals 184–5, 219
fuels 178
    alkanes as 180–1

biofuels 182–3, 246–7
hydrogen 70
functional groups 170, 225

gases, reaction volumes 132–3
giant lattices 65–7, 68–9
glycolysis 74
graphene 66
graphite 66
groups in the Periodic Table 24
    group 2 92–103
    group 7 104–13

Haber process 255, 274–5
haemoglobin 264
half-equations 84–5, 109–10, 129
halogenoalkanes 198–203, 205
halogens 104–8
    compounds 109–13
    reactions 155, 189–90, 191–2, 205
helium 22, 28
Hess's law 239–41
heterogeneous system 272–3
heterolytic fission 191
homogeneous system 272–3
homologous series 170–1
homolytic fission 184
hydration 63, 189–90
hydrocarbons 166–7, 218–19
    see also alkanes; alkenes
hydrogen
    bonds 42–3, 56–8, 62, 63–4
    electronic structure 22, 28
    fuel 70
    reactions 266–7
hydrogen halides 61–2, 113, 190, 191
hydrogenation 189
hydrogencarbonates 159
hydrolysis 199, 200–1
hydroxides 96–7, 158

indicators 141
industry
    and the atom economy 152–3
    catalysts used in 212, 259
    reversible reactions in 274–5
infrared spectroscopy 222–5
intermolecular interactions 55–64
iodine 66, 266–7
ionic equations 84–5, 128–9
ionisation energies 20–3, 28–9, 92–3
ions 40–1
    bonding 38–41, 68
    lattices 65, 68–9
    radii 39–40
    shapes 51–2
isomerism 175–7
isotopes 10–11, 12–13

IUPAC system 172–4

ketones 206

life cycle analysis 197
London forces 55–6

margarine 189
mass concentration 135, 136
mass number 10, 12
mass spectrometry 11, 12–15
    deducing structures from 220–1
    in organic compounds 218–19
mass to charge ratio ($m/z$ value) 218–19, 220–1
Maxwell–Boltzmann distribution curves 256–7
melting temperatures 27, 36–7, 104–5
metals
    bonding 36–7, 68
    giant lattices 65, 68–9
    reactions 154–5, 158
methane 184–5
molar concentration 136
molar mass 123
molar volumes 134–5
molecular formulae 122–4, 168–9, 225
molecules 48
    intermolecular interactions 55–64
    lattices 66–7, 69
    models 171
    polar and non-polar 53–4
    shapes 51–2
moles 122, 125–6

names
    formulae for 127
    organic compounds 172–4
    systematic 82
nanotechnology 260–1
natural gas 183
neutralisation 97, 237–8
neutrons 10–11
nitrates 99, 114
nitrogen oxides 181
non-polar molecules 53–4
nucleophiles 199
nucleophilic reactions 202–3

octet rule 49
organic compounds 166–7
    formulae 167, 168–9
    IR spectra 224–5
    mass spectrometry in 218–19
    names 172–4
    purification 208–11
oxidation 76–7, 80
    alcohols 206–7

alkenes 190
    group 2 elements 94
oxidation numbers 78–83, 85
oxides 96–7, 158, 181
oxidising agents 77

p orbitals 17
Periodic Table 24–9, 296
    patterns and trends 22–3, 26–9
periods in the Periodic Table 24, 27
photosynthesis 74
pi ($\pi$) bonds 42–3, 187
polar bonds 46
polar molecules 53–4
pollution 181
polymers/polymerisation 194–5, 197
precipitation reactions 156–7
precision 144–5
pressure
    and equilibrium position 269–70
    and reaction rate 254–5
primary standards 138–9
protons 10–11
purification 208–11

quantum shells 16–19

rate of reaction 252–9
    hydrolysis 200–1
    in industry 274–5
reacting masses 130–1
reaction mechanisms 184–5, 191–3, 202–3
reactivity, trends in 93, 105
recycling 196
redox reactions 74–85, 106–8, 109
reducing agents 77
reduction 76–7, 80
reflux 207
reforming 179
relative atomic mass 12
relative molecular mass 13–14, 122
reversible reactions 266–7, 274–5

s orbitals 16–17
SI units 123
sigma ($\sigma$) bonds 42–3, 186
silver nitrate test 112, 200
skeletal formulae 60, 168
sodium chloride 38, 63
solubility 96, 97
solutions
    concentrations 136–7
    standard 138–9
solvent extraction 210
solvents 62–4
standard solutions 138–9
standard temperature and pressure (STP) 233

state symbols 127
stereoisomerism 176–7
steric hindrance 253
structural formulae 49, 50, 168, 174
structural isomerism 175–6
substitution reactions 184–5, 202–3
sulfamic acid 138–9
sulfates 97, 156
sulfur oxides 181
sulfuric acid 109–11, 275
surface area, and reaction rate 255

temperature
    and equilibrium position 270
    and reaction rate 257
    *see also* boiling temperatures;
    melting temperatures
thermal stability 98–100
titrations 140–1, 142–3

waste disposal 196–7
water 62, 63–4
    reactions 95, 107, 113
wavenumbers 223, 224–5

yield 150–1